# Applied Structural
# Steel Design

Steel framing for the Joseph J. Bulmer Telecommunications and Computations Center on the campus of Hudson Valley Community College.

THIRD EDITION

# Applied Structural Steel Design

Leonard Spiegel, P.E.
Consulting Engineer

George F. Limbrunner, P.E.
Hudson Valley Community College

**PRENTICE HALL**

Upper Saddle River, New Jersey    Columbus, Ohio

*Library of Congress Cataloging-in-Publication Data*

SPIEGEL, LEONARD.
   Applied structural steel design / Leonard Spiegel, George F.
Limbrunner.—3rd ed.
    p.   cm.
  Includes bibliographical references and index.
  ISBN 0-13-381583-8
  1. Building, Iron and steel.  2. Structural design.
I. Limbrunner, George F.  II. Title.
TA684.S656  1997                   96-31114
624.1′821—dc20                 CIP

Cover Photo: Ewing Galloway
Editor: Ed Francis
Production Editor: Christine M. Harrington
Production Coordinator: Kathleen M. Lafferty/Roaring Mountain Editorial Services
Design Coordinator: Jill E. Bonar
Cover Designer: Russ Maselli
Production Manager: Pat Tonneman
Marketing Manager: Danny Hoyt

This book was set in Times Ten by Bi-Comp, Inc.,
and was printed and bound by R. R. Donnelley & Sons Company.
The cover was printed by Phoenix Color Corp.

 © 1997, 1993, 1986 by Prentice-Hall, Inc.
Simon & Schuster/A Viacom Company
Upper Saddle River, New Jersey 07458

Printed in the United States of America

10  9  8  7  6  5  4  3  2  1

ISBN: 0-13-381583-8

Prentice-Hall International (UK) Limited, *London*
Prentice-Hall of Australia Pty. Limited, *Sydney*
Prentice-Hall of Canada, Inc., *Toronto*
Prentice-Hall Hispanoamericana, S.A., *Mexico*
Prentice-Hall of India Private Limited, *New Delhi*
Prentice-Hall of Japan, Inc., *Tokyo*
Simon & Schuster Asia Pte. Ltd., *Singapore*
Editora Prentice-Hall do Brasil, Ltda., *Rio de Janeiro*

## NOTICE TO THE READER

The information contained in this book has been prepared in accordance with recognized engineering principles and is for general information only. Although it is believed to be accurate, this information should not be used for any specific application without competent professional examination and verification of its accuracy, suitability, and applicability by a licensed professional engineer, architect, or designer. The authors and publisher of this book make no warranty of any kind, expressed or implied, with regard to the material contained in this book nor shall they be liable for any special, consequential, or exemplary damages resulting, in whole or in part, from the reader's use of or reliance upon this material.

# Contents

## CHAPTER 9          OPEN WEB STEEL JOISTS AND METAL DECK          361

## CHAPTER 10          CONTINUOUS CONSTRUCTION
## AND PLASTIC DESIGN          377

## CHAPTER 11          STRUCTURAL STEEL DETAILING: BEAMS          395

# Preface

The primary objective of the third edition of *Applied Structural Steel Design* has remained unchanged since its first edition: to furnish the reader with a basic understanding of the strength and behavior of structural steel members and their interrelationships in simple structural systems.

The emphasis of this edition remains on the analysis and design of structural steel elements in accordance with the American Institute of Steel Construction (AISC) Specification for Structural Steel Buildings—Allowable Stress Design (ASD) and the AISC Manual of Steel Construction—ASD, 9th edition.

Allowable stress design has been the dominant design method for structural steel during the twentieth century. A modern design method called *Load and Resistance Factor Design* (LRFD) was officially introduced in 1986 when AISC published the first edition of the Manual of Steel Construction—Load and Resistance Factor Design and the LRFD Specification for Structural Steel Buildings.

Both design methods are currently being used, and although most engineering professionals agree that LRFD will become the dominant method in the future, the traditional ASD method still remains popular and practical and is the most widely used method. We therefore see this edition as a transitional text that bridges the two methods. ASD is utilized throughout the first 12 chapters. In these chapters we continually reference the AISC Manual of Steel Construction—ASD, 9th edition, and its use as a ready reference and companion publication to our text is strongly recommended. The last two chapters furnish a simplified (but comprehensive) introduction to the LRFD method. Chapter 13 deals with structural members, and Chapter 14 covers basic connections.

In this third edition, our text has undergone format revisions as well as content revisions reflecting changes in the ASD Manual, 9th edition, that have been made since it was first published in 1989 (changes such as column base plate design.) The LRFD chapters reflect the latest publication of the LRFD Manual (2nd edition) and LRFD Specification (December 1, 1993). Additionally, in this edition, many sections have been rewritten, new homework problems have been added, and many existing problems have been edited.

With relevant structural steel research and literature increasing at an exponential rate, it remains the intent of this book to translate this vast amount of information and data into an integrated source that reflects the latest information available. It is not intended to be a comprehensive theoretical treatise of the subject, because we believe that such a document could easily obscure the fundamentals that we strive to emphasize in engineering technology programs. In addition, we are of the opinion that adequate comprehensive books on structural steel design do exist for those who seek the theoretical background, the research studies, and more rigorous applications.

The text content has remained primarily an elementary, noncalculus, practical approach to the design and analysis of structural steel members, using numerous example problems and a step-by-step solution format. In addition, chapters on structural steel detailing of beams and columns are included in an effort to convey to the reader a feeling for the design-detailing sequence.

The book has been thoroughly tested over the years in our engineering technology programs and should serve as a valuable design guide and source for technologists, technicians, and engineering and architectural students. Additionally, it will aid engineers and architects preparing for state licensing examinations for professional registration.

As in the past, we are indebted to our students and colleagues who, with their questions, helpful criticisms, suggestions, and enthusiastic encouragement, have provided input for this edition. We also wish to express our appreciation to our wives and families for their continued support, tolerance, and encouragement during this time of preoccupation.

*Leonard Spiegel*
*George F. Limbrunner*

# CHAPTER 1

# Introduction to Steel Structures

## 1-1

## STEEL STRUCTURES

The material steel, as we know it today, is a relatively modern human creation. Its forerunners, cast iron (which may have been invented in China as early as the fourth century B.C.) and wrought iron, were used in building and bridge construction from the mid-eighteenth century to the mid-nineteenth century. In the United States, however, the age of steel began when it was first manufactured in 1856. The first important use of steel in any major construction project was in the still-existing Eads Bridge at St. Louis, Missouri, which was begun in 1868 and completed in 1874. This was followed in 1884 by the construction of the first high-rise steel-framed building, the 10-story (later, 12-story) Home Insurance Company Building in Chicago. The rapid development of steel-framed buildings in the Chicago area at that time seems to have resulted from that city's position as the commercial center for the booming expansion of the Midwest's economy. The rapid expansion caused an increased demand for commercial building space. This demand resulted in soaring land prices that, in turn, made high-rise buildings more cost-effective.

Since those beginnings, steel has been vastly improved in both material properties and in methods and types of applications. Steel structures of note at present include the Humber Estuary suspension bridge in England, the main span of which stretches 4626 ft; a guyed radio mast in Poland with a height of 2120 ft; and the Sears Tower in Chicago, with 109 stories, which rises to 1454 ft. Each of these structures owes its notability to the strength and quality of the steel that supports it.

This is not to say that steel offers the builder an answer to all structural problems. The other major common building materials (concrete, masonry, and wood) all have their place and in many situations will offer economies that will dictate their use. But for building applications where the ratio of strength to weight (or the strength per unit weight) must be kept high, steel offers feasible options.

Steels used in construction are generally *carbon steels*, alloys of iron and carbon. The carbon content is ordinarily less than 1% by weight. The chemical composition of the steel is varied, according to the properties desired, such as strength and corrosion resistance, by the addition of other alloying elements, such as silicon, manganese, copper, nickel, chromium, and vanadium, in very small amounts. When a steel contains a significant amount of any of such alloying elements, it is referred to as an *alloy steel*. Steel is not a renewable resource, but it can be recycled, and its primary component, iron, is plentiful.

Among the advantages of steel are uniformity of material and predictability of properties. Dimensional stability, ease of fabrication, and speed of erection are also beneficial characteristics of this building material. One may also list some disadvantages, such as susceptibility to corrosion (in most but not all steels) and loss of strength at elevated temperatures. Steel is not combustible, but it should be fireproofed to have any appreciable fire rating.

Some of the common types of steel structures are shown in Figure 1-1.

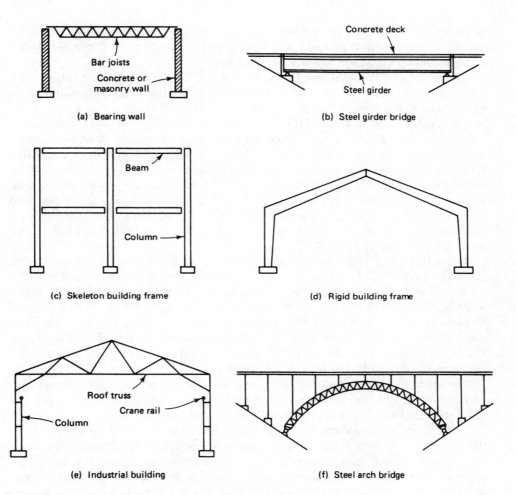

**FIGURE 1-1**  Types of steel structures.

## HANDBOOKS AND SPECIFICATIONS

Structural steel is a manufactured product and is available in various grades, sizes, and shapes. The use of standard handbooks is absolutely essential to anyone working in any phase of steel construction.

The American Institute of Steel Construction (AISC) is a nonprofit trade association that represents and serves the steel construction industry. Its objective is to improve, advance, and promote the use of structural steel. As a part of its many activities, AISC publishes and promotes an extensive collection of manuals, guides, supplements, specifications and codes, handbooks, proceedings, and special publications, including electronic publications and software.

Primary among the AISC publications are the *ASD Manual of Steel Construction, 9th edition* [1] (which we refer to as the *ASDM*); and the *LRFD Manual of Steel Construction, 2nd edition* (which we refer to as the *LRFDM*).

The ASDM contains the *AISC Specification for Structural Steel Buildings— Allowable Stress Design,* a set of guidelines covering various facets of steel design and construction based on the historic and traditional method of analysis and design for structural steel. The content of the first 12 chapters of this text is based on this allowable stress design (ASD) method. The ASDM should be a companion book, for it will be referenced repeatedly. We refer to the *AISC ASD Specification* as the *ASDS.* AISC publishes a supplement to the ASDM, *Volume II—Connections,* which is not referenced in this text because it covers only those applications not specifically treated in the ASDM.

The LRFDM is composed of two volumes. Volume II deals primarily with connections. Volume I contains the *AISC Load and Resistance Factor Design Specification for Structural Steel Buildings,* a modern specification first published by AISC in 1986 and one that will eventually replace the ASD Specification. The Load and Resistance Factor Design (LRFD) method is the topic of Chapters 13 and 14 of this text, and we refer to the AISC LRFD Specification as the *LRFDS.* Sufficient tables and charts from the LRFDM have been reproduced in this text for the purposes of our brief introductory treatment.

The manuals contain a wealth of information on products available, design aids and examples, erection guidelines, and other applicable specifications. They are indispensable tools for structural steel analysis and design.

The first AISC Specification for buildings was issued in 1923. Over the years, the specifications have been revised many times to reflect new product developments, new philosophies, improved analysis methods, and research results. It is generally recognized that the current specifications reflect reasonable sets of requirements and recommendations on which to base structural steel analysis and design. When a specification such as one of the current AISC Specifications is incorporated into the building code of a state or municipality (as it usually is), it becomes a legal document and is part of the law governing steel construction in a particular area.

## 1-3

## STEEL PROPERTIES

A knowledge of the various properties of steel is a requirement if one is to make intelligent choices and decisions in the selection of particular members. The more apparent mechanical properties of steel in which the designer is interested may be determined by a tension test. The test involves the tensile loading of a steel sample

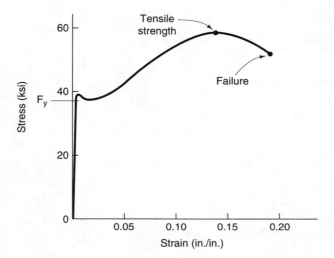

**FIGURE 1-2**   Typical stress-strain diagram for structural steel.

and the simultaneous measuring of load and elongation from which stress and strain may be calculated using

$$\text{stress*} = f_t = \frac{P}{A}$$

$$\text{strain} = \epsilon = \frac{\Delta L_0}{L_0}$$

where

$f_t$ = computed tensile stress (ksi)

$P$ = applied tensile load (kips)

$A$ = cross-sectional area of the tensile specimen (in.$^2$); this value is assumed constant throughout the test; decrease in cross-sectional area is neglected

$\epsilon$ = unit strain, elongation (in./in.)

$\Delta L_0$ = elongation or the change in length between two reference points on the tensile specimen (in.)

$L_0$ = original length between two reference points (may be two punch marks) placed longitudinally on the tensile specimen before loading (in.)

The sample is loaded to failure. The results of the test are displayed on a *stress-strain diagram.* Figure 1-2 is a typical diagram for commonly available structural steels. Upon loading, the tensile sample initially exhibits a linear relationship between stress and strain. The point at which the stress-strain relationship becomes

---

* Stress and strain, as referred to in this text, would be more precisely defined as *unit stress* and *unit strain*.

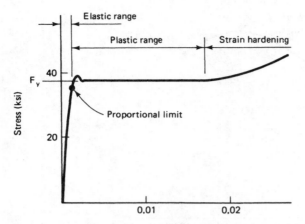

**FIGURE 1-3**  Partial stress-strain diagram for structural steel.

nonlinear is called the *proportional limit*. This is noted in Figure 1-3, which is the left part of Figure 1-2 shown to a large strain scale. The steel remains elastic (meaning that upon unloading it will return to its original length) as long as it is not stressed past a value slightly higher than the proportional limit called the *elastic limit*. The proportional limit and the elastic limit are so close that they are often considered to be the same value. Continuing the loading, a point will be reached at which the strain in the specimen increases rapidly at constant stress. The stress at which this occurs is called the *yield stress, $F_y$*, noted in Figures 1-2 and 1-3. This striking characteristic is typical of the commonly used structural steels. Also note in Figure 1-3 that $F_y$ is the stress value of the horizontal plateau region of the curve. The slightly higher stress that exists just after the proportional limit (sometimes called the *upper yield*) exists only instantaneously and is unstable. That part of the curve from the origin up to the proportional limit is called the *elastic range*. At present, most structural steel is designed so that actual stresses in the structural members do not exceed *allowable stresses,* well below $F_y$. This method of design is called the *allowable stress design* (ASD) method. In this design method, it is only the extreme left portion of the stress-strain diagram that is of immediate importance to the designer. There is still a vast range of stress and strain through which the steel will pass before it ultimately fails by tensile rupture, however.

In Figure 1-3, once the steel has been stressed past the proportional limit, it passes into the *plastic range* and strains under essentially constant stress ($F_y$). As the steel continues to strain, it reaches a point at which its load-carrying capacity increases. This phenomenon of increasing strength is termed *strain hardening*. The maximum stress to which the test specimen is subjected is called the *tensile strength $F_u$* (with units taken as ksi), noted in Figure 1-2. Although allowable stress design is still the most widely accepted design method, another design method, called *plastic design,* allows small but definite areas of members to be stressed to $F_y$ and strained into the plastic range. Plastic design is discussed further in Chapter 10.

For all practical purposes, in structural steel design, it is only the elastic range and the plastic range that need be of interest since the strains in the strain-hardening range are of such magnitude that the *deformation* of the structure would be unacceptable. Figure 1-4 shows an idealized diagram for structural steel that is sufficient for purposes of illustrating the steel stress-strain relationships. In Figure 1-4, the strain at the upper limit of the plastic range, $\epsilon_p$, is approximately 10 to 15 times the strain at the yield point, $\epsilon_y$.

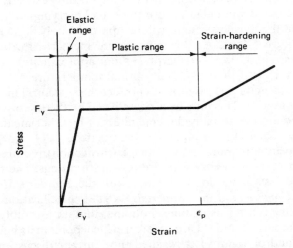

**FIGURE 1-4**   Idealized stress-strain diagram for structural steel.

Two other properties become apparent from the stress-strain diagram. The first is *modulus of elasticity E* (or Young's modulus), which is the proportionality constant between stress and strain *in the elastic range*. It is also the *slope* of the stress-strain curve in the elastic range:

$$E = \frac{\text{stress}}{\text{strain}} = \frac{f_t}{\epsilon}$$

Young's modulus $E$ is reasonably constant for structural steel. AISC recommends the value of $E$ to be 29,000 kips/in.[2]. The second property of note is that of *ductility,* the ability to undergo large deformations before failure. Ductility is frequently the reason that a steel frame will remain standing after portions of its members have been stressed far beyond the design allowable stresses. The deformation of parts of the structure will serve to transfer loads to other less heavily loaded parts and thus will keep the structure from collapsing even though all or parts of it may have deformed to the point where the structure is no longer usable. This "forgiving nature," the ductility of steel, is important in the safeguarding of lives and property in unknown and uncertain loading situations such as earthquakes.

Thus far in our discussion of steel properties, the implicit assumption has been that the steel is at a "normal" temperature. This is the usual case for steel used in buildings and bridges. Under normal conditions, the steel is within a fairly narrow range of temperature (commonly taken as $-30°F$ to $+120°F$), and the yield strength, tensile strength, and modulus of elasticity remain virtually constant. Steel is, nonetheless, a material that is significantly temperature-dependent. When structural steel members are subjected to the elevated temperatures of a fire, a substantial decrease in the steel strength and rigidity will develop over time.

Tests on those steels permitted by the ASDS indicate that the tensile strength of the steel does not significantly decrease until the temperature exceeds about 700°F. At about 800°F, the steels will begin to lose their load-carrying ability, and both the yield strength and the tensile strength will be approximately 80% of their values at normal temperature (70°F). As the temperature increases further, the yield strength and tensile strength drop off sharply, and at approximately 1600°F, both will be approximately 15% of their values at normal temperature.

To prevent such loss of strength, it is essential to protect the structural members against excessive temperature rise. This may be accomplished by the use of various insulating materials applied directly to individual members or by other means such as the use of a membrane ceiling in floor and roof construction.

Most building codes specify minimum fire resistance (protection) requirements and recognize performance tests as the basis for fire resistance ratings. The ratings are based on standard fire tests made in accordance with the *Standard Methods of Fire Tests of Building Construction and Materials,* ASTM Specification E119. Minimum ratings are specified in terms of time, in hours, that an assembly must be able to withstand exposure to a standard fire before a critical point in its behavior (such as loss of strength and/or rigidity) is reached. For further discussion, see "Effect of Heat on Structural Steel" in the ASDM.

## 1-4

## PRODUCTS AVAILABLE

In this text we consider only structural members made of hot-rolled steel. This process involves forming the steel to a desired shape by passing it between rolls while it is in the red-hot condition. Steel, as a material, is available in many different *types.* The primary property of interest is strength, in terms of $F_y$, since most allowable stresses are based on $F_y$. Also of interest may be the tensile strength, resistance to corrosion, and suitability for welding. Steels are usually specified according to ASTM (American Society for Testing and Materials) number. The ASDM, Part 1, Table 1,* contains a list of ASTM structural steels. Note that minimum $F_y$ ranges from 32 to 100 ksi and tensile strength $F_u$ ranges from 58 to 130 ksi.† Those interested in more detail concerning these steels are referred to

---

* Note that all references to the ASDM in this text are to the *9th edition.*

† The determination of $F_y$ is discussed further in Section 2-2 of this text.

the applicable ASTM literature [2]. Among these steels, the most widely used is the structural carbon steel designated A36. Its yield stress $F_y$ is 36 ksi except for cross sections in excess of 8 in. thick.

Grades A588 and A242 are weathering steels. These steels are popular for use in bridges and exposed building frames. They are generally used with no protective coating in all but the most corrosive environments and are obtainable with an atmospheric corrosion resistance anywhere from four to six times that of carbon steel.

Weathering steel develops a closely grained and tightly adherent oxide coating that acts as a barrier to moisture and oxygen and effectively prevents further corrosion of the steel. The development of this coating, however, occurs over a period of time. During this time, runoff water from the steel must be controlled and channeled and must be kept from contacting any surface on which a rust-colored stain would be undesirable.

Uncoated weathering steels are not recommended for exposure to concentrated industrial fumes, for exposure in marine locations where salt can be deposited on the steel, or where the steel is either buried in soil or submerged in water.

Currently, AISC reports that work is under way on the development of a new 50-ksi yield strength steel specification that will replace ASTM A36 as the industry

**PHOTO 1-1**    Rolling of a 36-in.-deep wide-flange shape. (Courtesy of Nucor-Yamato Steel Company.)

base standard. This new steel will be designed for improved performance through better-defined strength and material limits. The shift to the 50-ksi base material as the preferred material is intended to simplify and improve design practice.

Various standardized structural products are available in the different types of steels. The ASDM, Part 1, Tables 1 and 2, groups these products into *shapes* and *plates and bars.* Plates and bars are simply steel products of various widths and thicknesses. See the ASDM, Bars and Plates, for detailed information. Also note in Table 1 how, for any given type of steel, the properties of yield stress and tensile strength decrease as plate thickness increases. This is generally due to the larger number of passes through the rolls required for the thinner plates. Structural shapes, also called *sections,* are products that have had their cross sections tailored to one or more specific needs. The most common is the *wide-flange* shape, designated as a *W shape.*

The development of the wide-flange shape began around 1830 when wrought-iron rails were being rolled in England. By 1849, French engineers had somewhat improved the cross section for use in bending members with the development of the rolled wrought-iron I section (see Figure 1-5). The object, of course, was to provide an increased resistance to bending (greater moment of inertia) per unit weight of the cross section.

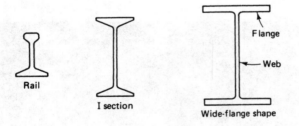

**FIGURE 1-5**    Development of the wide-flange shape.

The wide-flange shape, very efficient for bending applications, further improved on its forerunner by widening the flanges and thinning the web, thus obtaining an even better moment of inertia–to–weight ratio. The wide-flange beam is the invention of Henry Grey, a native of England who emigrated to the United States in 1870. He perfected the production method in 1897 but could find no company in the United States interested in manufacturing his new beam. His invention met with better acceptance in Europe, however. In 1902, in a German-owned steel mill in Differdange, in the Duchy of Luxembourg, the production of steel wide-flange shapes began. Shortly thereafter, Grey's beam captured the interest of Charles Schwab, president of Bethlehem Steel Company, and in 1908 the first wide-flange beam manufactured in the United States was rolled at Bethlehem, Pennsylvania [3].

Figure 1-6 summarizes the structural shapes that are included in the ASDM, Part 1.

*W shapes* are used primarily as *beam and column members* (see Figure 1-1). An example of the standard designation for a W shape is W36 × 300. This indicates

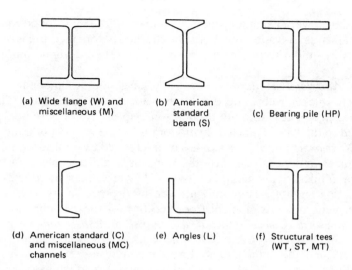

(a) Wide flange (W) and
miscellaneous (M)

(b) American
standard
beam (S)

(c) Bearing pile (HP)

(d) American standard (C)
and miscellaneous (MC)
channels

(e) Angles (L)

(f) Structural tees
(WT, ST, MT)

**FIGURE 1-6**    Hot-rolled steel shapes.

a W shape that is nominally (approximately) 36 in. deep from outside of flange to outside of flange and weighs 300 lb/ft. Note the range of W36 shapes in the ASDM, Part 1, Dimensions and Properties. Shapes in the set that runs from W36 × 135 to W36 × 256 are produced by the same set of rolls at the rolling mill. The weight difference is created by varying web thickness and flange width *and* thickness, as shown in Figure 1-7. The W40 is the deepest shape rolled in the United States. The ASDM, Part 1, indicates (by shading) the generally deeper and heavier shapes that are available only from nondomestic producers. This situation is changing, however, as new domestic steel-making facilities begin production. The deeper shapes, which will offer economies in some applications, are discussed further in Chapter 5.

   *M shapes* are *miscellaneous shapes* and have cross sections that appear to be exactly like W shapes, but cannot be strictly classified as W shapes, however. These

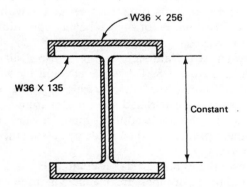

**FIGURE 1-7**    Cross-sectional variations.

shapes are rolled by fewer or smaller producers and therefore may not be commonly available in certain areas. Availability should be checked before their use is specified. Their application is similar to that of the W shape, as is their designation (e.g., M8 × 6.5).

*S shapes* are *American standard beams.* They have sloping inner faces on the flanges, relatively thicker webs, and depths that are mostly full inches. They are infrequently used in construction, but do find some application where heavy point loads are applied to the flanges, such as in monorails for the support of hung cranes.

*HP shapes* are *bearing pile shapes* and are characterized by a rather square cross section with flanges and webs of nearly the same or equal thickness (so that the web will withstand pile-driving hammer blows). They are generally used as driven piles for foundation support. They may also be used occasionally as beams and columns but are generally less efficient (more costly) for these applications.

*C and MC shapes* are *American Standard Channels* and *miscellaneous channels.* Examples of their designations are C15 × 50 and MC18 × 58, where the first number indicates the depth in inches and the second indicates the weight in pounds per foot. As with M shapes, MC shapes are generally not readily available and are produced by only a limited number of mills. The channel shapes are characterized by short flanges that have sloping (approximately $16\frac{2}{3}$% or 1:6) inner surfaces and depths to full inches. Their applications are usually as components of built-up cross sections, bracing and tie members, and members that frame openings.

*Angles* are designated by the letter L, leg length, and thickness. They may be equal leg or unequal leg angles. For unequal leg angles, the longer leg is stated first. For example, L9 × 4 × $\frac{1}{2}$ indicates an angle with one leg 9 in. long and one leg 4 in. long, both having a thickness of $\frac{1}{2}$ in. Note that the designation *does not* provide the unit weight of the angle as has been the case with all shapes discussed thus far. The weight in pounds per foot is tabulated. Angles are commonly used singly or in pairs as bracing members and tension members. They are also used as brackets and connecting members between beams and their supports. Light trusses and open web steel joists may also utilize angles for component parts.

*Structural tees* are shapes that are produced by splitting the webs of W, M, or S shapes. The tees are then designated WT, MT, or ST, respectively. For example, a WT18 × 105 (nominally 18 in. deep, 105 lb/ft) is obtained from a W36 × 210. This can be verified by checking the dimensional properties of the tee with those of the wide-flange shape. Tees are used primarily for special beam applications and as components in connections and trusses.

In addition to the foregoing *shapes,* structural steel *tubing* is also available to the designer and may be observed in Figure 1-8. Tubes are covered by different ASTM material specifications as reflected in the ASDM, Part 1, Table 3. One production method involves the forming of flat-rolled steel into a round profile. The edges are then welded together by the application of high pressure and an electric current. Square and rectangular profiles are then made by passing the round profile through another series of forming rolls.

The ASDM does provide dimensions and properties for commonly used round, square, and rectangular tubes. Round tubes, or pipes, are available in three weights, with wall thickness being the influencing factor. With reference to the ASDM, Part

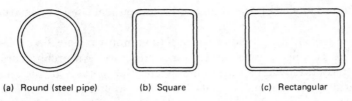

(a) Round (steel pipe)          (b) Square                (c) Rectangular

**FIGURE 1-8**  Structural steel tubing.

1, Pipe Dimensions, a standard-weight pipe section of 4 in. *nominal* outside diameter would be designated Pipe 4 Std. The same section in *extra strong* and *double extra strong* weights would be designated Pipe 4 X-Strong and Pipe 4 XX-Strong, respectively. Note the variation in wall thicknesses. Examples of structural steel tubing designations are TS5 $\times$ 5 $\times$ $\frac{1}{4}$ for a 5-in. square tube with $\frac{1}{4}$-in. wall thickness and TS12 $\times$ 8 $\times$ $\frac{1}{2}$ for a 12 in. $\times$ 8 in. rectangular tube with $\frac{1}{2}$-in. wall thickness. Tubes make excellent compression members, although the connections usually involve some welding. The most common use of tubes is in compression and tension members, but they are also used as beams in some situations. Structural members made of tubes are easier to clean and maintain than are their wide-flange counterparts.

All the foregoing product shapes and sections have advantages and disadvantages in particular structural applications. To make a reasonable decision concerning which to use, one should be familiar with the various properties of the available products. Availability will also play a role, and an alternative or substitute product may frequently be the economical choice. How the structure will be fabricated and erected must also be considered so that every possible economy be realized. Minimum weight is not always the least expensive.

The reader should be aware that many producers of the various structural steel shapes and tubes (such as Bethlehem Steel) publish design aids and technical information concerning their products. This is in addition to the information in the ASDM, which does not contain information on *all* available products. Availability of structural steel shapes should be checked with the producers or with local fabricators.

**1-5**

## THE BUILDING PROJECT

Building a structure involves many steps. The actual sequence of events, and the presence or absence of certain events, depends heavily on the size, scope, type of construction, and chosen method of managing the project. At the risk of oversimplifying the procedure, assuming that a low-rise steel frame building is to be constructed, a typical progression of events (primarily with respect to the structural aspect) may be broadly categorized into a design sequence and a construction sequence.

In the *design sequence,* a person or group (commonly called the *owner*) determines that there is a need, or a desire, for a structure. An architectural firm is contacted to study the owner's requirements and the features of the site, and to investigate various systems and layouts that offer solutions. Design calculations may be very rough and sketchy at this point, but the main building dimensions and the locations of the principal members are determined. Together with a structural engineer, the architect recommends a design from the various possible solutions investigated. This phase may be called the *planning* or *preliminary design phase.* Upon the owner's approval, as well as a municipality approval, the final design begins. As the architectural, mechanical, and electrical designs are progressing, the structural design portion of the project proceeds through the stages of firming up the layout and selecting or designing all the structural members. Contract drawings and specifications are also prepared.

In the *construction sequence,* the project is advertised and contractors are invited to submit bids. Based on bids submitted, a contractor is selected and a contract is signed. The contractor then engages the services of a *steel fabricator* (a subcontractor). The steel fabricator utilizes the contract drawings and specifications to prepare *shop drawings* of all the steel members, showing precise dimensions and details and including all the details of the connections. Economy is gained in this way since the fabricator is well versed in structural practice and knows the strengths (and weaknesses) of his or her own shop. The shop drawings are submitted to the architect/engineer for approval before actual *fabrication* is begun. The fabrication shop work includes the actual cutting to length of the members, punching, drilling, shaping, grinding, welding, painting, and sometimes preassembling parts of the structure. Following fabrication, the steel is transported to the site and *erection* of the steel commences.

In summary, the foregoing is only one way in which various events in a building project may take place. Any particular project may be completely different. One example is the "fast-tracking" method, whereby construction is begun based on a sketchy and incomplete design. Design and construction then progress together, generally under the guidance of a professional construction manager. Under this arrangement, the construction manager acts as the owner's agent to direct both the design and the construction of a building. The economic benefits of early completion and occupancy outweigh the increased cost of construction brought about by the various unknown factors created by the incomplete design. Traditionally, however, steel construction projects proceed generally as described.

## 1-6

## DESIGN CONSIDERATIONS

Above all other considerations, the structural designer must be concerned with the safety and well-being of the general public, who will become the users of the structure. Economy, beauty, functionality, maintainability, permanence, and the like are all secondary considerations when compared with the safety and well-being of the users. The competence of the designer is of utmost importance. Not only must codes be followed, but their requirements (and the recommendations of the various specifications) must be tempered and applied with sound judgment. The blind following of a code will not relieve the designer of ultimate responsibility. Codes and specifications, because of their bureaucratic nature, normally lag the state of the art. It therefore behooves designers and practitioners to keep abreast of current developments and happenings through all the trade, technical, and professional channels available.

In the consideration of safety, a decision must be made as to just how safe a structure should be. The expression of safety is normally made in terms of a *factor of safety*. The factor of safety may be defined in many ways, but in a broad sense it is the ratio of (1) the load (or stress) that causes failure to (2) the maximum load (or stress) actually allowed in the structure. In *allowable stress design* the attainment of yield stress in a member is considered to be analogous to failure. Although the steel will not actually fail (rupture) at yield, significant and unacceptable deformations are on the verge of occurring, which may render the structure unusable. The maximum stress (allowable stress) to be used in the proportioning of the member is specified. This stress will exist, as a limit, in the member. The factor of safety is

**PHOTO 1-2**    Rolled beams of nominal 40-in. depth and 102-ft span being placed for a bridge span. Note shear studs on the top flange for composite action between concrete deck and steel girders. Ditgesbaach Valley, Luxembourg. (Coutesy of Trade Arbed, Inc., New York.)

then a factor of safety *against yielding.* As an example, assume that a member composed of a steel having yield stress $F_y$ has a specified allowable stress of $0.66F_y$. The factor of safety (F.S.) against yielding would then be

$$\text{F.S.} = \frac{\text{``failure stress''}}{\text{maximum stress}} = \frac{\text{yield stress}}{\text{maximum stress}} = \frac{F_y}{0.66F_y} = 1.5$$

Another way of considering this case is to think of the member as having a 50% *reserve of strength* against yielding in this particular application.

The factors of safety recommended by the various specifications and codes depend on many things. Danger to life and property as a result of the collapse of a particular type of structure, confidence in the analysis methods, confidence in the prediction of loads, variation in material properties, and possible deterioration during the design life of the structure are all possible considerations. Recommended factors of safety are the result of cumulative pooled experience and history and are the minimum values that have been traditionally accepted as good practice.

The designer must also consider *loadings.* All the forces produced by loads that act on a structure must be transmitted through the structure to the underlying foundation. The designer must determine, based on the code requirements, judgment, and experience, just which loads are applicable and what their magnitudes will be. Codes and specifications again give guidelines in terms of minimum loads to be used based on general occupancy categories. The designer must decide if the minimum loads are satisfactory and, if not, make a better estimate.

Loads can be broadly grouped into dead loads and live loads. *Dead loads* are static loads that produce vertical forces due to gravity and include the weight of the steel framework and all materials permanently attached to it and supported by it. Reasonable estimates of the weight of the structure can usually be made based on the preliminary design work. *Live loads* include all vertical loads that may be either present on or absent from the structure. Generally, lateral loads are considered live loads whether they are permanent or not. Examples of live loads are the following:

Snow
People and furniture
Stored materials
Vehicles (on bridges and in warehouses)
Cranes
Wind
Lateral pressure due to earth or stored liquids
Earthquakes

State, municipal, or other applicable building codes normally provide minimum loads for a designer's use in a particular area. In the absence of such requirements, the reader is referred to Reference 4. A detailed treatment of load calculations and theory is beyond the scope of this book. For an excellent discussion on theory and determination of loads and forces for structural design, the reader is referred to Reference 5.

## 1-7
## NOTATION AND CALCULATIONS

The U.S. Customary System of weights and measures is the primary unit system being used in this text. Eventually, the metric system (*Système international d' unités,* or SI) will supersede the U.S. Customary System. As a means of introducing the SI metric system, Appendix B in this text furnishes various tables for familiarization purposes. SI base units, derived units, and prefixes are all tabulated, along with a table of conversion factors applicable to the structural design field.

For the many calculations that will be part of this text, nomenclature and symbols of the ASDM are used. The reader is referred to the table of symbols furnished immediately prior to the Index in the ASDM. Also, additional nomenclature applicable to beam diagrams and formulas are found in Part 2. In some cases, the AISC nomenclature and symbols are supplemented.

With regard to numerical accuracy of calculations, one should keep in mind the accuracy of the starting data in each calculation. Numbers resulting from the calculation need not be more accurate than the original data. The following guidelines, with respect to numerical accuracy in structural steel design calculations, are suggested; they are partially adopted from other sources [6]:

1.  Loads to the nearest 1 psf; 10 lb/ft; 100-lb concentration
2.  Span lengths and dimensions to 0.1 ft (for design)
3.  Total loads and reactions to 0.1 kip or three-figure accuracy
4.  Moments to the nearest 0.1 ft-kip or three-figure accuracy

The practice of presenting numerical data with many more digits than warranted by the intrinsic uncertainty of the data (referred to by some as "superdigitation") is common. To help guard against superdigitation, several rules of thumb are advocated. One is based on the premise that engineering data are rarely known to an accuracy of greater than 0.2%. This would be equivalent to a possible error of 100 lb in a load of 50 kips (1 kip = 1000 lb). Therefore, loads in the range of 50 kips should be represented no more precisely than 0.1 kip.

Traditionally, three-significant-digit accuracy has been sufficient and acceptable for structural engineering calculations, although four digits may be used for numbers beginning with 1. (This tradition is rooted in the not-so-ancient widespread use of the 10-in. slide rule.) More accuracy is generally not warranted for structural design calculations. Therefore, the following representations would be common:

|        |         |
|-------:|---------|
|  4.78  | 1.742   |
|  32.1  | 0.00932 |
|  728   | 0.1781  |
| 88,300 |         |

We have attempted to follow this tradition in the problem solutions in this text by rounding off *intermediate* and *final* numerical solutions as described. For the text presentation, we have used the intermediate solution(s), *as shown,* in the subsequent calculations. When working on a calculator, however, one would normally maintain *all* digits and round *only* the final answer. For this reason, the reader may frequently obtain numerical results that differ slightly from those given in the text. This should not cause undue concern.

## REFERENCES

[1]  *Manual of Steel Construction,* 9th ed., American Institute of Steel Construction, Inc., 1 East Wacker Drive, Suite 3100, Chicago, IL 60601-2001.

[2]  American Society for Testing and Materials, 1916 Race Street, Philadelphia, PA 19103.

[3]  Robert Hessen, *Steel Titan, The Life of Charles M. Schwab* (New York: Oxford University Press, 1975), pp. 172–173.

[4]  *Minimum Design Loads for Buildings and Other Structures,* ASCE 7-93 (Formerly ANSI 58.1), 1993, American Society of Civil Engineers, 345 East 47th Street, New York, NY 10017.

[5]  William McGuire, *Steel Structures* (Englewood Cliffs, NJ: Prentice-Hall, Inc., 1968), Chapter 3.

[6]  *CRSI Handbook,* Concrete Reinforcing Steel Institute, 933 North Plum Grove Road, Schaumburg, IL 60173-4758.

## PROBLEMS

**1-1.**   List the various types of hot-rolled shapes for which properties are given in the ASDM. Draw a freehand sketch of each shape.

**1-2.**   Define the following symbls: $A$, $A_g$, $I_x$, $L$, $T$, $r_T$, ksi.

**1-3.**   List the titles of the seven parts of the ASDM.

**1-4.** Determine the unit weight (lb/ft) for the following steel shapes:
  **(a)** W21 × 62.
  **(b)** MC10 × 25.
  **(c)** WT12 × 88.
  **(d)** L6 × 6 × 1.
  **(e)** TS16 × 12 × $\frac{1}{2}$.

**1-5.** Sketch and completely dimension (use design dimensions) the cross sections of the following:
  **(a)** W36 × 210.
  **(b)** W14 × 500.
  **(c)** HP13 × 60.
  **(d)** MC13 × 50.
  **(e)** L6 × 4 × $\frac{1}{2}$.
  **(f)** WT12 × 52.
  **(g)** TS7 × 5 × $\frac{1}{2}$.

**1-6.** Calculate the weight (in pounds) of a 1-ft length of a steel built-up member having a cross section as shown.

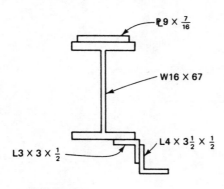

**PROBLEM 1-6**

**1-7.** How many of each of the following shapes are included in the ASDM?
  **(a)** W33.
  **(b)** W14.
  **(c)** W8.
  **(d)** Equal leg angles.

**1-8.** Sketch approximate stress-strain diagrams of $\frac{1}{2}$-in.-thick plate and 5-in.-thick plate, both of A441 steel. Superimpose the two diagrams. Label the numerical values for the properties of minimum yield stress and tensile strength for each plate thickness (refer to ASDM, Part 1, Table 1). Show how the modulus of elasticity can be determined.

**1-9.** Using the conversion factors from Appendix B-3, convert all the dimensions for a W30 × 148 from tabulated values (U.S. Customary System) to SI values. Sketch the results.

**1-10.** Using the conversion factors from Appendix B-3, convert all the properties for a W30 × 148 from the tabulated values (U.S. Customary System) to SI values.

# CHAPTER 2

# Tension Members

## 2-1

### INTRODUCTION

The proportioning of tension members is among the simpler of the problems that face the structural engineer. Although easy to proportion, however, tension members, and structures in which the main load-carrying members are in tension, require great care in the design and *detailing* of their connections. Some catastrophic structural failures have been directly attributed to poor tension member connection details. Tension members do not have the inherent stability problems of beams and columns. A tensile load applied along the longitudinal axis of the member tends to hold the member in alignment, thereby making instability a minor concern.

Tension members therefore do not generally require the *bracing* usually associated with beams and columns. The resulting tension member structures are less *redundant,* and the potential for sudden failure exists if there is any inadequacy present, such as a weakness in a connection.

Of most concern in the *selection* of tension members is the choice of the configuration of the cross section so that the connections will be simple and efficient. Also, the connection should transmit the load to the member with as little eccentricity as possible.

Examples of tension members may be found in many structures. They include hangers for catwalks and storage bins, truss web and chord members, cables for direct support of roofs, sag rods, tie rods, and various types of braces. Most of the common hot-rolled structural steel shapes may be used as tension members. Small bracing members may be circular threaded steel rods or flexible members such as wire ropes or cables. Single angles, double angles, tees, and channels may also be used for bracing purposes, as well as light truss tension members. Large truss members may consist of W shapes as well as an almost infinite variety of built-up members. Suspension bridges at one time were supported on chains composed of long plates of rectangular cross section having enlarged ends through which large-

**PHOTO 2-1**   This frame is braced in both the vertical and horizontal planes with structural steel angles. The angles are bolted to gusset plates which are welded to the frame. (Courtesy of the American Hot Dip Galvanizers Association.)

diameter steel pins were inserted as connectors. These are the *pin-connected members* referred to in the ASDS, Section D1. These "eye-bar chains" have been superseded by more efficient multiwire steel cables that are either spun in place or prefabricated.

## 2-2

### TENSION MEMBER ANALYSIS

The direct stress formula is the basis for tension member analysis (and design). It may be written for *stress,*

$$f_t = \frac{P}{A}$$

or for *tensile capacity,*

$$P_t = F_t A$$

where

$f_t$ = computed tensile stress

$P$ = applied axial load

$P_t$ = axial tensile load capacity (or maximum allowable axial tensile load)

$F_t$ = allowable axial tensile stress

$A$ = cross-sectional area of axially loaded tension member (either gross area $A_g$, net area $A_n$, or *effective* net area $A_e$)

Gross area $A_g$ is the original, unaltered cross-sectional area of the member. $A_n$, net area, is illustrated in Figure 2-1 and is logically the cross-sectional area actually available to be stressed in tension. Net area may be visualized by imagining that

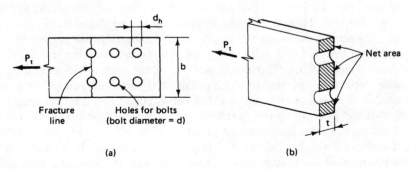

**FIGURE 2-1**   Net area.

the tension member (a plate in this case) fractures along the line shown in Figure 2-1a. The net area, shown crosshatched in Figure 2-1b, is then calculated:

$$A_n = A_g - \text{(area of holes)}$$

$$= bt - 2(d_h t)$$

where $b$ and $t$ are plate width and thickness and $d_h$ is hole diameter for analysis purposes.

The ASDS, Section B2, directs that for net area computations, the hole diameter $d_h$ be taken as $\frac{1}{16}$ in. greater than the actual nominal diameter of the hole. Hole diameters are normally punched (or drilled) $\frac{1}{16}$ in. larger than the diameter of the fastener. Therefore, for purposes of analysis and design, hole diameters are taken as the fastener diameter plus $\frac{1}{8}$ in. The effective net area $A_e$ is a function of the end connection of the tension member and is discussed in Section 2-3 of this chapter.

The direct stress formula applies directly to homogeneous axially loaded tension members. Its use is based on the assumption that the tensile stress is uniformly distributed over the net section of the tension member, despite the fact that high stress concentrations are known to exist (at working loads) around the holes in a tension member. The commonly used structural steels are sufficiently ductile so that they undergo yielding and stress redistribution. This will result in a uniform stress distribution at ultimate load.

The allowable tensile stress $F_t$ takes into consideration two types of failure. First, the member may rupture on the least net area as shown in Figure 2-1. This is the classical and historical approach to tension member analysis. For this type of failure, the ASDS, Section D1, states that $F_t = 0.50F_u$ on the net area, where $F_u$ is the specified minimum tensile strength of the steel. Second, the tension member may undergo uncontrolled yielding of its gross area *without rupture*. Excessive elongation of a tension member is undesirable in that it normally results in deformation of the structure and can lead to failure in other parts of the structural system. For this type of failure, the ASDS, Section D1, establishes an allowable tensile stress $F_t = 0.60F_y$ on the gross cross-sectional area of the member. In determining $F_t$, the yield stress $F_y$ must be known. The best source for the value of $F_y$ is the ASDM, Part 1, Tables 1 and 2. For a known *plate* thickness, determine the $F_y$ opposite the specified steel type in Table 1. For a known *shape,* first determine the appropriate *group* from Table 2. Then determine $F_y$ opposite the specified steel type in Table 1. These allowable stresses do not apply to pin-connected members (such as eye bars or plates connected with relatively large pins, as discussed in the ASDS, Section D3), threaded steel rods, or flexible tension members such as cables and wire rope.

Another type of failure that must be considered for tension members is depicted in Figure 2-2. This is a tearing failure that can occur at end connections along the perimeter of welds or along the perimeter of a group of bolt holes. Depending on the end connection, the failure could occur in either the tension member itself or in the member to which it is attached (e.g., a gusset plate). It is characterized by a combination of shear failure along a plane through the welds or bolt holes and a simultaneous tension failure along a perpendicular plane. Due to the shape of

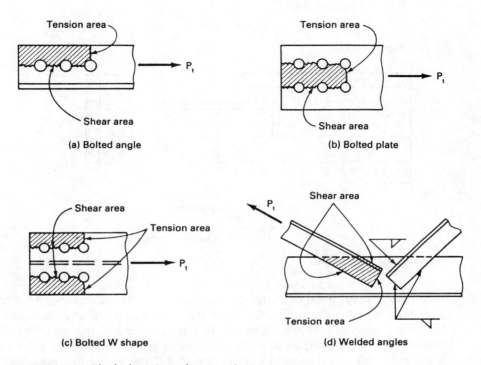

**FIGURE 2-2**   Block shear in end connections.

the element that is torn from the tension member, this failure mode is called *block shear,* and it is discussed in the ASDS, Section J4. Reference 1 contains some background.

Block shear strength is calculated from the summation of net shear area $A_v$ times the allowable shear stress $F_v$ and net tension area $A_t$ times the allowable tensile stress $F_t$, where $F_v = 0.30F_u$ and $F_t = 0.50F_u$.

Mathematically, this is stated as

$$P = A_v F_v + A_t F_t$$

For purposes of block shear strength calculations for bolted connections, hole diameters for the net area determination are taken as the fastener diameter plus $\frac{1}{8}$ in. This is the same approach as that used for tensile net area calculations.

Connections play an important role in the strength of tension members. We consider connection analysis and design in Chapters 7, 8, and 11. The *geometry* of bolted connections is instrumental in net area and block shear calculations, however. Therefore, a few definitions are in order.

Figure 2-3 shows a tension member composed of a single steel angle with a 4-bolt connection. The tensile load $P$ is assumed to be applied parallel to and coincident with the longitudinal axis of the member. The bolt holes are located on *gage*

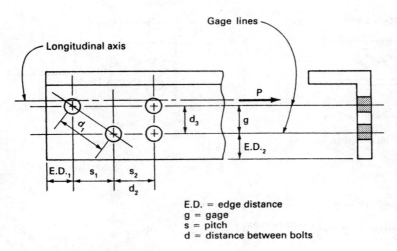

E.D. = edge distance
g = gage
s = pitch
d = distance between bolts

**FIGURE 2-3**   Definitions.

*lines* that are also parallel to the longitudinal axis. The dimension *g* between the gage lines is called the *gage*. The dimension *s* parallel to the gage line and taken between centers of bolt holes is called the *pitch* (or the *bolt spacing*). The *distance between bolts* is a straight line distance between any two bolts. The *edge distance* is the perpendicular distance from the *center of a hole* to the nearest edge.

The analysis of a tension member involves the determination of the individual allowable loads based on the various failure modes. Once these have been determined, the axial tensile load capacity of the member is taken as the smallest value.

**Example 2-1** _____

Find the axial tensile load capacity $P_t$ of the lapped, bolted tension member shown in Figure 2-4. Bolts are $\frac{3}{4}$-in. diameter, and the plate material is A36

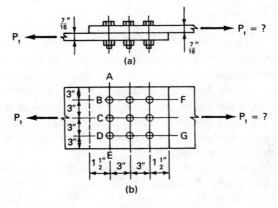

**FIGURE 2-4**   Tension member analysis.

steel ($F_u = 58 - 80$ ksi from the ASDM, Part 1, Table 1). Assume that the fasteners are adequate and do not control the tensile capacity. Pitch, gage, and edge distance are as shown.

**Solution:**

$$P_t = F_t A_n \quad \text{or} \quad F_t A_g$$

1.  Based on gross area,

$$P_t = 0.60 F_y A_g$$

$$= 0.60(36)\left(\frac{7}{16}\right)(12) = 113.4 \text{ kips}$$

2.  Based on net area, visualizing a transverse fracture along line *ABCDE* in Figure 2-4b,

$$A_n = A_g - A_{\text{holes}}$$

$$= \left(\frac{7}{16}\right)(12) - 3\left(\frac{3}{4} + \frac{1}{8}\right)\left(\frac{7}{16}\right) = 4.10 \text{ in.}^2$$

$$P_t = 0.50 F_u A_n$$

Using the lower limit of the $F_u$ range (conservative),

$$P_t = 0.50(58)(4.10) = 118.9 \text{ kips}$$

3.  Next, check the block shear strength in accordance with the ASDS, Section J4. Two possible cases are investigated. The block shear strength is written

$$P_t = A_v F_v + A_t F_t = A_v(0.30 F_u) + A_t(0.50 F_u)$$

where $A_v$ and $A_t$ are the *net* shear area and *net* tension area, respectively. Note that the hole diameter is taken as

$$\frac{3}{4} + \frac{1}{8} = 0.875 \text{ in.}$$

*Case I—failure line FBCDG (see Figure 2-4b):*

$$A_v = 2\left(\frac{7}{16}\right)[7.5 - 2.5(0.875)] = 4.65 \text{ in.}^2$$

$$A_t = \frac{7}{16}[6 - 2(0.875)] = 1.859 \text{ in.}^2$$

$$P_t = 4.65(0.30)(58) + 1.859(0.50)(58) = 134.8 \text{ kips}$$

*Case II—failure line ABCDG* (*see Figure 2-4b*):

$$A_v = \left(\frac{7}{16}\right)[7.5 - 2.5(0.875)] = 2.32 \text{ in.}^2$$

$$A_t = \left(\frac{7}{16}\right)[9 - 2.5(0.875)] = 2.98 \text{ in.}^2$$

$$P_t = 2.32(0.30)(58) + 2.98(0.50)(58)$$

$$= 126.8 \text{ kips (564 kN)}$$

Therefore, the capacity $P_t$ of this tension member is 113.4 kips as controlled by general yielding of the gross area.

In Example 2-1, the critical net area on which fracture could logically be expected was easy to visualize. In some cases the fasteners will be arranged so that the controlling fracture line will be something other than transverse, as shown in Figure 2-5. This situation can occur when fasteners are staggered to accommodate a desired size or shape of connection. Note in Figure 2-5 that there are two possible failure lines across the width of the plate. These may be defined as lines *ABCD* and *ABE*. For large values of *s*, line *ABE* will be the more critical failure line (smaller net area). For small values of *s*, line *ABCD* will be more critical. Actually, both gage and pitch will affect the problem. A combination of shear and tensile stresses acts on the sloping line *BC* of failure line *ABCD*. The interaction of these stresses presents a rather complicated theoretical problem. The ASDS uses a simplified method of analysis in this situation that is based on the studies and observations of V. H. Cochrane and T. A. Smith in the early part of the twentieth century. Reference 2 contains some background. The ASDS, Section B2, stipulates that where a fracture line contains within it a diagonal line, the net width of the part

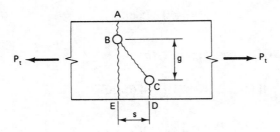

**FIGURE 2-5**  Staggered holes.

should be obtained by deducting from the gross width the diameters of all the holes along the fracture line and adding, for each diagonal line, the quantity

$$\frac{s^2}{4g}$$

where $s$ and $g$ are as previously defined. An expression for *net width* $w_n$ may be written

$$w_n = w_g - \Sigma d_h + \Sigma \frac{s^2}{4g}$$

where $w_g$ represents gross width and $d_h$ represents the hole diameter to be used for design. The foregoing formula for $w_n$ is convenient to use with members of uniform thickness. If the formula is multiplied by thickness $t$, it becomes

$$w_n t = w_g t - \Sigma d_h t + \Sigma \frac{s^2 t}{4g}$$

Or, since $w_n t = A_n$ and $w_g t = A_g$,

$$A_n = A_g - \Sigma d_h t + \Sigma \frac{s^2 t}{4g}$$

The latter formula for $A_n$ is the more useful since it provides net area directly and is also applicable with members that do not have uniform thickness (i.e., channels). In a determination of critical net area where multiple possible failure lines exist, the critical net area is, of course, the least net area.

**Example 2-2** _____

Determine the critical net width $w_n$ for the plate shown in Figure 2-6. Fasteners will be 1-in.-diameter bolts.

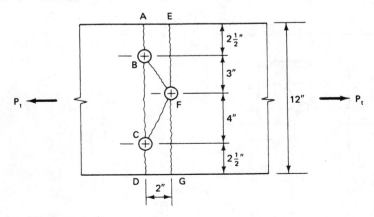

**FIGURE 2-6**  Net width calculation.

**Solution:**

Use the formula for net width:

$$w_n = w_g - \Sigma d_h + \Sigma \frac{s^2}{4g}$$

For illustrative purposes, all possible failure lines will be checked. Practically, some may be seen by inspection to be noncritical. Note how failure lines have been designated with letters. For *analysis purposes,* the diameter of the holes $d_h = 1\frac{1}{8}$ in. = 1.13 in.

| Line | Net width |
|------|-----------|
| ABCD | $12 - 2(1.13) + 0 = 9.74$ in. |
| EFG | $12 - 1(1.13) + 0 = 10.87$ in. |
| ABFG | $12 - 2(1.13) + \frac{2^2}{4(3)} = 10.07$ in. |
| EFCD | $12 - 2(1.13) + \frac{2^2}{4(4)} = 9.99$ in. |
| ABFCD | $12 - 3(1.13) + \frac{2^2}{4(3)} + \frac{2^2}{4(4)} = 9.19$ in. (controls) |

Therefore, the critical net width for this plate is 9.19 in. The critical net area would be found by multiplying the critical net width by the plate thickness.

To shorten the calculations, as a general rule, the transverse section having the greatest number of holes should be checked first. Follow this by checking every zigzag line (section) that has more holes than the initially checked transverse section. With reference to Example 2-2, it is seen that this procedure would have eliminated the checking of noncritical lines *EFG, ABFG,* and *EFCD.*

Finally, the failure line should be considered in a logical way with regard to the transmission of the load through it. In Figure 2-7 the tension member carries a load

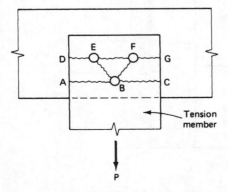

**FIGURE 2-7**   Three-bolt tension member connection.

$P$. Each of the three bolts may be assumed to transmit $P/3$ into the supporting member. Failure lines *ABC, ABFG, DEBC,* and *DEBFG* would each carry the full load $P$. Failure line *DEFG,* however, would carry only $\frac{2}{3}P$ since the bolt at $B$ transmits the other $P/3$.

A zigzag failure line sometimes occurs in a member that has more than one element making up its cross section, such as an angle. This situation is shown in Figure 2-8. The analysis can proceed exactly as before if the angle is visualized as "flattened out" as shown in Figure 2-8c. The ASDS, Section B2, provides that, in this case, the gage $g$ for holes in opposite adjacent legs shall be the sum of the gages from the back of the angle less the thickness. Therefore,

$$g = 2.50 + 1.75 - 0.25 = 4.00 \text{ in.}$$

Failure lines may then be investigated as previously shown.

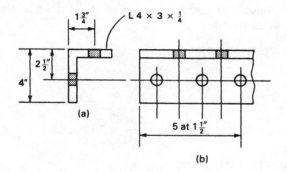

(a)

(b)

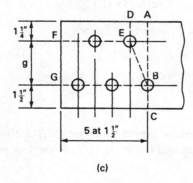

(c)

**FIGURE 2-8**  Zigzag failure lines in an angle.

**Example 2-3** _____

Calculate the tensile capacity $P_t$ of the angle shown in Figure 2-8. Assume that the member to which the angle is connected (and the bolts) do not govern capacity. Assume A36 steel and $\frac{3}{4}$-in.-diameter bolts.

**Solution:**

1.    Calculate $P_t$ based on yielding of the gross area. The gross width of the angle is determined from

$$w_g = 4 + 3 - \tfrac{1}{4} = 6.75 \text{ in.}$$

The gross area is determined from

$$A_g = 6.75(0.25) = 1.69 \text{ in.}^2$$

Therefore,

$$P_t = A_g F_t = A_g(0.60F_y)$$
$$= 1.69(0.60)(36) = 36.5 \text{ kips}$$

2.    Next, calculate $P_t$ based on tensile fracture of the net area. Two net width values are determined:

Line $ABC$ (transverse section):

$$w_n = 6.75 - 0.875 = 5.88 \text{ in.}$$

Line $DEBC$:

$$w_n = 6.75 - 2(0.875) + \frac{1.5^2}{4(4)} = 5.14 \text{ in.}$$

The zigzag line controls, therefore,

$$A_n = 5.14(0.25) = 1.285 \text{ in.}^2$$

The tensile load capacity $P_t$ based on net area is

$$P_t = A_n F_t = A_n(0.5F_u)$$
$$= 1.285(0.50)(58) = 37.3 \text{ kips}$$

3.    Last, check block shear.

*Case I:*   In the block shear failure line defined by *FEBG*, the diagonal line is considered to be tension area and the $s^2/4g$ term is used as previously. The hole diameter is taken as

$$\frac{3}{4} + \frac{1}{8} = 0.875 \text{ in.}$$

$$P_t = A_v F_v + A_t F_t = A_v(0.30F_u) + A_t(0.50F_u)$$
$$A_v = 0.25[9(1.5) - 4(0.875)] = 2.50 \text{ in.}^2$$
$$A_t = 0.25\left[4 - (0.875) + \frac{1.5^2}{4(4)}\right] = 0.816 \text{ in.}^2$$

Therefore,

$$P_t = 2.50(0.30)(58) + 0.816(0.50)(58) = 67.2 \text{ kips}$$

*Case II:*   Use the block shear failure line defined by *FEBC,* where *EB* and *BC* are considered tension areas and *FE* is a shear area.

$$A_v = 0.25[6.0 - 1.5(0.875)] = 1.172 \text{ in.}^2$$

$$A_t = 0.25[1.5 - 0.5(0.875)] + 0.25\left[4.00 - 0.875 + \frac{1.5^2}{4(4)}\right] = 1.082 \text{ in.}^2$$

from which

$$P_t = 1.172(0.30)(58) + 1.082(0.50)(58) = 51.8 \text{ kips}$$

Failure line *DEBG* could also be checked. It would be found to have a tensile capacity slightly greater than line *FEBC.* Therefore, the tensile capacity $P_t$ of the angle is 36.5 kips (162 kN), controlled by yielding of the gross area.

## 2-3

## EFFECTIVE NET AREA

For some tension members, such as rolled shapes, that do not have all elements of the cross section connected to the supporting members, the failure load is less than would be predicted by the product $A_n F_u$. The phenomenon to which this situation is generally attributed is called *shear lag* and is illustrated in Figure 2-9. Note that the angle is connected along only one leg. This leads to a concentration of stress along that leg and leaves part of the unconnected leg unstressed or stressed very little. Studies have shown that the shear lag effect diminishes as the length of the connection is increased.

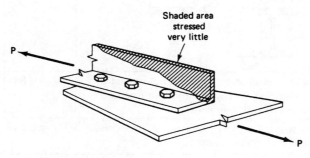

**FIGURE 2-9**   Shear lag.

The ASDS, Section B3, accounts for the effect of shear lag through the use of an effective net area, which is a function of how the tension member is connected at its ends.

When the load is transmitted directly to *each* of the cross-sectional elements by either bolts or welds, the effective net area $A_e$ is equal to the net area $A_n$. When the load is transmitted by *bolts* through some, but not all, of the cross-sectional elements of the member, the effective net area $A_e$ shall be computed from

$$A_e = UA_n \qquad \text{ASDS Eqn. (B3-1)}$$

where

$A_n$ = the net area of the member (in.$^2$)

$U$ = reduction coefficient (see Table 2-1)

When the load is transmitted by *welds* through some, but not all, of the cross-sectional elements of the member, the effective net area $A_e$ shall be computed from

$$A_e = UA_g \qquad \text{ASDS Eqn. (B3-2)}$$

where

$A_g$ = the gross area of the member (in.$^2$)

**TABLE 2-1   Values for Reduction Coefficient, $U$**

| Case I | W, M, S shapes or their tees. Connection is to the flanges. Minimum of three bolts per line in the direction of stress. | $U = 0.90$ |
|---|---|---|
| Case II | All shapes and built-up cross sections not meeting the requirements of case I. Minimum of three bolts per line in the direction of stress. | $U = 0.85$ |
| Case III | All members whose connections have only two bolts per line in the direction of stress. | $U = 0.75$ |

Note that a $U$ value from Table 2-1 is the same for a welded connection as that for a bolted connection except that Case III is not applicable and the condition as to the number of bolts in Cases I and II does not apply.

With respect to welded end connections, the ASDS, Section B3, furnishes effective net area criteria for two special considerations:

(a)  When a load is transmitted by transverse welds to some, but not all, of the cross-sectional elements of W, M, or S shapes and structural tees cut from these shapes, the effective net area $A_e$ shall be taken as that area of the *directly connected* elements.

(b)  When a load is transmitted by longitudinal welds used alone along both edges of a flat bar (or plate) axially loaded tension member, the length of each weld shall not be less than the width of the plate. The effective net area $A_e$ shall be computed by ASDS Equation (B3-2) using the reduction coefficient $U$ shown in Table 2-2,

where

$\ell$ = weld length (in.)

$w$ = plate width (distance between welds) (in.)

■■■ **TABLE 2-2  Values of $U$ (Longitudinal Welds on a Flat Bar or Plate)**

| Condition | $U$ |
|---|---|
| $\ell > 2w$ | 1.0 |
| $2w > \ell > 1.5w$ | 0.87 |
| $1.5w > \ell > w$ | 0.75 |

These reduction coefficients $U$ furnished in the ASDS are average values based on an empirical equation as shown in the ASDS Commentary, Section B3. It is the authors' preference to use the average values for both analysis and design rather than to utilize the equations to compute a $U$ value.

Additionally, for relatively short connection fittings such as splice plates, gusset plates, and beam-to-column fittings subjected to tensile force, the effective net area shall be taken as the actual net area except that it shall not be taken as greater than 85% of the gross area. Therefore, for these short plates and fittings subjected to tension, $U$ does not apply, and

$$A_e = A_n \text{ (not to exceed } 0.85A_g)$$

**Example 2-4** _____

A tension member in a truss is to be composed of a W8 $\times$ 24 and will be connected with two lines of $\frac{3}{4}$-in.-diameter bolts in each flange as shown in Figure 2-10. Assume three bolts per line, 3-in. pitch, $1\frac{1}{2}$-in. edge distance, and A36 steel. Find the tensile load capacity $P_t$. Assume that the gusset plates and the bolt capacities are satisfactory.

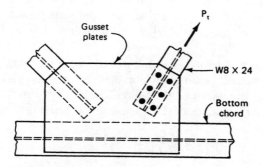

**FIGURE 2-10**   Truss connection.

**Solution:**

Properties of the W8 $\times$ 24:

$$A_g = 7.08 \text{ in.}^2$$

$$d = 7.93 \text{ in.}$$

$$b_f = 6.495 \text{ in.}$$

$$t_f = 0.40 \text{ in.}$$

1.   Based on gross area:

$$P_t = F_t A_g$$

$$P_t = 0.60 F_y A_g$$

$$= 0.60(36)(7.08) = 152.9 \text{ kips}$$

2.   Based on effective net area:

$$P_t = F_t A_e$$

$$P_t = 0.50 F_u A_e$$

$$A_e = U A_n$$

$$A_n = A_g - 4\left(\frac{7}{8}\right)(0.40) = 7.08 - 1.40 = 5.68 \text{ in.}^2$$

For $U$ evaluation, the member is covered by case I (Table 2-1) if $b_f \geq \frac{2}{3}d$:

$$\frac{2}{3}d = 0.67(7.93) = 5.31 \text{ in.}$$

$$b_f = 6.495 \text{ in.} > 5.31 \text{ in.} \qquad\qquad \textbf{O.K.}$$

Therefore, $U = 0.90$, and

$$P_t = 0.50 F_u U A_n$$

$$= 0.50(58)(0.90)(5.68) = 148 \text{ kips}$$

3.   The block shear consideration involves four "blocks," two in each flange, as shown in Figure 2-11.

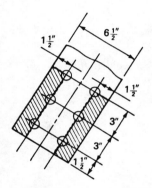

**FIGURE 2-11**   Flange block shear.

The hole diameter is taken as

$$\frac{3}{4} + \frac{1}{8} = 0.875 \text{ in.}$$

$$P_t = A_v F_v + A_t F_t = A_v(0.30 F_u) + A_t(0.50 F_u)$$

$$A_v = 4(0.40)(7.5 - 2.5(0.875)) = 8.50 \text{ in.}^2$$

$$A_t = 4(0.40)\left(1.5 - \frac{0.875}{2}\right) = 1.700 \text{ in.}^2$$

Therefore,

$$P_t = 8.50(0.30)(58) + 1.700(0.50)(58) = 197.2 \text{ kips}$$

For this member, $P_t = 148$ kips as controlled by a rupture failure based on the least net area.

**Example 2-5**

Find the tensile load capacity $P_t$ for the double-angle tension member shown in Figure 2-12. All structural steel is A36 ($F_u = 58$ ksi). Assume that the welds are adequate and do not control the tensile capacity.

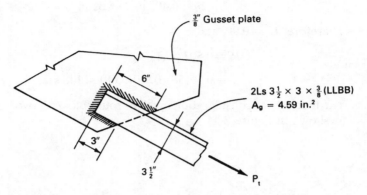

**FIGURE 2-12**   Double-angle tension member welded end connection.

**Solution:**

1. Based on gross area:

$$P_t = F_t A_g$$
$$= 0.60 F_y A_g$$
$$= 0.60(36)(4.59) = 99.1 \text{ kips}$$

2. Based on *effective* net area (since only one leg of each angle is connected to the gusset plate):

$$P_t = F_t A_e$$
$$= 0.50 F_u A_e$$

where $A_e = U A_g$ and $U = 0.85$ from Case II in Table 2-1. Therefore,

$$P_t = 0.50(58)(0.85)(4.59) = 113.1 \text{ kips}$$

3. Check block shear in the gusset plate along the perimeter of the welds.

$$P_t = A_v F_v + A_t F_t = A_v(0.30 F_u) + A_t(0.50 F_u)$$
$$A_v = 0.375(6 + 3) = 3.38 \text{ in.}^2$$
$$A_t = 3.5(0.375) = 1.313 \text{ in.}^2$$

Therefore,

$$P_t = 3.38(0.30)(58) + 1.313(0.50)(58)$$
$$= 96.9 \text{ kips}$$

For this member, $P_t = 96.9$ kips as controlled by block shear failure in the gusset plate.

## 2-4

## LENGTH EFFECTS

As mentioned earlier in this chapter, tension members do not suffer from the problems of instability and buckling that compression members and beams do. Therefore, *length* plays a minor role. The ASDS, Section B7, suggests *upper limits* for the *slenderness ratios* of tension members. Recall that the slenderness ratio is the ratio of the member's unbraced length to its least radius of gyration. Staying within the recommended limits is not essential for the structural integrity of the members. If the slenderness ratio is within the recommended limit, however, there will be some resistance to undesirable vibrations as well as some resistance to bending and deformation during shipping and erection handling. The recommended upper limit for slenderness ratio $\ell/r$ is 300 for all tension members.

The recommended upper limit on $\ell/r$ is *preferred,* but is not mandatory, and applies to tension members *other than* steel rods and cables.

### Example 2-6

In Example 2-4, assume that the W8 $\times$ 24 tension member is 20 ft long. Determine whether the member's slenderness ratio is within the ASDS recommendations.

### Solution:

Maximum preferred $\ell/r = 300$. Calculate actual $\ell/r$. Use the least radius of gyration.

$$\frac{\ell}{r} = \frac{20(12)}{1.61} = 149 < 300 \qquad \textbf{O.K.}$$

Note that the slenderness ratio is unitless. Since $r$ is tabulated in units of inches, $\ell$ must be converted to inches.

**2-5**

## DESIGN OF TENSION MEMBERS

The design, or selection, of adequate tension members involves provision of the following:

1.   Adequate gross area $(A_g)$.
2.   Adequate radius of gyration $r$ to meet the preferred $\ell/r$ limits.
3.   Adequate net area $(A_n$ or $A_e)$.
4.   Adequate block shear strength.
5.   A cross-sectional shape such that the connections can be simple.

The minimum required properties as governed by the first two foregoing items involve strength and slenderness and are easily calculated using the principles already discussed in this chapter. The third required property (*net* area) is also easily calculated. Sections, however, are tabulated on the basis of *gross* area. The relationship between gross area and net area depends in part on the thickness of the material, which is unknown at this point. The fourth item (block shear strength) depends on the material thickness, the size and type of fasteners, and the geometry of the connection. If the connection details can be established or approximated, a required thickness can be calculated. Alternatively, the block shear strength can be checked once the member has been selected and the connection designed. The fifth consideration involves the way the member will fit into and be affected by the structure of which it is a part. It could involve a considerable amount of judgment on the part of the designer. The selection of a tension member (particularly in trusses) must be based on *assumed* end connections. After a member is selected and the end connections have been designed, *it may be necessary to revise the selection.*

Since block shear plays such an important role, a preliminary understanding of simple connections for tension members is essential. We will limit this discussion for the present to the essentials of bolted connections. Chapters 7, 8, and 11 present an in-depth discussion on both bolted and welded connections.

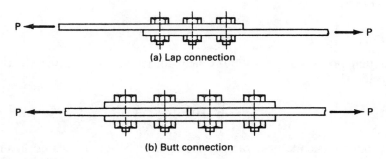

**FIGURE 2-13**   Types of connections.

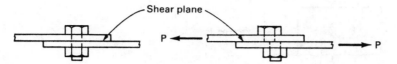

**FIGURE 2-14**   Bolt in single shear.

The two types of connections that we will consider here, *lap* and *butt* connections, are depicted in Figure 2-13.

In most structural connections, the bolt is required to prevent the movement of the connected material in a direction perpendicular to the length of the bolt as shown in Figure 2-14.

In this case, the bolt is said to be loaded in *shear*. In the connection shown, the bolt has a tendency to shear off along the single contact plane of the two plates. Since the bolt is resisting the tendency of the plates to slide past one another along the contact surface and is being sheared on a single plane, the bolt is said to be in *single* shear. In a butt connection, such as that shown in Figure 2-13b, there are two contact planes. Therefore, the bolt is offering resistance along two planes and is said to be in *double* shear.

Our later discussion will detail the rationale behind the determination of the strength of a single bolt in single or double shear. Table 2-3 provides these strengths. This table represents an extreme oversimplification of Table I-D from the ASDM, Part 4, in that it provides bolt strengths for only a very narrow range of conditions and connection types. For the purposes of this chapter, however, it is adequate for estimating a required number of bolts for a connection when designing tension members. Note that there are two types of bolts shown: A325 and A490. These designations are ASTM material designations. The notations under "Loading" refer to single shear (S) and double shear (D).

In addition to considering shear failure *in* the bolts, we must also be concerned with the members being connected (the plates of Figure 2-13) where they bear *on* the bolts. If a material is overly *thin,* the hole will elongate into an oval shape and the connection will be said to have failed in *bearing*. This condition is shown in

■■■■■  **TABLE 2-3   Shear Allowable Load per Bolt (kips)**

|  |  | Diameter | | |
| --- | --- | --- | --- | --- |
| **Bolt** | **Loading** | **3/4 in.** | **7/8 in.** | **1 in.** |
| A325 | S | 7.51 | 10.2 | 13.4 |
| A325 | D | 15.0 | 20.4 | 26.7 |
| A490 | S | 9.28 | 12.6 | 16.5 |
| A490 | D | 18.6 | 25.3 | 33.0 |

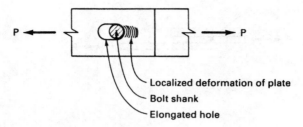

**FIGURE 2-15**   Bearing failure.

Figure 2-15. In later chapters we will discuss this type of failure in depth and how the bearing strength of a bolt is determined. Table 2-4 shows these strengths. This table is an oversimplification of Table I-E in the ASDM, Part 4, and applies to only A36 steel.

**■■■■■ TABLE 2-4   Bearing Allowable Load per Bolt (kips)**

| Material thickness (in.) | Bolt diameter | | |
|:---:|:---:|:---:|:---:|
| | 3/4 in. | 7/8 in. | 1 in. |
| 1/4 | 13.1 | 15.2 | 17.4 |
| 5/16 | 16.3 | 19.0 | 21.8 |
| 3/8 | 19.6 | 22.8 | 26.1 |
| 7/16 | 22.8 | 26.6 | 30.5 |
| 1 | 52.2 | 60.9 | 69.6 |

Pitch and edge distance were discussed in Section 2-2 of this chapter. The ASDS recommends that the minimum distance between bolts be taken as $2\frac{2}{3}$ times the bolt diameter (and that 3 times the bolt diameter is preferred). Table 2-5 is a condensed version of ASDS Table J3-5 showing minimum edge distances for various

**■■■■■ TABLE 2-5   Minimum Edge Distance (in.)**

| Bolt diameter (in.) | At sheared edges | At rolled, gas cut, or saw-cut edges |
|:---:|:---:|:---:|
| $\frac{3}{4}$ | $1\frac{1}{4}$ | 1 |
| $\frac{7}{8}$ | $1\frac{1}{2}$ | $1\frac{1}{8}$ |
| 1 | $1\frac{3}{4}$ | $1\frac{1}{4}$ |

bolt sizes and edge conditions. Recall that edge distance is the distance from the center of a bolt hole to the nearest edge. Using the information provided, a reasonable estimate can be made as to number of fasteners required and connection geometry for purposes of tension member design.

**Example 2-7** _____

Select the lightest double-angle tension member for member $BC$, a web member in a light truss, as shown in Figure 2-16. The tensile load will be 46 kips. Use A36 steel. The length is 13.4 ft. Fasteners will be $\frac{3}{4}$-in.-diameter A325 bolts and will connect the double-angle member to a $\frac{3}{8}$-in.-thick gusset plate. Assume that the strength of the gusset plate will not control.

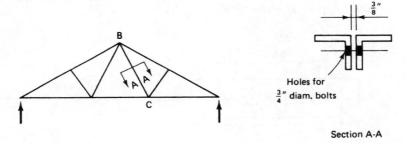

**FIGURE 2-16**   Tension member design.

**Solution:**

The required gross area, based on general yielding of the member, is calculated from

$$\text{required } A_g = \frac{P}{F_t} = \frac{P}{0.60F_y} = \frac{46}{0.60(36)} = 2.13 \text{ in.}^2$$

The minimum required radius of gyration is calculated from

$$\text{maximum } \frac{\ell}{r} = 300$$

$$\text{minimum required } r = \frac{\ell}{300} = \frac{13.4(12)}{300} = 0.54 \text{ in}$$

Remaining to be considered are (a) minimum required net area based on tensile fracture and (b) block shear, both of which depend on the thickness of the member and on connection details. We will assume one row (one gage line) of $\frac{3}{4}$-in.-diameter A325 bolts. A check of double-angle properties (ASDM, Part 1) shows that angles of $\frac{1}{4}$-in. thickness could provide the required $A_g$ of

2.13 in.$^2$. Proceeding farther with the $\frac{1}{4}$-in.-thick angles, compare the allowable load per bolt for shear and bearing from Tables 2-3 and 2-4. We see that the bearing allowable load of 26.2 kips (on two $\frac{1}{4}$-in. thicknesses of angle) is less critical than the shear allowable load of 15 kips. Therefore, we calculate the required minimum number of bolts from

$$N = \frac{P}{\text{bolt allowable}} = \frac{46}{15.0} = 3.07 \text{ bolts} \qquad \textbf{Use 4 bolts.}$$

Assuming a minimum pitch of

$$2.67d_b = 2.00 \text{ in.}$$

and a minimum edge distance of $1\frac{1}{4}$ in. (from Table 2-5), the connection detail would appear as shown in Figure 2-17.

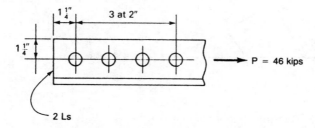

**FIGURE 2-17** Connection detail.

The reader should note that this approximation neglects other factors that determine bolt strength. Final connection design is covered in detail in Chapters 7 and 8. Changes in the angle thickness or changes in the connection, which affect strength, may yet occur. For instance, if the pitch or edge distance changes in the final connection, the block shear strength will be affected.

We could, at this point, check the block shear strength using the assumed approximate connection details and an *assumed angle thickness*. Rather than do that, we will determine the *required minimum angle thickness* based on the required block shear strength, which is equal to the applied tensile load. The hole diameter is taken as

$$\frac{3}{4} + \frac{1}{8} = 0.875 \text{ in.}$$

$$P_t = A_v F_v + A_t F_t$$

$$A_v = 2t(7.25 - 3.5(0.875)) = 8.38t$$

$$A_t = 2t(1.25 - 0.875/2) = 1.625t$$

Substituting,

$$P_t = 8.38t(0.3)F_u + 1.625t(0.5)F_u$$

from which

$$\text{minimum required } t = \frac{46}{8.38(0.3)(58) + 1.625(0.5)(58)} = 0.238 \text{ in.}$$

Assuming an angle thickness of $\frac{1}{4}$ in., calculate the minimum required gross area based on the required effective net area $A_e$:

$$P_t = F_t A_e = 0.5 F_u U A_n$$
$$= 0.5 F_u U (A_g - A_{\text{holes}})$$

Therefore,

$$\text{required } A_g = \frac{P}{U(0.5)F_u} + A_{\text{holes}}$$

From Table 2-1, a reduction coefficient $U$ of 0.85 is selected, from which

$$\text{required } A_g = \frac{46}{0.85(0.5)(58)} + 2\left(\frac{3}{4} + \frac{1}{8}\right)\left(\frac{1}{4}\right) = 2.30 \text{ in.}^2$$

We now select a double-angle member based on the preceding requirements. If unequal leg angles are selected, they will be oriented LLBB (this more nearly balances the radius of gyration values).

A member composed of 2L3 $\times$ 2 $\times \frac{1}{4}$ (LLBB) meets the requirements. The following properties are noted:

$$A_g = 2.38 \text{ in.}^2 \qquad \qquad \textbf{O.K.}$$

$$r_y = 0.891 \text{ in.} \qquad \qquad \textbf{O.K.}$$

$$t = \frac{1}{4} \text{ in.} \qquad \qquad \textbf{O.K.}$$

Therefore, use 2L3 $\times$ 2 $\times \frac{1}{4}$. Note that 2L2$\frac{1}{2} \times$ 2$\frac{1}{2} \times \frac{1}{4}$ would also be satisfactory.

---

## 2-6

## THREADED RODS IN TENSION

Rods of circular cross section are commonly used for tension members. These are sometimes called tie rods or sag rods, depending on the application. The connection to other structural members is made by threading the end of the rod and installing a nut. The thread reduces the cross-sectional area available to carry tension. Historically, design of threaded rods has been based on the cross-sectional area (called the *root area*) at the base of the threads. In Figure 2-18, the root area has a diameter $K$. More recently, threaded rod design was based on an area greater than the root area but less than the cross-sectional area of the unthreaded body of the rod. This area was called the *tensile stress area*. In each case, an allowable tensile stress was determined as a fraction of the yield stress (i.e., $0.60F_y$ applied to the tensile stress area).

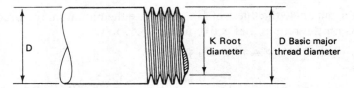

FIGURE 2-18   Threaded rod—partial view.

Rods with enlarged ends, called *upset rods* (shown in Figure 2-19), have been used to circumvent the problem of reduced cross-sectional area under the threads. The enlarged end is proportioned so that the strength at the threads is greater than the strength of the unthreaded body of the rod. The forging process used to create the enlarged end is expensive. Therefore, upset rods are less common in construction applications than they are in machine applications.

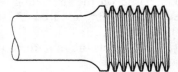

**FIGURE 2-19**   Upset rod.

Under the current ASDS, the allowable tensile stress for threaded rods, other than upset rods, is taken as $0.33F_u$, and this is applied to the *unthreaded nominal body area* of the rod (area of diameter $D$ in Figure 2-18). Refer to the ASDS, Table J3.2. For upset rods, the threaded upset end must be proportioned so that its capacity is *greater than* the capacity of the unthreaded body of the rod. The capacity of the threaded upset end is calculated from

$$P_t = F_t A_D = 0.33F_u A_D$$

where $A_D$ is the gross area of the upset end. The capacity of the unthreaded body of the rod is calculated from

$$P_t = F_t A_D = 0.60F_y A_D$$

where $A_D$ is the gross area of the body of the rod. Therefore, the capacity of a properly proportioned upset end rod may be calculated from the latter formula.

Circular rods are available in most steels commonly used in construction. For useful data on threads, the reader is referred to the threaded fastener data contained in the ASDM, Part 4.

In the selection of threaded rods, it should be noted that there are no slenderness ratio recommendations by the ASDS. A common rule of thumb is to use a rod diameter not less than $\frac{1}{500}$ of the rod length. The minimum size of rod should be limited to $\frac{5}{8}$ in. in diameter since smaller rods are easily damaged during construction. Also, the minimum design load for a threaded fastener (and, therefore, for a threaded rod) is 6 kips, as per the ASDS, Section J1.6.

## Example 2-8

A hanging storage bin weighing 30 tons will be supported by three circular threaded rods. Use A36 steel. Determine the required rod diameter and specify the required threads.

**Solution:**

Each rod will carry 10 tons (or 20 kips). The required unthreaded nominal body area (gross $A_D$) is

$$\text{required } A = \frac{P}{0.33 F_u} = \frac{20}{0.33(58)} = 1.04 \text{ in.}^2$$

From the ASDM, Part 4, threaded fastener data, basing the selection on gross area $A_D$, use a $1\frac{1}{4}$-in.-diameter rod. Using the standard thread designation, the required thread will be **$1\frac{1}{4}$–7 UNC 2A.**

## Example 2-9

Rework Example 2-8 using upset rods. Determine the required rod size and the required diameter of the upset end. Specify the required thread.

**Solution:**

The required nominal body area (gross area $A_D$) of the rod before upsetting is

$$\text{required } A = \frac{P}{0.60 F_y} = \frac{20}{0.60(36)} = 0.93 \text{ in.}^2$$

Use a $1\frac{1}{8}$-in.-diameter rod ($A_D = 0.994$ in.$^2$). The selection of the upset end will be based on the capacity of the nominal body area, which is

$$0.60(F_y)(0.994) = 0.60(36)(0.994) = 21.5 \text{ kips}$$

The required area of the upset end (gross area $A_D$) is

$$\text{required } A = \frac{P}{0.33 F_u} = \frac{21.5}{0.33(58)} = 1.12 \text{ in.}^2$$

Use a $1\frac{1}{4}$-in.-diameter upset end ($A_D = 1.227$ in.$^2$). The required thread is, as before, **$1\frac{1}{4}$–7 UNC 2A.**

## REFERENCES

[1]   S. Hardash and R. Bjorhovde, "New Design Criteria for Gusset Plates in Tension," *AISC Engineering Journal,* Vol. 22, No. 2, 2nd Qtr., 1985.

[2]   William McGuire, *Steel Structures* (Englewood Cliffs, NJ: Prentice-Hall, Inc., 1968), pp. 308–318.

## PROBLEMS

**2-1.**   Compute the tensile capacity for the $\frac{1}{2}$-in.-thick plate shown. The bolts are $\frac{3}{4}$-in.-diameter high-strength bolts. The steel is A36.

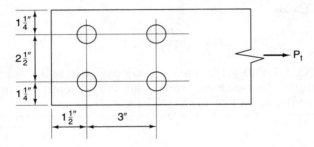

**PROBLEM 2-1**

**2-2.**   Compute the tensile capacity for the angles and conditions shown. Use A36 steel and standard angle gages.

**(a)**  L4 × 4 × $\frac{1}{2}$, connected with $\frac{3}{4}$-in.-diameter high-strength bolts.

**(b)**  L6 × 4 × $\frac{5}{8}$, short leg connected with $\frac{7}{8}$-in.-diameter high-strength bolts.

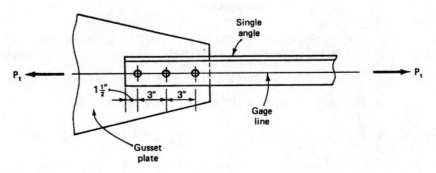

**PROBLEM 2-2**

**2-3.** Compute the tensile capacity for the angles and conditions shown. Use A36 steel and standard angle gages. The gusset plate is $\frac{3}{8}$ in. thick.

**(a)** L5 × 3 × $\frac{3}{8}$, long leg connected with $\frac{7}{8}$-in.-diameter high-strength bolts.

**(b)** L6 × 4 × $\frac{5}{8}$, long leg connected with $\frac{3}{4}$-in.-diameter high-strength bolts.

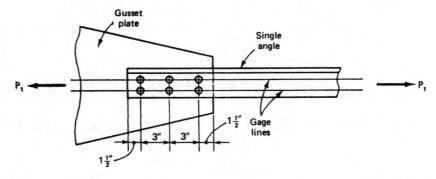

**PROBLEM 2-3**

**2-4.** Compute the tensile capacity for the double-angle member shown. The bolts are $\frac{7}{8}$-in. diameter and the steel is A36. Use standard gages. Assume that the plate does not control.

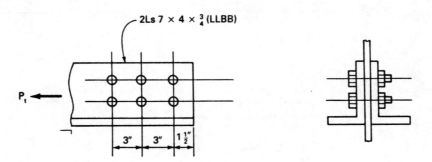

**PROBLEM 2-4**

**2-5.** Compute the tensile capacity for the connection shown. (Neglect bolt shear and bearing on the plates.) The bolts are $\frac{7}{8}$-in. diameter and the steel is A36.

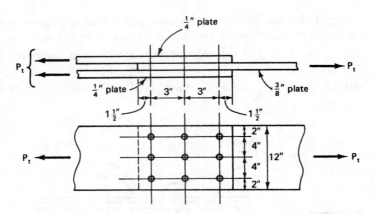

**PROBLEM 2-5**

**2-6.** Compute the net area for the plates shown. Assume $\frac{3}{4}$-in. bolts.

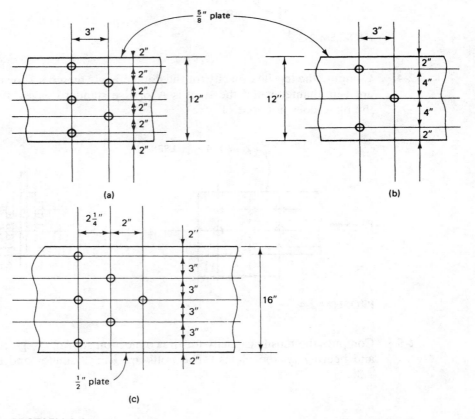

(a)                                                                 (b)

(c)

**PROBLEM 2-6**

**2-7.** Compute the tensile capacity for the A36 steel plate shown. Bolts are $\frac{7}{8}$-in. diameter. Neglect block shear.

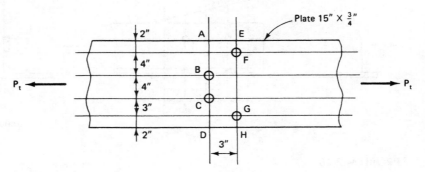

**PROBLEM 2-7**

**2-8.** A truss tension member is to be composed of a W6 $\times$ 15 and will be connected with two gage lines of $\frac{3}{4}$-in.-diameter bolts in each flange, as shown in Figure 2-10. Assume three bolts per gage line and A36 steel. Compute the tensile capacity $P_t$. Edge distance is $1\frac{1}{4}$ in. typical. Pitch is 3 in.

**2-9.** Compute the tensile capacity for the double-angle member shown. The bolts are $\frac{7}{8}$-in. diameter and the steel is A36.

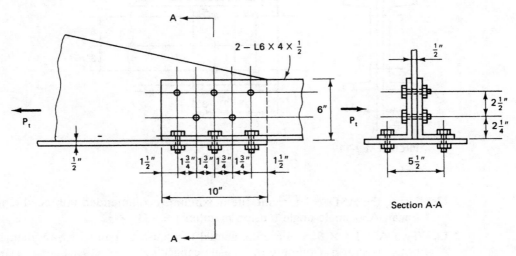

**PROBLEM 2-9**

**2-10.** Compute the tensile capacity for the double-angle member shown. The bolts are $\frac{3}{4}$-in. diameter and the steel is A441. Use standard gages as per the ASDM.

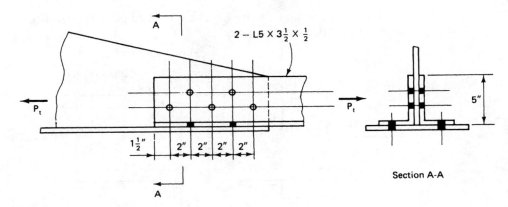

**PROBLEM 2-10**

**2-11.** Compute the tensile capacity for the double-channel member (two C10 × 15.3) shown. The channels are $\frac{1}{2}$ in. back-to-back and are 16 ft–0 in. long. Check the slenderness ratio. The steel is A36.

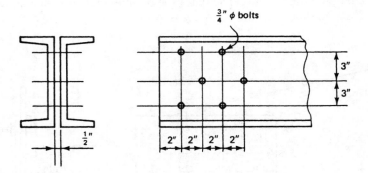

**PROBLEM 2-11**

**2-12.** Using the ASDS, determine the maximum recommended unbraced length for an A36 single-angle tension member L6 × 3$\frac{1}{2}$ × $\frac{1}{2}$.

**2-13.** Two A36 L5 × 3$\frac{1}{2}$ × $\frac{1}{2}$ are connected to a gusset plate with $\frac{3}{4}$-in.-diameter bolts, as shown. Compute the tensile capacity $P_t$.

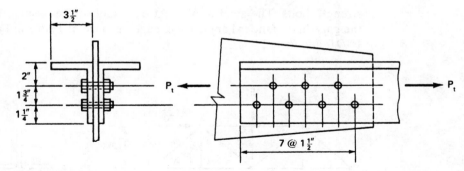

**PROBLEM 2-13**

**2-14.** Rework Example 2-4 (Figure 2-10). Assume a W12 × 40, four $\frac{3}{4}$-in.-diameter bolts per line, $1\frac{1}{4}$-in. edge distance, and 3-in. pitch.

**2-15.** Using the ASDS, compute the maximum recommended unbraced length for the A36 tension members having cross sections as shown.

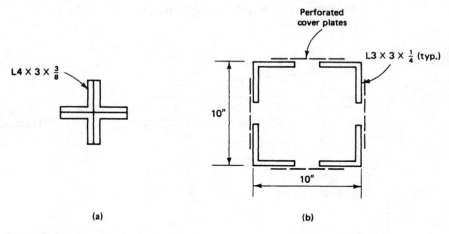

(a)                                    (b)

**PROBLEM 2-15**

**2-16.** A welded tension member is fabricated using one S5 × 14.75 and two C5 × 6.7 channels. The shapes are joined at the flange tips with the channel flanges turned in to form a box shape. The steel is A36 and the member length is 32 ft–0 in. Compute the tensile load capacity and check the slenderness ratio. Neglect block shear.

**2-17.** For the truss shown, member *DE* is composed of 2L4 × 3 × $\frac{1}{4}$. All joints are assumed to be pin-connected with one gage line of $\frac{3}{4}$-in.-diameter high-strength bolts. The steel is A36. At each end of $DE$ assume four bolts on the gage line, standard angle gage, edge distance of $1\frac{1}{4}$ in., and a $2\frac{1}{4}$-in. pitch. Is $DE$ adequate?

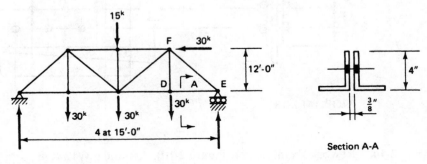

Section A-A

**PROBLEM 2-17**

**2-18.** Determine whether the A36 double-angle tension member shown can resist a tensile load of 64 kips. The bolts are $\frac{3}{4}$-in. diameter. Assume a standard angle gage, $2\frac{1}{2}$-in. pitch, and $1\frac{1}{4}$-in. end edge distance.

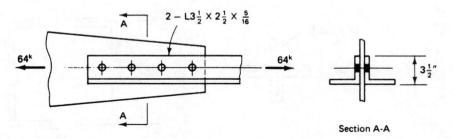

Section A-A

**PROBLEM 2-18**

**2-19.** Select a W10 shape to resist a tensile load of 125 kips. The member is to be A36 steel and 27 ft–0 in. long. Assume end connections consisting of two gage lines of $\frac{3}{4}$-in.-diameter bolts per flange. Assume three bolts per gage line, edge distance of $1\frac{1}{4}$ in., and a 3-in. pitch. Check block shear in the W10 shape but assume that block shear in the gusset plates will not control.

**2-20.** Select a W14 shape to support a tensile load of 400 kips. The member is to be A36 steel and will be 20 ft long. Assume end conditions consisting of two gage lines of 1-in.-diameter bolts in each flange. Assume four bolts per gage line, edge distance of $1\frac{1}{2}$ in., and a 3-in. pitch. Check block shear in the W14 shape but neglect block shear in the end connection gusset plates.

**2-21.** Select a W shape to resist a tensile load of 250 kips. Use A36 steel. The member length is 25 ft–0 in. Assume end connections consisting of two gage lines of $\frac{3}{4}$-in.-diameter bolts per flange. Assume five bolts per gage line, $1\frac{1}{4}$-in. edge distances, and a 2-in. pitch. Check block shear in the W shape but assume that block shear in the gusset plates will not control.

**2-22.** Select the lightest A36 single-angle tension member using $\frac{7}{8}$-in.-diameter high-strength bolts. Assume an end connection consisting of one gage line with four bolts. Use standard angle gages, $1\frac{1}{2}$-in. end edge distance, and $3\frac{1}{4}$-in. pitch.

**(a)** Tensile load = 55 kips, length = 10 ft–6 in.

**(b)** Tensile load = 42 kips, length = 12 ft–0 in.

**2-23.** Rework Problem 2-22 but select a double-angle member, placing angles $\frac{3}{8}$ in. back to back. The cross section is similar to that shown for Problem 2-18.

**2-24.** In the truss of Problem 2-17, assume that the loads applied along the bottom chord have increased to 40 kips. Using the information given in Problem 2-17, select the lightest weight double-angle tension member for *DF* and *DE*. Assume five bolts on the gage line for each member with $1\frac{1}{2}$-in. edge distance and $2\frac{1}{2}$-in. bolt pitch.

**2-25.** Select a square tube to serve as a tension member for the condition shown. Use A501 steel with $F_y$ = 36 ksi. The tensile load is 75 kips and the length of the member is 21 ft–0 in. Assume welded end connections.

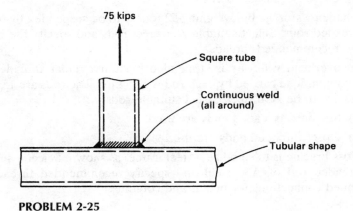

**PROBLEM 2-25**

**2-26.** Select a double-angle tension member like that shown to support a load *P* = 180 kips. Use A36 steel and $\frac{7}{8}$-in.-diameter high-strength bolts. The length of the member is 15 ft–0 in. Use standard angle gages, $1\frac{1}{2}$-in. end edge distance, and 3-in. pitch. Assume a total of eight fasteners.

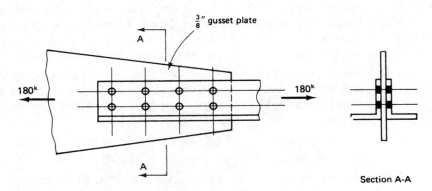

Section A-A

**PROBLEM 2-26**

**2-27.** Select the lightest pair of American Standard channels to support a tensile load of 390 kips. The channels are to be placed 12 in. back-to-back with the flanges turned in. The member is to be A36 steel, 33 ft long, and welded to $\frac{3}{8}$-in. gusset plates at each end. Assume a reduction coefficient $U$ of 0.87.

**2-28.** Select an A36 threaded tie rod to resist a tensile load of 21 kips.

**(a)** Use standard threaded ends.

**(b)** Use upset threaded ends.

**2-29.** A hanging storage bin weighing 52 tons is to be suspended by three circular threaded rods. Select suitable A36 steel rods and specify the rod diameter and recommended threads.

**2-30.** An overhead walkway is suspended by four circular threaded rods. The maximum load carried by each rod is 17.5 kips. The rods are 15 ft–0 in. long and are to be A36 steel. Select suitable rods if

**(a)** Standard threaded ends are used.

**(b)** Upset threaded ends are used.

**2-31.** Cross-bracing is to stabilize a steel frame, as shown. Select a suitable round threaded rod of A36 steel and specify recommended threads. (Assume pinned connections for beams and columns.)

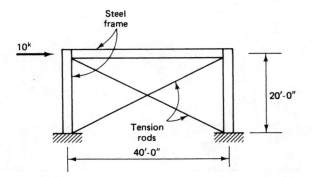

**PROBLEM 2-31**

**2-32.** Member *AB* is an A36 steel tie-back rod with threaded ends. Determine the required diameter and specify the threads.

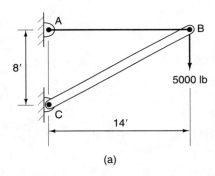
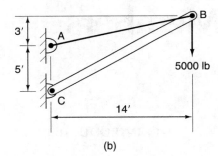

(a)                               (b)

**PROBLEM 2-32**

# CHAPTER 3

# Axially Loaded Compression Members

## 3-1

### INTRODUCTION

Structural members that carry compressive loads are sometimes given names that identify them as to their function. Compression members that serve as bracing are commonly called *struts*. Other compression members may be called *posts* or *pillars*. Trusses are composed of members that are in compression and members that are in tension. These may be either chord or web members. Of primary interest in this chapter are the main vertical compression members in building frames, which are called *columns*. Additionally, double-angle compression members are discussed.

**PHOTO 3-1**   Erecting a three-story structural steel wide-flange column.

Columns are compression members that have their length dimension considerably larger than their least cross-sectional dimension.

In this chapter we consider members that are subjected to *axial* (concentric) *loads;* that is, the loads are coincident with the longitudinal centroidal axis of the member. This is a special case and one that exists rarely, if at all. Where *small* eccentricities exist, however, it may be assumed that an appropriate factor of safety will compensate for the eccentricity and the column may be designed as though it were axially loaded. Columns may support varying amounts of axial load and bending moment. If we consider the range of possible combinations of load and moment supported on columns, then at one end of the range is the axially loaded column. This column carries no moment. At the other end of the range is the member that carries only moment with no (or very little) axial load. (As a moment-carrying member, it could be considered a *beam*). When a column carries both axial load and moment, it is called a *beam-column*. Beam-columns are considered in Chapter 6.

Commonly used cross sections for steel compression members include most of the rolled shapes. These and other typical cross sections are shown in Figure 3-1. For the W shapes (Figure 3-1b), the cross sections usually used are those that are rather square in shape and that have nominal depths of 14 in. or less. These shapes are more efficient than others for supporting compressive loads (the deeper shapes are more efficient when used as bending members). For larger loads it is common to use a *built-up* cross section. In addition to providing increased cross-sectional area, the built-up sections allow a designer to tailor to specific needs the radius of gyration ($r$) values about the $x$-$x$ and $y$-$y$ axes. The dashed lines shown on the cross sections of Figure 3-1f and g represent tie plates, lacing bars, or perforated cover plates and do not contribute to the cross-sectional properties. Their functions

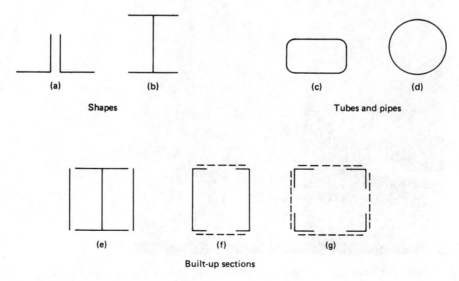

**FIGURE 3-1**   Compression member cross sections.

are to hold the main longitudinal components of the cross section in proper relative position and to make the built-up section act as a single unit.

In dealing with compression members, the problem of *stability* is of great importance. Unlike tension members, where the load tends to hold the members in alignment, compression members are very sensitive to factors that may tend to cause lateral displacements or buckling. The situation is similar in ways to the lateral buckling of beams. The buckling problem is intensified and the load-carrying capacity is affected by such factors as eccentric load, imperfection of material, and initial crookedness of the member. Also, *residual stresses* play a role. These are the variable stresses that are "locked up" in the member as a result of the method of manufacture, which involves unequal cooling rates within the cross section. The parts that cool first (such as the flange tips) will have residual compression stresses, while parts that cool last (the junction of the flange and web) will have residual tension stresses. Residual stresses may also be induced by nonuniform plastic deformation caused by cold working, such as in the straightening process.

## 3-2

## IDEAL COLUMNS

The development of the theory of elastic column behavior took place long ago. Leonard Euler (1707–1783), a Swiss mathematician, is credited with many significant contributions in the field of Newtonian mechanics, among them the derivation of the elastic column buckling formula. His theory was presented in 1759 and is still the basis for the analysis and design of slender columns. The buckling of a long or slender column may be demonstrated by loading an ordinary wooden yardstick in compression. The yardstick will buckle (constituting failure) as shown in Figure 3-2c, but will not fracture if the load is not increased beyond the buckling load.

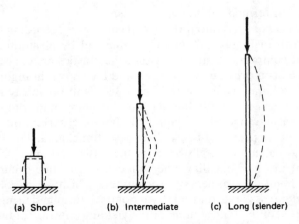

(a) Short          (b) Intermediate          (c) Long (slender)

**FIGURE 3-2**  Columns types and failure modes.

Upon release of the load, it will return to its original shape. Simply stated, Euler's formula gives the buckling load $P_e$ for a pin-ended, homogeneous, initially straight, long column of an elastic material that is concentrically loaded. This is considered to be the ideal column. The Euler buckling load is expressed as

$$P_e = \frac{\pi^2 EI}{\ell^2}$$

where

$P_e$ = concentric load that will cause initial buckling

$\pi$ = mathematical constant (3.1416)

$E$ = material modulus of elasticity

$I$ = least moment of inertia of the cross section

$\ell$ = length of the column from pin end to pin end

Tests have verified that Euler's formula accurately predicts buckling *load*, where the buckling *stress* is less than (approximately) the proportional limit of the material and adherence to the basic assumptions is maintained. Since the buckling stress must be compared with the proportional limit, Euler's formula is commonly written in terms of stress. This may easily be derived from the preceding buckling load formula, recognizing that $I = Ar^2$:

$$f_e = \frac{\pi^2 E}{(\ell/r)^2}$$

where

$f_e$ = uniform compressive stress at which initial buckling occurs

$r$ = least radius of gyration of the cross section $\sqrt{I/A}$, where $A$ is the cross-sectional area

It will be recalled that $\ell/r$ is termed the *slenderness ratio*.

It is convenient to classify columns into three broad categories according to their modes of failure, as shown in Figure 3-2. Columns that fail by elastic buckling, where buckling occurs at compressive stresses within the elastic range, are called *long columns*. These have been previously discussed and are shown in Figure 3-2c. A very short and stocky column, as shown in Figure 3-2a, will obviously not fail by elastic buckling. It will crush or squash due to general yielding, and compressive stresses will be in the inelastic range. If yielding is the failure criterion, the failure load may be determined as the product of $F_y$ and cross-sectional area. This column is called a *short column*. A column that falls between these two extremes, as shown in Figure 3-2b, will fail by inelastic buckling when a localized yielding occurs. This will be initiated at some point of weakness or crookedness. This type of column is called an *intermediate column*. Its failure strength cannot be determined using either the elastic buckling criterion of the long column or the yielding criterion of the short column. It is designed and analyzed using empirical formulas based on extensive test results.

As a review of column behavior as treated in strength of materials, ideal columns will first be analyzed and designed using Euler's formula. This is not to be construed as the modern and practical approach to column analysis and design, but merely as a basis for the more modern methods. The AISC approach will follow later in this chapter.

**Example 3-1** _____

Determine the Euler buckling load ($P_e$) for an axially loaded W14 × 22 shown in Figure 3-3. The column has pinned ends. Assume A36 steel with a proportional limit of 34 ksi. The column length is 12 ft.

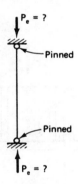

**FIGURE 3-3**  Euler column analysis.

**Solution:**

Properties of the W14 × 22 are

$$A = 6.49 \text{ in.}^2$$

$$r_y = 1.04 \text{ in.}$$

$$I_y = 7.00 \text{ in.}^4$$

Solve for the buckling stress:

$$f_e = \frac{\pi^2 E}{(\ell/r)^2} = \frac{\pi^2(29{,}000)}{[(12 \times 12)/1.04]^2} = 14.93 \text{ ksi}$$

14.93 ksi < 34 ksi (Euler's formula applies)

$$P_e = f_e A = 14.93(6.49) = 96.9 \text{ kips}$$

or

$$P_e = \frac{\pi^2 EI}{\ell^2} = \frac{\pi^2 (29{,}000)(7.00)}{(12 \times 12)^2} = 96.6 \text{ kips } (430 \text{ kN})$$

## 3-3

## EFFECTIVE LENGTHS

Euler's formula gives the buckling load for a column that has pinned ends. A practical column, in addition to being nonperfect in other aspects, may have end conditions (end supports) that provide restraint of some magnitude and will not allow the column ends to rotate freely. Figure 3-4a and b shows column supports

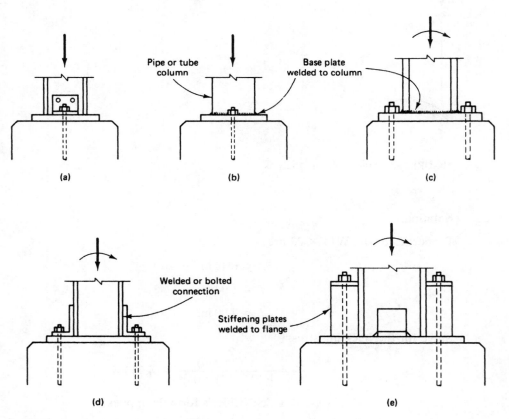

**FIGURE 3-4**   Typical column supports.

that offer very little resistance to column end rotations. These are essentially pinned end supports. The supports shown in Figure 3-4c, d, and e provide significant resistance to column end rotation. The use of Euler's formula may be extended to columns having other than pinned ends through the use of an *effective length*. This concept is illustrated in Figure 3-5, where a column having rigid (or fixed) ends is shown. The deflected shape of the buckled column is shown in Figure 3-5a. Inflection points exist at the quarter points of the column length. These are points of zero moment and may be theoretically replaced with pins without affecting the equilibrium or the deflected shape of the column. If the central portion of the fixed-ended column is considered separately, as in Figure 3-5b, it is seen that it behaves as a pin-ended column of length $\ell/2$. The Euler critical load for the fixed-ended column is then seen to be the same as for a pin-ended column of length $\ell/2$. The length $\ell/2$ is said to be the effective length of the fixed-ended column, and the *effective length factor K* is $\frac{1}{2}$ or 0.50. The effective length is written as $K\ell$, where $\ell$ is the actual length of the column. Euler's formula may be rewritten with the inclusion of the effective length as

$$P_e = \frac{\pi^2 EI}{(K\ell)^2}$$

and

$$f_e = \frac{\pi^2 E}{(K\ell/r)^2}$$

For the fixed-ended column just discussed,

$$P_e = \frac{\pi^2 EI}{(0.5\ell)^2} = \frac{4\pi^2 EI}{\ell^2}$$

It is seen that the buckling load is increased by a factor of 4 when rigid end supports are furnished for a column.

Other combinations of column end conditions are covered in the ASDS Commentary. Table C-C2.1 provides theoretical $K$ values for six idealized conditions in

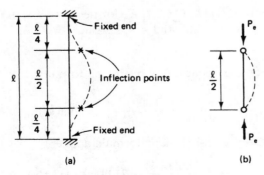

**FIGURE 3-5**  Column effective length.

which joint rotation and translation are either fully realized or nonexistent. Since there is no perfectly rigid column support and no perfect pin support, the referenced table also provides recommended design values for $K$ where ideal conditions are approximated. These values are slightly higher than the ideal values and therefore are conservative (the predicted $P_e$ will be on the low side). The reader should carefully study the end condition criteria and the buckled shapes of the columns as shown in Table C-C2.1.

Euler's formula for buckling load may also be adapted to result in an expression for an allowable compressive load capacity. This will be termed, for convenience, $P_a$. A factor of safety (F.S.) is introduced (see Section 1-6 of this text for a discussion of factor of safety):

$$P_a = \frac{P_e}{\text{F.S.}}$$

$$P_a = \frac{\pi^2 EI}{(K\ell)^2(\text{F.S.})}$$

**Example 3-2** _____

A W10 $\times$ 49 column of A36 steel has end conditions that approximate the fixed-pinned condition (fixed at the bottom, pinned at the top, no sidesway). Assume a proportional limit of 34 ksi.

(a)  If the length of the column is 26 ft, find the allowable compressive load capacity, $P_a$, using Euler's formula and a factor of safety of 2.0.

(b)  What is the minimum length of this column at which the Euler formula would still be valid?

**Solution:**

From the ASDS Commentary, Table C-C2.1, $K$ (for design) = 0.80. For the W10 $\times$ 49, $A$ = 14.4 in.$^2$, $I_y$ = 93.4 in.$^4$, and $r_y$ = 2.54 in.

(a)  Find $f_e$ first and check the applicability of Euler's formula:

$$f_e = \frac{\pi^2 E}{(K\ell/r)^2} = \frac{\pi^2(29,000)}{[0.80(26 \times 12)/2.54]^2} = 29.6 \text{ ksi}$$

29.6 ksi < 34 ksi (Euler's formula applies)

$$P_a = \frac{P_e}{\text{F.S.}} = \frac{f_e A}{\text{F.S.}} = \frac{29.6(14.4)}{2.0} = 213 \text{ kips (947 kN)}$$

(b)   Find the length at which $f_e$ equals the proportional limit:

$$f_e = \frac{\pi^2 E}{(K\ell/r)^2}$$

$$\ell = \sqrt{\frac{\pi^2 E}{f_e(K/r)^2}} = \sqrt{\frac{\pi^2(29,000)}{34(0.8/2.54)^2}} = 291 \text{ in.} = 24.3 \text{ ft } (7.41 \text{ m})$$

## Example 3-3

Use Euler's formula to select a W-shape column to support an axial load of 50 kips. The length is 12 ft and the ends are pinned. Use A36 steel with a proportional limit assumed to be 34 ksi. Check the applicability of Euler's formula. Assume a factor of safety = 3.0 (*Note:* This is *not* the AISC method of column selection.)

### Solution:

Select a column that has an allowable compressive load capacity $P_a$ of at least 50 kips. Assume that Euler's formula applies.

$$P_a = \frac{\pi^2 EI}{(K\ell)^2(\text{F.S.})}$$

$$\text{required } I = \frac{P_a(K\ell)^2(\text{F.S.})}{\pi^2 E} = \frac{50(1.0 \times 12 \times 12)^2(3.0)}{\pi^2(29,000)} = 10.9 \text{ in.}^4$$

Try W6 × 20:

$$I_y = 13.3 \text{ in.}^4$$

$$A = 5.87 \text{ in.}^2$$

$$r_y = 1.50 \text{ in.}$$

Check the applicability of Euler's formula and the capacity of the W6 × 20:

$$f_e = \frac{\pi^2 E}{(K\ell/r)^2} = \frac{\pi^2(29,000)}{(144/1.50)^2} = 31.06 \text{ ksi}$$

31.06 ksi < 34 ksi (Euler's formula applies)

Calculating the buckling load, we have

$$P_e = 31.06(5.87) = 182 \text{ kips}$$

from which the allowable compressive load capacity is

$$P_a = \frac{P_e}{\text{F.S.}} = \frac{182}{3.0} = 60.8 \text{ kips} > 50 \text{ kips}$$                    **O.K.**

## 3-4

## ASDS ALLOWABLE STRESSES
## FOR COMPRESSION MEMBERS

In the preceding sections, Euler's formula was used to analyze and design columns. In each case, the applicability of the approach was checked. In each case, Euler's formula did result in $f_e$ less than the proportional limit. In effect, all the columns were slender columns. Practical columns, however, generally do not fall into this category. The practical analysis/design method must concern itself with the entire possible range of slenderness ratio $K\ell/r$. Theoretical formulas are not applicable for intermediate and short columns because of many material and geometric uncertainties. The strength of intermediate and short columns cannot be predicted accurately theoretically; therefore, the results of extensive testing and experience must be utilized.

Figure 3-6 shows a plot of the failure stresses of columns versus their $K\ell/r$ ratios as determined by testing. Since no two practical columns are identical, the failure stresses are expected to fall within a *range* of values for a particular $K\ell/r$ value. Columns with $K\ell/r$ values to the right of line A-A have their failure stresses closely predicted by Euler's formula. They are subject to elastic buckling where the buckling occurs at a stress less than the proportional limit. Columns with $K\ell/r$ values to the left of line A-A fail by inelastic buckling (yielding occurs), and a departure of the test data from the curve that represents Euler's formula is noted.

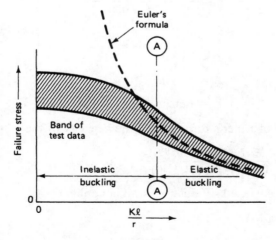

**FIGURE 3-6**   Column test data.

The ASDS allowable stresses for compression members, as found in Section E2, may be shown diagrammatically as in Figure 3-7. The maximum $K\ell/r$ is preferably limited to 200 for compression members. In the region labeled *elastic buckling*

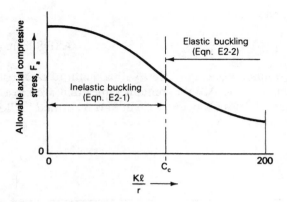

**FIGURE 3-7**  ASDS $F_a$ versus $\dfrac{K\ell}{r}$.

in Figure 3-7, the shape of the curve closely follows the shape of the Euler formula curve of Figure 3-6. It is essentially the same curve with a factor of safety applied. Allowable axial compressive stress on the gross section is denoted $F_a$. The value of $K\ell/r$ that separates elastic buckling from inelastic buckling has been arbitrarily established as that value at which the Euler buckling stress ($f_e$) is equal to $F_y/2$. This $K\ell/r$ value is denoted as $C_c$. Its value may be determined as follows:

$$f_e = \frac{\pi^2 E}{(K\ell/r)^2}$$

Let $f_e = F_y/2$ and let $K\ell/r = C_c$; then

$$\frac{F_y}{2} = \frac{\pi^2 E}{(C_c)^2}$$

from which

$$C_c = \sqrt{\frac{2\pi^2 E}{F_y}}$$

Table 4 in the Numerical Values section of the ASDS lists values of $C_c$ for various values of $F_y$.

For column $K\ell/r$ values less than $C_c$, $F_a$ is determined by

$$F_a = \frac{\left[1 - \dfrac{(K\ell/r)^2}{2C_c^2}\right]F_y}{\dfrac{5}{3} + \dfrac{3(K\ell/r)}{8C_c} - \dfrac{(K\ell/r)^3}{8C_c^3}} \qquad \text{ASDS Eqn. (E2-1)}$$

For column $K\ell/r$ values greater than $C_c$,

$$F_a = \frac{12\pi^2 E}{23(K\ell/r)^2} \qquad\qquad \text{ASDS Eqn. (E2-2)}$$

This is the familiar Euler formula for buckling stress with a factor of safety of 23/12 or 1.92 incorporated. Fortunately, the ASDM contains tables that are most useful in the determination of $F_a$; see Tables C-36 and C-50 in Part 3 entitled "Allowable Stress for Compression Members."

## 3-5

## ANALYSIS OF COLUMNS (ASDS)

Several examples demonstrate the analysis method using the ASDS and available tables. It should be noted that, in accordance with the ASDS, all column analysis and design is based on the *gross* cross-sectional area of the column.

### Example 3-4

Find the allowable compressive load capacity $P_a$ for a W12 × 120 column that has a length of 16 ft. Use A36 steel. The ends are pinned.

**Solution:**

For the W12 × 120,

$$A = 35.3 \text{ in.}^2$$

$$r_y = 3.13 \text{ in.}$$

$$\frac{K\ell}{r} = \frac{1.0(16)(12)}{3.13} = 61$$

The $K\ell/r$ value has been rounded to the nearest whole number for table use. Interpolation is not considered to be warranted. From the ASDM, Part 3, Table C-36, $F_a = 17.33$ ksi. Therefore,

$$P_a = F_a A = 17.33(35.3) = 612 \text{ kips (2720 kN)}$$

### Example 3-5

A W10 × 68 column of A36 steel is to carry an axial load of 300 kips. The length is 20 ft. Determine if the column is adequate if

(a)   The ends are pinned.

(b)   The ends are fixed.

**Solution:**

For the W10 × 68,

$$A = 20.0 \text{ in.}^2$$

$$r_y = 2.59 \text{ in.}$$

(a)   $K = 1.0$ from the ASDS, Table C-C2.1; therefore

$$\frac{K\ell}{r} = \frac{1.0(20)(12)}{2.59} = 93$$

and from Table C-36, $F_a = 13.84$ ksi:

$$P_a = F_a A = 13.84(20.0) = 277 \text{ kips}$$

$$277 \text{ kips} < 300 \text{ kips} \qquad\qquad \textbf{N.G.}$$

(b)   $K = 0.65$ from the ASDS, Table C-C2.1; therefore

$$\frac{K\ell}{r} = \frac{0.65(20)(12)}{2.59} = 60$$

and from Table C-36, $F_a = 17.43$ ksi:

$$P_a = F_a A = 17.43(20.0) = 349 \text{ kips}$$

$$349 \text{ kips} > 300 \text{ kips} \qquad\qquad \textbf{O.K.}$$

---

**Example 3-6** _____

Find the compressive axial load capacity for a built-up column that has a cross section as shown in Figure 3-8. The steel is A36, the length is 18 ft, and the ends are assumed to be fixed-pinned (totally fixed at bottom; rotation free, translation fixed at top).

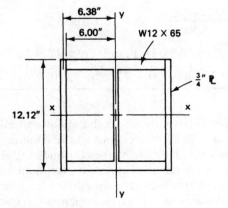
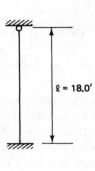

**FIGURE 3-8**   Built-up column.

**Solution:**

Properties of the W12 × 65 are

$$A = 19.1 \text{ in.}^2$$

$$d = 12.12 \text{ in.}$$

$$b_f = 12.00 \text{ in.}$$

$$I_x = 533 \text{ in.}^4$$

$$I_y = 174 \text{ in.}^4$$

Determine the least moment of inertia for the built-up cross section:

$$I = \Sigma I_c + \Sigma Ad^2$$

$$I_x = 533 + 2\left(\frac{1}{12}\right)(0.75)(12.12)^3 = 756 \text{ in.}^4$$

$$I_y = 174 + 2(0.75)(12.12)(6.38)^2 = 914 \text{ in.}^4$$

The x-x axis controls since its moment of inertia is smaller. Notice in the $I_x$ calculation that the $Ad^2$ terms are zero for both the W shape and the plates since the centroidal axes of these component parts coincide with the composite centroidal axis. In the $I_y$ calculation the $I_c$ terms for the plates have been neglected since they are very small. Calculating the radius of gyration, we have

$$\text{total } A = 19.1 + 2(12.12)(0.75) = 37.28 \text{ in.}^2$$

$$r_x = \sqrt{\frac{I_x}{A}} = \sqrt{\frac{756}{37.28}} = 4.50 \text{ in.}$$

With an effective length factor K of 0.80, the capacity may be calculated as usual:

$$\frac{K\ell}{r_x} = \frac{0.8(18)(12)}{4.50} = 38.4 \qquad \text{(use 38.0)}$$

Reference to Table C-36 of the ASDM, Part 3, gives $F_a = 19.35$ ksi:

$$P_a = AF_a = 37.28(19.35) = 721 \text{ kips (3210 kN)}$$

---

Columns are sometimes braced differently about the major and minor axes, as shown by column AB in Figure 3-9. If all connections to the column are assumed to be simple (pinned) connections, the deflected shapes for buckling about the two axes will be as shown. Note that the column is braced so that the unbraced length for weak axis buckling is less than the unbraced length for buckling about the strong axis. In this situation either axis may control, depending on which has the

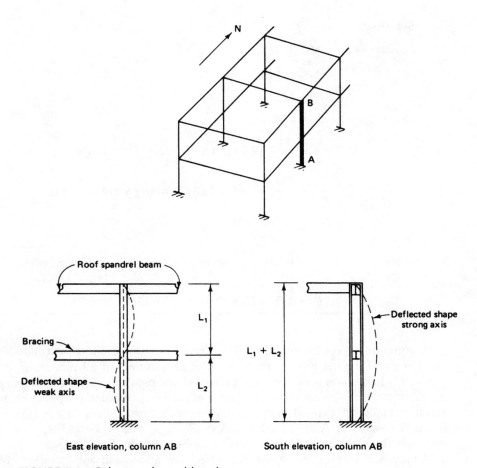

**FIGURE 3-9**   Column unbraced lengths.

associated larger $K\ell/r$ ratio. Naturally, if there is no reasonable certainty that the bracing will not be removed, the columns should be designed with the bracing neglected.

**Example 3-7** _____

Find the allowable compressive axial load capacity for a W10 × 88 that has an unbraced length of 24 ft with respect to axis $x$-$x$ and 12 ft with respect to axis $y$-$y$. Assume an A36 steel member, pin-connected at the top and fixed at the bottom. (Assume that the column is pin-connected at mid-height.)

**Solution:**

For the W10 × 88,

$$A = 25.9 \text{ in.}^2$$

$$r_y = 2.63 \text{ in.}$$

$$r_x = 4.54 \text{ in.}$$

$$\frac{K\ell_y}{r_y} = \frac{1(12)(12)}{2.63} = 54.8 \text{ (top part of column)}$$

$$\frac{K\ell_y}{r_y} = \frac{0.8(12)(12)}{2.63} = 43.8 \text{ (bottom part of column)}$$

$$\frac{K\ell_x}{r_x} = \frac{0.8(24)(12)}{4.54} = 50.8$$

Since 54.8 is the controlling slenderness ratio, $F_a = 17.90$ ksi (using $K\ell/r = 55.0$). Thus

$$P = AF_a = 25.9(17.9) = 463.6 \text{ kips (2060 kN)}$$

A common type of column used in one-story commercial construction is the unfilled circular steel pipe column. With equal stiffness in all directions, it is an efficient compression member. Connections to the pipe column may require special considerations, however. Three categories of pipe for structural purposes are manufactured: standard, extra strong, and double-extra strong. Steel pipe may be manufactured to ASTM A501 or ASTM A53 (Type E and S, Grade B). For design purposes the yield stress for each steel may be taken as $F_y = 36$ ksi.

Square and rectangular structural tubing are also commonly used as building columns. The tubing is manufactured to $F_y = 46$ ksi under ASTM A500 Grade B. The tubular members are also relatively efficient and have an advantage in that end-connection details are simpler than with the pipe columns.

**Example 3-8** _____

Find the allowable compressive load capacity for a 10-in. standard steel pipe column that has an unbraced length of 15 ft. Ends are pin-connected ($K = 1$), and the steel is A501 ($F_y = 36$ ksi).

**Solution:**

For the 10-in. standard steel pipe (ASDM, Part 1),

$$A = 11.9 \text{ in.}^2$$

$$r = 3.67 \text{ in.}$$

$$\frac{K\ell}{r} = \frac{1(15)(12)}{3.67} = 49$$

$$F_a = 18.44 \text{ ksi}$$

$$P_a = AF_a = 11.9(18.44) = 219 \text{ kips (974 kN)}$$

**Example 3-9**

Find the allowable compressive load capacity for an 8-in. double-extra-strong steel pipe column that has an unbraced length of 20 ft. The ends are pin-connected, and the steel is A501 ($F_y = 36$ ksi).

**Solution:**

For the 8-in. double-extra-strong steel pipe (ASDM, Part 1),

$$A = 21.3 \text{ in.}^2$$

$$r = 2.76 \text{ in.}$$

$$\frac{K\ell}{r} = \frac{1(20)(12)}{2.76} = 87$$

$$F_a = 14.56 \text{ ksi}$$

$$P_a = AF_a = 21.3(14.56) = 310 \text{ kips (1379 kN)}$$

**Example 3-10**

Find the allowable compressive load capacity for a rectangular structural tubing, TS8 $\times$ 4 $\times \frac{5}{16}$, that has an unbraced length of 13 ft. The ends are pin-connected, and the steel is A500 Grade B ($F_y = 46$ ksi).

**Solution:**

For the TS8 $\times$ 4 $\times \frac{5}{16}$ (ASDM, Part 1),

$$A = 6.86 \text{ in.}^2$$

$$r_y = 1.62 \text{ in.}$$

$$r_x = 2.80 \text{ in.}$$

$$\frac{K\ell}{r} = \frac{1(13)(12)}{1.62} = 96.3$$

$F_a$ cannot be determined directly using the ASDM since $F_a$ tables are available only for $F_y = 36$ ksi and 50 ksi. Tables 3 and 4 of the Numerical Values section of the ASDS may be used, however.

From Table 4, for $F_y = 46$ ksi, $C_c = 111.6$. Therefore, $K\ell/r < C_c$, and Table 3 can be used to determine $F_a$. From Table 3, enter with the ratio

$$\frac{K\ell/r}{C_c} = \frac{96.3}{111.6} = 0.86$$

and select $C_a = 0.330$.

$F_a$ may then be determined by

$$F_a = C_a F_y$$

$$= 0.330(46) = 15.18 \text{ ksi}$$

$$P_a = AF_a = 6.86(15.18) = 104 \text{ kips } (463 \text{ kN})$$

Also note that ASDS Eq. (E2-1) could be used for $F_a$.

---

**Example 3-11** _____

Find the allowable compressive load capacity for a W8 × 40 with an unbraced length equal to 26 ft. The member is used in a wind-bracing system and is pin-connected. Use A36 steel.

**Solution:**

For the W8 × 40,

$$A = 11.7 \text{ in.}^2$$

$$r_y = 2.04 \text{ in.}$$

$$\frac{K\ell}{r} = \frac{1(26)(12)}{2.04} = 153$$

$$F_a = 6.38 \text{ ksi}$$

$$P_a = AF_a = 11.7(6.38) = 74.6 \text{ kips } (332 \text{ kN})$$

---

## 3-6

## DESIGN OF AXIALLY LOADED COLUMNS

The selection of cross sections for columns is greatly facilitated by the availability of design aids. We have seen that the allowable axial stress $F_a$ depends on the effective slenderness ratio $K\ell/r$ of the column provided. Therefore, there is no direct solution for a required area or moment of inertia. If ASDS Equation (E2-2) were known to control, a required least $r$ could be calculated. This is not a practical solution, however.

Most structural steel columns are composed of W shapes, structural tubing, and/ or pipes. The ASDM, Part 3, contains allowable axial load tables (referred to as the "column load tables") for the popular column shapes. Note that the W shapes include only those of nominal depth of 14 in. and less. Although deeper shapes can be used as columns, they are less efficient than square shapes. Deeper shapes are used for beam applications. Allowable loads ($P_a$) are tabulated as a function of $KL$ (in feet) and cover the common length ranges. The actual column length $L$ for columns in building frames is normally taken as the floor-to-floor distance. The effective length factor $K$ may be determined using the aids discussed in Section 3-3 of this text.

The tables may be used for analysis as well as for design. For instance, in Example 3-4, the allowable compressive load capacity of a W12 × 120 of A36 steel was computed to be 612 kips. From the ASDM, Part 3, column load table for the W12 × 120, with $KL = 1.0 \times 16$ ft, the allowable load of 611 kips may be obtained directly. (Note that the unshaded areas are for $F_y = 36$ ksi and the shaded areas are for $F_y = 50$ ksi.) The General Notes at the beginning of the load tables discuss use and limitations.

The tabular values of allowable loads are with respect to the members' minor (or weak) axis. Although the column load tables are indispensable for the selection of the types of cross sections noted, if built-up sections are required (see Figure 3-1), or a section is desired for which a column load table is not available, a trial-and-error calculation approach will have to be used.

## Example 3-12

Select the lightest W shape for a column that will support an axial load $P$ of 200 kips. The length of the column will be 20 ft and the ends may be assumed to be pinned. Use A36 steel.

### Solution:

Using the ASDM, Part 3, column load tables, with $KL = 1.0 \times 20$ ft and $P = 200$ kips, the following W shapes are observed to be adequate ($P_a \geq P$):

$$W14 \times 61 \quad (P_a = 237 \text{ kips})$$
$$W12 \times 53 \quad (P_a = 209 \text{ kips})$$
$$W10 \times 54 \quad (P_a = 217 \text{ kips})$$
$$W8 \times 67 \quad (P_a = 221 \text{ kips})$$

The W12 × 53 is selected since it is the lightest shape with adequate capacity.

**Example 3-13** _____

Select the lightest W10 for column $AB$ shown in Figure 3-9, $P = 160$ kips. The overall length $(L_x)$ is 30 ft. The weak axis is braced at midheight $(L_y = 15$ ft$)$. Assume pinned ends $(K = 1.0)$ for both axes and A36 steel.

**Solution:**

Assume that the weak axis $(y\text{-}y$ axis$)$ will control. Select the column using the ASDM, Part 3, column load tables, and then check whether the assumption is correct:

$$KL_y = 15 \text{ ft} \qquad P = 160 \text{ kips}$$

Select a W10 × 39 $(P_a = 162$ kips based on weak axis buckling$)$. For the W10 × 39, $r_x = 4.27$ in., $r_y = 1.98$ in.,

$$\frac{K\ell_x}{r_x} = \frac{1.0(30)(12)}{4.27} = 84.3$$

$$\frac{K\ell_y}{r_y} = \frac{1.0(15)(12)}{1.98} = 90.9$$

The larger $K\ell/r$ controls and the assumption of the weak axis controlling was correct. A W10 × 39 will be adequate.

---

Under different conditions, it is possible that the strong axis will control and be the buckling axis for the column of Example 3-13. The column load tables then cannot be used directly. Once an initial section has been selected (based on the assumption that the $y\text{-}y$ axis controls), however, a very rapid analysis check can be made using the tabulated properties at the bottom of the column load tables. The procedure is as follows:

1. Divide the strong-axis effective length $(KL_x)$ by the $r_x/r_y$ ratio.
2. Compare with $KL_y$. The larger of the two values becomes the controlling $KL$.
3. With the controlling $KL$ value, find $P_a$ in the appropriate column load table.

**Example 3-14** _____

Rework Example 3-13 except this time, the weak axis is braced at the third points so that $L_y = 10$ ft. $L_x$ remains at 30 ft.

**Solution:**

As previously, select on the basis of the weak axis controlling:

$$KL_y = 10 \text{ ft} \qquad P = 160 \text{ kips}$$

Try a W10 × 33. $P_a = 167$ kips (based on weak axis controlling) and $r_x/r_y = 2.16$:

$$\frac{KL_x}{r_x/r_y} = \frac{30 \text{ ft}}{2.16} = 13.89 \text{ ft}$$

This is an *equivalent weak-axis length* (i.e., column length based on weak-axis buckling that results in the same capacity as does the 30-ft strong-axis buckling length). *Since 13.89 ft $> L_y$, the strong axis controls.* Rounding the $KL$ of 13.89 ft to 14 ft and entering the column load table for the W10 × 33 gives $P_a = 142$ kips, and

$$142 \text{ kips} < 160 \text{ kips} \qquad\qquad \textbf{N.G.}$$

Try W10 × 39:

$$\frac{KL_x}{r_x/r_y} = \frac{30 \text{ ft}}{2.16} = 13.89 \text{ ft} \approx 14 \text{ ft}$$

$$13.89 \text{ ft} > L_y$$

Therefore, the strong axis controls. From the column load table, $P_a = 170$ kips $> 160$ kips. Therefore, **use a W10 × 39.**

---

## 3-7

## DOUBLE-ANGLE MEMBERS

In single-plane trusses, both tension and compression members are frequently composed of double-angle members. Generally, the two angles are not in contact with each other but are separated by the thickness of a gusset plate used at each end of the member for connection purposes. The two angles must be connected at intervals along their length in accordance with ASDS requirements, however. This is usually accomplished through the use of *connectors* (or separators) made up of *filler plates* in combination with bolts or welds.

The ASDM, Part 1, contains tables of properties of double-angle members with the two angles in contact or separated. In addition, the ASDM, Part 3, contains special tables for double-angle compression members. These tables furnish an allowable concentric compressive load and are of significant value for the analysis and design of double-angle compression members. Their use will be illustrated shortly.

The allowable load tables in Part 3 furnish an allowable concentric compressive load based on an effective length in feet ($KL$) with respect to both the *x-x* and *y-y* axes. The tabulated loads with respect to the *y-y* axis assume a $\frac{3}{8}$-in. spacing back-to-back of the angles. Since double-angle members are commonly used as truss members, it is usual practice to assume $K = 1.0$. The table is limited to compression members with a slenderness ratio $K\ell/r$ of 200 or less.

The allowable loads with respect to the $x$-$x$ axis have been computed in accordance with the column equations of Section E2 of the ASDS, modified, where necessary, by local buckling considerations due to an excessive width-thickness ratio of the angle legs.

The allowable loads with respect to the $y$-$y$ axis have been computed in a similar manner. In addition to the local buckling consideration, however, a flexural-torsional buckling consideration is introduced into the column equations of Section E2 of the ASDS. This is accomplished by the introduction of an *effective slenderness ratio* with respect to the $y$-$y$ axis that, in turn, is based on a critical flexural-torsional buckling stress $F_e$. The use of the allowable load tables of Part 3 removes the need for the complex procedure of determining the critical flexural-torsional buckling stress.

Flexural-torsional buckling may be described as a form of instability of the compression member that involves a combination of bending and twisting of the member. In this sense, it resembles the lateral buckling of unbraced beams as is discussed in Chapter 4 of this text.

At this point, with regard to compression members, it should be noted, that flexural-torsional buckling applies to all shapes except those that are doubly symmetric such as rolled wide-flange sections, tubular sections, and pipe sections. For the singly symmetric shapes such as double-angles and tees, the column capacity *may* be controlled by flexural-torsional buckling. In fact, a singly symmetric shape will buckle in one of two modes: flexural-torsional or simple flexural buckling. One mode or the other will govern the strength of the member. For a more in-depth discussion, see Reference 1.

With respect to double-angle members, where the angles are not in contact, the ASDS, Section E4, requires that the connectors must be spaced along the length of the member so that the local slenderness ratio $a/r_z$ of the individual angle does not exceed $\frac{3}{4}$ times the governing slenderness ratio of the overall member. In addition, at least two intermediate connectors must be used to provide for adequate shear transfer. The connectors must be welded or fully tightened high-strength bolts must be used (high-strength bolts are discussed in Chapter 7). Compliance with the spacing requirements will prevent shearing of the connectors. Note that

$a$ = spacing between connectors (in.)

$r_z$ = radius of gyration with respect to the $z$-$z$ axis of a single angle

**Example 3-15** _____

Determine the capacity of an axially loaded, A36 steel, double-angle compressive truss member. The member is composed of 2 L8 × 4 × $\frac{1}{2}$ with long legs $\frac{3}{8}$ in. back-to-back. The unbraced length of the member is 20 ft. Assume that $K = 1.0$.

**Solution:**

From the double-angle column load tables of the ASDM, Part 3, with $KL_x = 20$ ft and $KL_y = 20$ ft,

$$P_x = 151 \text{ kips} \qquad P_y = 63 \text{ kips}$$

Therefore, the allowable load is the smaller of the two values: $P_a = 63$ kips (280 kN).

---

**Example 3-16**

Determine the capacity of an axially loaded double-angle compression member if the member is composed of 2 L6 $\times$ 4 $\times$ $\frac{5}{8}$ with long legs $\frac{3}{8}$ in. back-to-back. Use A36 steel and $K = 1.0$. The unbraced length of the member is 16 ft. In addition, calculate the required number and spacing of intermediate connectors.

**Solution:**

From the double-angle column load tables of the ASDM, Part 3, with $KL_x = 16$ ft and $KL_y = 16$ ft,

$$P_x = 150 \text{ kips} \qquad P_y = 124 \text{ kips}$$

Therefore, the $y$-$y$ axis controls and the allowable load $P_a$ is 124 kips (552 kN).
    The slenderness ratio with respect to the $y$-$y$ axis is

$$\frac{K\ell_y}{r_y} = \frac{1(16)(12)}{1.67} = 115$$

The maximum slenderness ratio allowed for the individual angle is calculated from

$$\frac{a}{r_z} \le 0.75(115) = 86$$

Therefore,

$$\text{maximum } a = 86(0.864) = 74 \text{ in.}$$

from which

$$\frac{16(12)}{74} = 2.6 \text{ spaces} \qquad \text{(use three spaces)}$$

The spacing for the connectors is then calculated from

$$\frac{16(12)}{3} = 64 \text{ in.} < 74 \text{ in.}$$

This also satisfies the requirement that at least two intermediate connectors must be used.

---

**Example 3-17** _____

Design an unequal leg double-angle compression member for a truss. The unbraced length $L$ is 12 ft. Assume $K$ to be 1.0. Axial load $P$ is 80 kips. Use A36 steel and long legs back-to-back. The member is connected to a $\frac{3}{8}$-in.-thick gusset plate at each end. Calculate the required number and spacing of the intermediate connectors.

**Solution:**

The unbraced length with respect to each axis ($KL_x$ and $KL_y$) is 12 ft. Enter the ASDM, Part 3, with these $KL$ values and load $P$ of 80 kips. Select the most economical member that can carry safely the applied load with respect to each axis.

| Angles | $P_x$ (kips) | $P_y$ (kips) | Weight (lb/ft) |
|---|---|---|---|
| $5 \times 3\frac{1}{2} \times \frac{1}{2}$ | 113 | 104 | 27.2 |
| $6 \times 4 \times \frac{3}{8}$ | 107 | 89 | 24.6 |

From the preceding possible members, select the lightest (most economical): 2 L6 $\times$ 4 $\times \frac{3}{8}$. Since the $y$-$y$ axis controls, the maximum slenderness ratio for the member is

$$\frac{K\ell_y}{r_y} = \frac{1.0(12)(12)}{1.62} = 89$$

The maximum slenderness ratio allowed for the individual angle is calculated from

$$\frac{a}{r_z} \leq 0.75(89) = 67$$

Therefore,

$$\text{maximum } a = 67(0.877) = 59 \text{ in.}$$

Using three equal spaces with two intermediate connectors gives

$$\text{spacing} = \frac{12(12)}{3} = 48 \text{ in.} < 59 \text{ in.} \qquad \textbf{O.K.}$$

This also satisfies the requirement that at least two intermediate connectors must be used.

## 3-8

## COLUMN BASE PLATES (AXIAL LOAD)

Columns are usually supported on concrete supports such as footings or piers. Since the steel of the column is a higher-strength material than the concrete, the column load must be spread out over the support. This is accomplished by use of a rolled-steel *base plate.*

Base plates may be square or rectangular. They must be large enough to keep the actual bearing pressure under the plate below an allowable bearing pressure $F_p$, which may be obtained from ASDS, Section J9, as follows.

For a plate covering the full area of concrete support,

$$F_p = 0.35 f'_c$$

For a plate covering less than the full area of concrete support,

$$F_p = 0.35 f'_c \sqrt{\frac{A_2}{A_1}} \le 0.7 f'_c$$

where

$f'_c$ = specified compressive strength of concrete (ksi)

$A_1$ = area of steel concentrically bearing on a concrete support (in.²)

$A_2$ = maximum area of the portion of the supporting surface that is geometrically similar to and concentric with the loaded area (in.²)

**PHOTO 3-2**   Detail of a base plate welded to a wide-flange column.

The base plate must be thick enough so that bending in the plate itself will not be critical. Two-way bending is involved since as the column pushes down on the base plate, the parts of the plate not directly under the column itself will tend to curl (or deflect) upward. For all but the smallest base plates, the required plate thickness may be determined by considering 1-in.-wide sections of the base plate to act as cantilever beams spanning in each of two directions, fixed at the edges of a rectangle whose sides are $0.80b_f$ and $0.95d$, as shown in Figure 3-10. In Figure 3-10 and in the following discussion, the notation is as follows:

$P$ = total column load (kips)

$A_1 = B \times N$, area of plate (in.$^2$)

$m, n$ = length of cantilever from assumed critical plane of bending, for thickness determination (in.)

$d$ = depth of the column section (in.)

$b_f$ = flange width of the column section (in.)

$F_p$ = allowable bearing pressure on concrete support (ksi)

$F_b$ = allowable bending stress in plate (ksi)

$f_p$ = actual bearing pressure on concrete support (ksi)

$f_c'$ = compressive strength of concrete (ksi)

$t_p$ = thickness of plate (in.)

The column load $P$ is assumed to be uniformly distributed over the top of the base plate within the described rectangle. In turn, it is assumed that the plate is sufficiently rigid and will distribute the applied load so that the pressure underneath the base plate is also uniformly distributed.

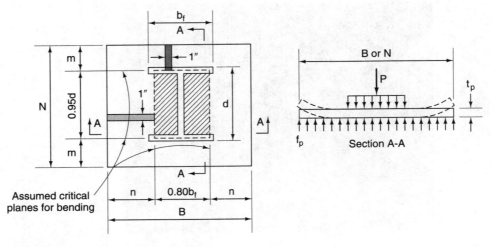

**FIGURE 3-10**   Column base plate design.

The most economical column base plate will result if the length $N$ and the width $B$ are selected so that $m = n$. This condition is approached when

$$N \approx \sqrt{\text{required } A_1} + \Delta$$

where

$$\Delta = 0.5(0.95d - 0.80b_f)$$

Once the length $N$ of the plate has been determined, the required width $B$ can be computed from

$$\text{required } B = \frac{\text{required } A_1}{N}$$

after which $m$ and $n$ can be determined.

The allowable bending stress $F_b$ in the plate is taken as $0.75F_y$ (ASDS, Section F2). The required thickness of the base plate can be computed from

$$t_p = 2m\sqrt{\frac{f_p}{F_y}} \quad \text{or} \quad t_p = 2n\sqrt{\frac{f_p}{F_y}}$$

These formulas are also applicable to beam bearing plates and are further discussed in Section 4-13 of this text. Note that the largest required plate thickness will result from the larger value of $m$ or $n$.

When the base plate is just large enough in area to accommodate the outside envelope of the column (dimensions $d$ and $b_f$), the values of $m$ and $n$ will be small. The resulting plate thickness will then be small also, and the assumption of a uniform bearing pressure under the plate is no longer valid. For *light loads* with this type of base plate, the column load is assumed to be distributed on an H-shaped area of the footing under the plate. This area is cross-hatched in Figure 3-11.

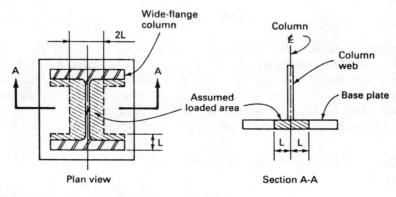

**FIGURE 3-11** Common load distribution.

For small base plates that are more heavily loaded, the required plate thickness may be taken as

$$t_p = 2n'\sqrt{\frac{f_p}{F_y}} \qquad\qquad \text{(ASDS, Part 3)}$$

where

$$n' = \frac{\sqrt{db_f}}{4}$$

$d$ = depth of column section (in.)

$b_f$ = flange width of column section (in.)

To provide for a smooth design transition between small plates that are heavily loaded and those which are lightly loaded, the following coefficients have been developed:

$$\lambda = \frac{2(1 - \sqrt{1 - q})}{\sqrt{q}} \le 1.0$$

$$q = \frac{4f_p db_f}{(d + b_f)^2 F_p} < 1.0$$

When $\lambda$ is less than 1.0 or when $q$ is less than 0.64, the design for lightly loaded base plates governs. If $q > 0.64$, take $\lambda$ as 1.0.

After computing $\lambda$, calculate $\lambda n'$ and determine the required plate thickness from

$$t_p = 2(\lambda n')\sqrt{\frac{f_p}{F_y}}$$

Since the expression for required $t_p$ is in the same form as the expression for $t_p$ for large plates, the largest dimension ($m$, $n$, or $\lambda n'$) will control and the expression may be rewritten as

$$\text{required } t_p = 2c\sqrt{\frac{f_p}{F_y}}$$

where $c$ represents the maximum value of ($m$, $n$, or $\lambda n'$).

The treatment of small base plates first appeared in the 8th edition of the ASDM. The small plate analysis theory was completely revised in the 9th edition and was, in turn, revised again in subsequent printings of the 9th edition. Reference 2 contains some interesting background on the current approach.

A step-by-step procedure for column base plate design is as follows:

1.   Determine the required base plate area:

$$\text{required } A_1 = \frac{P}{F_p}$$

2.  Select $B$ and $N$ so that $m$ and $n$ are approximately equal (if possible). Use

$$N \approx \sqrt{\text{required } A_1} + \Delta$$

where

$$\Delta = 0.5(0.95d - 0.80b_f)$$

$$\text{required } B = \frac{\text{required } A_1}{N}$$

Select $B$ and $N$ (usually to full inches) such that

$$B \times N \geq \text{required } A_1$$

3.  Calculate the actual bearing pressure under the plate:

$$f_p = \frac{P}{BN}$$

4.  Calculate $m$, $n$, and $n'$:

$$m = \frac{N - 0.95d}{2}$$

$$n = \frac{B - 0.80b_f}{2}$$

$$n' = \frac{\sqrt{db_f}}{4}$$

5.  Check for the case of a lightly loaded small base plate using

$$q = \frac{4f_p db_f}{(d + b_f)^2 F_p} < 1.0$$

and, if necessary (when $q < 0.64$),

$$\lambda = \frac{2(1 - \sqrt{1 - q})}{\sqrt{q}} \leq 1.0$$

If $q < 0.64$, the design for a lightly loaded base plate governs ($\lambda$ should be determined). If $q \geq 0.64$, then $\lambda = 1.0$. The upper limit ($\lambda = 1.0$) can always be used as a conservative assumption and will simplify the computations a bit.

6.  Calculate the required plate thickness $t_p$ using $c$ as the larger of ($m$, $n$, or $\lambda n'$):

$$\text{required } t_p = 2c\sqrt{\frac{f_p}{F_y}}$$

7.  Specify the base plate: width, thickness, and length.

The ASDM, Part 3, column base plate design procedure varies slightly from the preceding. In the ASDM procedure, it is shown that the particular size of the supporting area $A_2$ (which is assumed to be a variable) can be determined such that the allowable bearing pressure will be at its maximum value of $0.7F_y$. This will result in a minimum required base plate area.

In the examples that follow, it is assumed that the base plate will cover the full area of a concrete support. We therefore conservatively take $F_p$ to be $0.35f_c'$. This will simplify the computations.

**Example 3-18** _____

Design a rectangular base plate for a W14 × 74 column that is to carry an axial load of 350 kips. Assume that the base plate will cover the full area of a concrete pier of $f_c' = 3$ ksi. Use A36 steel.

**Solution:**

From the ASDS, Section J9,

$$F_p = 0.35f_c' = 0.35(3) = 1.05 \text{ ksi}$$

1.  The required area is

$$\text{required } A_1 = \frac{P}{F_p} = \frac{350}{1.05} = 333 \text{ in.}^2$$

2.  For the W14 × 74, $d = 14.17$ in. and $b_f = 10.07$ in. Referring to Figure 3-10, $B$ and $N$ may be selected as follows:

$$\Delta = 0.5(0.95d - 0.80b_f)$$

$$= 0.5[0.95(14.17) - 0.80(10.07)]$$

$$= 2.70 \text{ in.}$$

$$N \approx \sqrt{\text{required } A_1} + \Delta$$

$$= \sqrt{333} + 2.70 = 20.95 \text{ in.}$$

$$B = \frac{\text{required } A_1}{N} = \frac{333}{20.95} = 15.89 \text{ in.}$$

Round these required plate dimensions to the next whole inch:

$$N = 21 \text{ in.} \qquad B = 16 \text{ in.}$$

$$\text{area furnished} = 21(16) = 336 \text{ in.}^2 > 333 \text{ in.}^2 \qquad \textbf{O.K.}$$

3.  The actual bearing pressure under the plate is

$$f_p = \frac{P}{BN} = \frac{350}{16(21)} = 1.042 \text{ ksi} < 1.05 \text{ ksi} \qquad \textbf{O.K.}$$

4. Calculate $m$, $n$, and $n'$:

$$m = \frac{N - 0.95d}{2} = \frac{21 - 0.95(14.17)}{2} = 3.77 \text{ in.}$$

$$n = \frac{B - 0.80b_f}{2} = \frac{16 - 0.80(10.07)}{2} = 3.97 \text{ in.}$$

$$n' = \frac{\sqrt{db_f}}{4} = \frac{\sqrt{14.17(10.07)}}{4} = 2.99 \text{ in.}$$

5. Check for the case of a lightly loaded small base plate using

$$q = \frac{4f_p db_f}{(d + b_f)^2 F_p}$$

$$= \frac{4(1.042)(14.17)(10.07)}{(14.17 + 10.07)^2 1.05}$$

$$= 0.964 > 0.64$$

Therefore, take $\lambda = 1.0$. Alternatively, the upper limit ($\lambda = 1.0$) can always be used as a conservative assumption. If we were to make that assumption, it can be seen that in this case $\lambda n'$ would be 2.99 and would not exceed either $m$ or $n$ and, therefore, would not control.

6. Calculate the required plate thickness using $c$ as the larger of $m$, $n$, and $\lambda n'$:

$$\text{required } t_p = 2c \sqrt{\frac{f_p}{F_y}} = 2(3.97) \sqrt{\frac{1.042}{36}} = 1.351 \text{ in.}$$

7. Refer to the ASDM, Part 1, Bars and Plates—Product Availability, for information on plate thicknesses available. Select a thickness of $1\frac{3}{8}$ in. **Use a base plate 16 × $1\frac{3}{8}$ × 1′–9.**

---

## Example 3-19

Design an economical rectangular column base plate for a W12 × 50 column that is to carry an axial load of 65 kips. All steel is A36. Assume that the base plate will cover the full area of the concrete support. Here, $f_c' = 3.0$ ksi; therefore,

$$F_p = 0.35 f_c' = 0.35(3.0) = 1.05 \text{ ksi}$$

**Solution:**

1. The required area is

$$\text{required } A_1 = \frac{P}{F_p} = \frac{65}{1.05} = 61.9 \text{ in.}^2$$

2.  For the W12 × 50, $d = 12.19$ in. and $b_f = 8.08$ in. The area of the rectangular profile of the column is

$$b_f d = 8.08(12.19) = 98.5 \text{ in.}^2$$

$$98.5 \text{ in.}^2 > 61.9 \text{ in.}^2$$

Therefore, to accommodate the envelope of the W12 × 50, $B$ and $N$ must be selected to the full inch dimensions that are larger than $b_f$ and $d$. Use $N = 13$ in. and $B = 9$ in. This will furnish an area of 117 in.².

3.  The actual bearing pressure under the plate is

$$f_p = \frac{P}{BN} = \frac{65}{9(13)} = 0.556 \text{ ksi} < 1.05 \text{ ksi} \qquad \textbf{(O.K.)}$$

4.  Calculate $m$, $n$, and $n'$:

$$m = \frac{N - 0.95d}{2} = \frac{13 - 0.95(12.19)}{2} = 0.710 \text{ in.}$$

$$n = \frac{B - 0.80b_f}{2} = \frac{9 - 0.80(8.08)}{2} = 1.268 \text{ in.}$$

$$n' = \frac{\sqrt{db_f}}{4} = \frac{\sqrt{12.19(8.08)}}{4} = 2.48 \text{ in.}$$

5.  Check for a lightly loaded plate:

$$q = \frac{4f_p db_f}{(d + b_f)^2 F_p} = \frac{4(0.556)(12.19)(8.08)}{(12.19 + 8.08)^2(1.05)} = 0.508$$

When $q < 0.64$, the design for a lightly loaded plate governs. Since in this case $q$ is less than 0.64, calculate $\lambda$ and $\lambda n'$:

$$\lambda = \frac{2(1 - \sqrt{1 - q})}{\sqrt{q}} = \frac{2(1 - \sqrt{1 - 0.508})}{\sqrt{0.508}} = 0.838$$

and

$$\lambda n' = 0.838(2.48) = 2.08 \text{ in.}$$

6.  Calculate the required plate thickness using $c$ as the largest of $m$, $n$, and $\lambda n'$:

$$t_p = 2c\sqrt{\frac{f_p}{F_y}}$$

$$= 2(2.08)\sqrt{\frac{0.556}{36}}$$

$$= 0.517 \text{ in.}$$

Referring to ASDM, Part 1, select a plate thickness of $\frac{9}{16}$ in.

7.  **Use a base plate 9 × $\frac{9}{16}$ × 1′–1.**

The length and width of column base plates are usually selected in multiples of full inches and their thickness in multiples of $\frac{1}{8}$ in. if the plate thickness required is between 1 and 3 in. Since good contact between the column and the plate is a necessity, the ASDS requires that rolled steel bearing plates over 2 in. but not over 4 in. in thickness must be straightened by pressing or milling. Plates over 4 in. in thickness must be milled. The bottom surface of the plates need not be milled since a layer of grout will be placed between the plate and the underlying foundation to ensure full bearing contact. Plates 2 in. or less in thickness may be used without milling provided that a satisfactory contact bearing is obtained between the plate and the column.

Steel columns and their base plates are usually anchored to the foundation by steel anchor bolts embedded in the concrete. The anchor bolts pass through the base plate in slightly oversize holes. This allows for some misalignment of the bolts without redrilling the base plate holes or without taking out and resetting the anchor bolts. Angles may be used to bolt or weld the base plate to the column. If so, the anchor bolts will also pass through the angles. With the exception of base plates for larger columns, current practice is to omit the angles and shop weld the base plate to the column, thereby permitting the column and base plate to be shipped to the job site as a single unit. For a more thorough discussion on some of the practical aspects of column bases, see Reference 3.

## REFERENCES

[1] Cynthia J. Zahn and Nestor R. Iwankiw, "Flexural-Torsional Buckling and Its Implications for Steel Compression Member Design," *AISC Engineering Journal,* 4th Qtr., 1989.

[2] William A. Thornton, "Design of Base Plates for Wide-Flange Columns, A Concatenation of Methods," *AISC Engineering Journal,* 4th Qtr., 1990.

[3] David T. Ricker, "Some Practical Aspects of Column Base Selection," *AISC Engineering Journal,* 3rd Qtr., 1989.

## PROBLEMS

**3-1.** Two channels C12 × 25 serve as a 35-ft-long pin-ended column as shown. Find the Euler buckling load $P_e$. The proportional limit is 34 ksi.

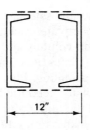

**PROBLEM 3-1**

**3-2.** Calculate $P_e$ for the pipe column shown for the following lengths ($E = 29,000$ ksi, and the proportional limit $= 34$ ksi):

**(a)** 15 ft.

**(b)** 20 ft.

**(c)** 30 ft.

**(d)** 40 ft.

**(e)** 50 ft.

Draw a graph of $f_e$ versus $\ell$ (ft) showing the results of the calculation above. Show the proportional limit. Assume pinned ends.

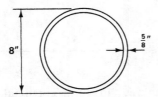

**PROBLEM 3-2**

**3-3.** A W8 × 40 is to serve as a pin-ended, 18-ft-long column. The proportional limit is 34 ksi. Find the allowable axial compressive load using Euler's formula and a factor of safety of 2.5.

**3-4.** Use Euler's formula and a factor of safety of 2.5 to design a W14 column to support a load of 350 kips. The length is 34 ft. The ends are pinned. The proportional limit is 34 ksi.

**3-5.** A solid square steel bar 2 in. × 2 in. in cross section is used as a pin-ended column. The proportional limit is 34 ksi. Find the allowable axial compressive load using Euler's formula and a factor of safety of 3.0 if the unbraced length is

**(a)** 4 ft.

**(b)** 8 ft.

**3-6.** A W10 × 49 is to serve as a pin-ended, 40-ft-long column. The column is braced at midheight with respect to its weak axis. The proportional limit is 34 ksi. Find the allowable axial compressive load using Euler's formula and a factor of safety of 2.5.

**3-7.** Find the allowable axial compressive load $P_a$ for the following compression members. ASTM A501 steel furnishes $F_y = 36$ ksi.

| Shape | K | L (ft) | Steel |
|-------|-----|------|-------|
| **(a)** W16 × 67 | 1.3 | 16.0 | A36 |
| **(b)** HP12 × 53 | 1.0 | 14.0 | A588 |
| **(c)** Pipe 8 X-Strong | 0.8 | 26.3 | A501 |

**3-8.** Find the allowable axial compressive load for a W12 × 50 column of A36 steel. End conditions and the unbraced length are as follows:

**(a)** Fixed ends, $L = 15$ ft.

**(b)** Pinned ends, $L = 15$ ft.

**(c)** One end pinned, one fixed, $L = 15$ ft.

**(d)** Pinned ends, $L = 30$ ft.

**(e)** Pinned ends, $L_x = 15$ ft, $L_y = 7.5$ ft.

**3-9.** Find the allowable axial compressive load for a W10 × 39 column of A36 steel. End conditions and the unbraced length are as follows:

**(a)** Pinned ends, $L = 18$ ft.

**(b)** Pinned ends, $L = 26$ ft.

**(c)** Pinned ends, $L_x = 26$ ft, $L_y = 13$ ft.

**3-10.** Find the allowable axial compressive load for a built-up column composed of a W12 × 50 with a 12 × $\frac{1}{2}$ plate bolted to each flange. The member is 30 ft long with pinned ends and is A36 steel.

**3-11.** Find the allowable axial compressive load for the following compression members:

**(a)** W8 × 58, fixed ends, $L = 16$ ft–6 in., A36 steel.

**(b)** W12 × 96, pinned ends, $L = 20$ ft, A36 steel.

**(c)** W14 × 211, pinned ends, $L = 20$ ft, A441 steel.

**(d)** W12 × 87, fixed at one end, pinned at one end, $L_x = 24$ ft, $L_y = 12$ ft, A36 steel

**(e)** W10 × 88, fixed at one end, pinned at one end, $L = 24$ ft–6 in., $F_y = 50$ ksi.

**(f)** W14 × 176, fixed ends, $L = 22$ ft, $F_y = 46$ ksi.

**(g)** 10-in. standard steel pipe, pinned ends, $L = 22$ ft, A36 steel.

**(h)** W10 × 54, pinned ends, $L_x = 22$ ft, $L_y = 11$ ft, A36 steel.

**(i)** Structural tubing TS6 × 4 × $\frac{3}{8}$, pinned ends, $L = 14$ ft, $F_y = 46$ ksi.

**3-12.** Find $P_a$ for the built-up column shown. Use A36 steel. The length is 18 ft and the ends are fixed.

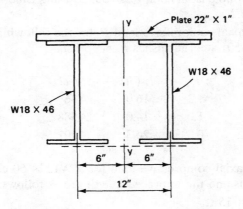

**PROBLEM 3-12**

**3-13.** Find $P_a$ for the built-up column shown. Use A36 steel. The length is 19 ft–6 in., and $K = 1.0$.
  **(a)** Find $P_a$ for the core W14 × 257.
  **(b)** Find $P_a$ for the built-up column.

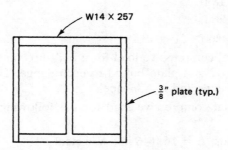

**PROBLEM 3-13**

**3-14.** Select the lightest shape indicated to serve as an axially loaded column. Use A36 steel (A500 Grade B for tubing, with $F_y = 46$ ksi).

|     | Shape | P | Ends | L (ft) |
|-----|-------|-----|-------------|--------|
| **(a)** | W | 100 | Pinned | 24 |
| **(b)** | W | 280 | Fixed | 20 |
| **(c)** | HP | 260 | Fixed/pinned | 12 |
| **(d)** | Tube | 200 | Pinned | 15 |

**3-15.** Select the lightest shape indicated to serve as an axially loaded column.

| | Shape | $P$ | Ends | $L$ ($ft$) | Steel |
|---|---|---|---|---|---|
| **(a)** | W | 600 | Fixed | 15 | A36 |
| **(b)** | W | 600 | Fixed | 15 | A441 |
| **(c)** | W | 100 | Fixed | $L_x = 30$, $L_y = 15$ | A36 |
| **(d)** | W | 100 | Pinned | 26 | A36 |
| **(e)** | W | 275 | Fixed/pinned | 10 | A36 |
| **(f)** | Std. pipe | 41 | Pinned | 12 | A501 |
| **(g)** | Tube | 50 | Pinned | 12 | A500 B |
| **(h)** | W | 2700 | Pinned | 15 | A36 |
| **(i)** | W | 2700 | Pinned | 15 | A441 |

**3-16.** Select the lightest W12 columns of A36 steel to support the axial loads as follows:

(a) $P = 400$ kips, pinned ends, $L_x = 24$ ft, $L_y = 18$ ft.

(b) $P = 400$ kips, pinned ends, $L_x = 24$ ft, $L_y = 12$ ft.

(c) $P = 400$ kips, pinned ends, $L_x = 24$ ft, $L_y = 8$ ft.

(d) $P = 700$ kips, pinned ends, $L_x = 31$ ft, $L_y = 16$ ft.

**3-17.** Design a typical interior column for the four-story braced frame shown. The frames are 32 ft on center. Assume that $K = 1.0$. Loads may be determined from the following: the roof snow load is 45 psf, the roof dead load is 30 psf, the floor live load is 100 psf and 15 psf for partitions, and the floors are composed of 5-in.-thick reinforced concrete slabs. Include 10 psf dead load for both roof and floor to account for the weight of beams and girders. Splice the columns just above the second-floor level. The steel is A441.

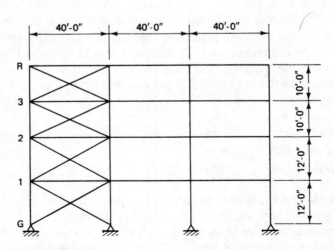

**PROBLEM 3-17**

**3-18.** An axially loaded column has its weak axis braced at the third points. $P = 600$ kips, the overall length is 27 ft, $K = 1.0$, and the steel is A36. Select the lightest W shape.

**3-19.** Select or design the most economical column to support an axial load of 550 kips. $K = 1.3$, and $L = 16$ ft. The column must fit into a 12 in. $\times$ 11 in. space. The column is to be A36 steel.

**3-20.** Compute the allowable axial compressive load for a double-angle strut consisting of 2L6 $\times$ 4 $\times$ $\frac{3}{4}$ long legs back-to-back, straddling a $\frac{3}{8}$-in. gusset plate. The strut is a truss member of A36 steel with an unbraced length of

   **(a)** $L = 15$ ft.

   **(b)** $L = 34$ ft.

**3-21.** Compute the allowable axial compressive load for a double-angle strut consisting of 2L7 $\times$ 4 $\times$ $\frac{3}{8}$, long legs back-to-back (LLBB), straddling a $\frac{3}{8}$-in. gusset plate. The strut is a truss member of A36 steel with an unbraced length $L = 8$ ft. Calculate the required number and spacing of intermediate connectors.

**3-22.** Determine the maximum preferable unbraced length for a truss compression member consisting of 2L6 $\times$ 4 $\times$ $\frac{1}{2}$, LLBB, straddling a $\frac{3}{8}$-in. gusset plate. The steel is A36.

**3-23.** Select the lightest double-angle truss compression member, LLBB, straddling a $\frac{3}{8}$-in. gusset plate for the conditions given. The steel is A36. Also calculate the required number and spacing of intermediate connectors.

   **(a)** $P = 100$ kips, $L = 10$ ft.

   **(b)** $P = 32$ kips, $L = 19$ ft.

   **(c)** $P = 50$ kips, $L = 8$ ft.

**3-24.** Design a *square* base plate for a W10 $\times$ 112 column that supports an axial load of 450 kips. The base plate is to be of A36 steel. The footing size (pile cap) is 8 ft square. The compressive strength of concrete $f'_c$ is 3000 psi.

**3-25.** Design the lightest base plates for the columns and loads given. The plate size is to be full inches, and the thickness is to be governed by the ASDM. Determine the *weight* of the plate in each case. Use A36 steel. $F_p = 1050$ psi. Note that the W18 $\times$ 119 is not commonly used as a column.

|     | Column | $P$ (kips) |
| --- | --- | --- |
| **(a)** | W12 $\times$ 45 | 75 |
| **(b)** | W14 $\times$ 132 | 650 |
| **(c)** | W18 $\times$ 119 | 700 |

**3-26.** An axially loaded W10 $\times$ 60 column is supported on a 1-in.-thick base plate having $B = 13$ in. and $N = 1$ ft–3 in. The concrete pedestal on which the base plate is set is an 18-in. square. All steel is A36, and $f'_c = 3500$ psi. Find the maximum allowable axial load that this base plate can support.

# CHAPTER 4

# Beams

## 4-1

## INTRODUCTION

Beams are among the most common members that one will find in structures. They are structural members that carry loads that are applied at right angles to the longitudinal axis of the member. These loads cause the beam to bend. In this chapter we consider beams that carry no axial force. Figure 1-1a, b, and c illustrates some typical examples of beam applications.

When visualizing a beam (or any structural member) for the purposes of analysis or design, it is convenient to think of the member in some idealized form. This idealized form represents as closely as possible the actual structural member, but it has the advantage that it can be dealt with mathematically. For instance, in Figure 4-1a the beam is shown with simple supports. These supports, a pin (knife-edge or hinge) on the left and a roller on the right, create conditions that are easily treated mathematically when it becomes necessary to find beam reactions, shears, moments, and deflections. Recall that the pin support will provide vertical and horizontal reactions (but no resistance to rotation), and the roller pin will provide only a vertical reaction. This is particularly significant for bridges, where provisions must be made for expansion and contraction due to temperature changes. In buildings, each support is generally capable of furnishing vertical and horizontal reactions. The beams, however, are still considered to be simply supported since the requirement of a simple support is to permit freedom of rotation. In Figure 4-1b the cantilever beam has a fixed support on the left side. This type of support provides vertical and horizontal reactions as well as resistance (or a reaction) to rotation. The one fixed support is sufficient for static equilibrium of the beam. Although the idealized conditions generally will not exist in the actual structure, the actual conditions will approximate the ideal conditions and should be close enough to allow for a reasonable analysis or design.

In the process of beam design, we will be concerned initially with the *bending moment* in the beam. The bending moment is produced in the beam by the loads it supports. Other effects, such as shear or deflection, may eventually control the

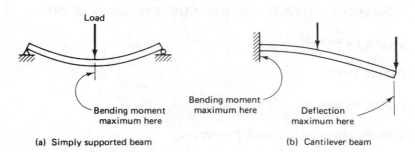

(a) Simply supported beam                    (b) Cantilever beam

**FIGURE 4-1**   Beam types.

design of the beam and will have to be checked. But usually moment is critical, and it is, therefore, of initial concern.

Beams are sometimes called by other names that are indicative of some specialized function(s):

Girder: a major, or deep, beam that often provides support for other beams

Stringer: a main longitudinal beam, usually in bridge floors

Floor beam: a transverse beam in bridge floors

Joist: a light beam that supports a floor

Lintel: a beam spanning an opening (a door or a window), usually in masonry construction

Spandrel: a beam on the outside perimeter of a building that supports, among other loads, the exterior wall

Purlin: a beam that supports a roof and frames between or over supports, such as roof trusses or rigid frames

Girt: generally, a light beam that supports only the lightweight exterior sides of a building (typical in preengineered metal buildings)

## 4-2

## THE MECHANICS OF BENDING

When the simply supported beam of Figure 4-2a is subjected to two symmetrically placed loads, it bends as shown by the deflected shape. The diagrams of the induced shear ($V$) and moment ($M$) are as shown in Figure 4-2b and c. These diagrams neglect the weight of the beam and consider the two concentrated loads ($P$) only. It is assumed that the reader is completely familiar with the development of shear and moment diagrams from strength of materials. We consider a section at midspan (or anywhere between the two concentrated loads) where the moment is maximum as shown in Figure 4-2c. The maximum stress due to flexure (bending) in the beam may be determined by use of the *flexure formula:*

$$f_b = \frac{Mc}{I} = \frac{M}{S}$$

where

$f_b$ = computed bending stress (maximum at top and/or bottom)

$M$ = maximum applied moment

$c$ = distance from the neutral axis to the extreme outside of the cross section

$I$ = moment of inertia of the cross section about the bending neutral axis

$S$ = section modulus ($S = I/c$) of the cross section about the bending neutral axis

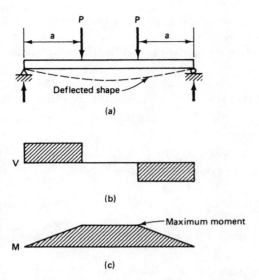

**FIGURE 4-2**   Load, shear, and moment diagrams for a simply supported beam.

The fundamental assumptions and the derivation of the flexure formula may be found in most textbooks on strength of materials.

The actual use of the flexure formula is straightforward, although the units must be carefully considered. Any conversion factors must be applied so that compatibility of units exists. For example, in the formula $f_b = M/S$, the *usual* units are

$$f_b: \text{ kips/in.}^2 \text{ or ksi (stress)}$$

$$M: \text{ ft-kips (moment)}$$

$$S: \text{ in.}^3 \text{ (section modulus)}$$

The *units* for the calculation of $M/S$ may be written

$$f_b = \frac{M}{S}: \quad \frac{\text{ft-kips}}{\text{in.}^3}$$

For compatibility of units, with stress $f_b$ resulting as ksi, a conversion factor of 12 in./ft must be used. For example,

$$\frac{50 \text{ ft-kips}}{30 \text{ in.}^3} \times 12 \frac{\text{in.}}{\text{ft}} = 20 \frac{\text{kips}}{\text{in.}^2} \quad (137.9 \text{ MPa})$$

For numerical problems worked out in this text, necessary conversion factors are shown without further explanation.

It should be noted that for *empirical formulas,* such as those discussed in Section 4-3, units are sometimes *not* compatible (e.g., $\sqrt{F_y}$). In these formulas numerical values with units precisely as defined in the ASDM must be used carefully.

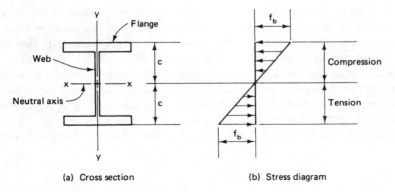

(a)  Cross section                              (b)  Stress diagram

**FIGURE 4-3**   Simple beam bending.

Assuming that the beam of Figure 4-2 is a typical wide flange (W shape), Figure 4-3 shows the cross section and depicts the resulting bending stress diagram. The shape of the diagram is typical for bending stress at any point along the beam. Several points should be noted.

1.   For wide-flange beams, the moment of inertia about the $x$-$x$ axis, $I_x$, is greater than the moment of inertia about the $y$-$y$ axis. The beam is oriented so that bending occurs about the $x$-$x$ axis. This is true except in very rare situations.

2.   In this case, due to symmetry, the neutral axis is at the center of the cross section. The $c$ distance is equal whether on the tension or compression side.

3.   The maximum stress occurs at the top and the bottom of the cross section. The beam of Figure 4-2 bends so that compression occurs above the neutral axis and tension occurs below the neutral axis (commonly, this is called *positive moment*).

4.   Generally, only the maximum bending stress is of interest. Therefore, unless otherwise stated, $f_b$ is assumed to be the maximum stress. The flexure formula may also be used to find the stress at any level in the cross section by substituting in place of $c$ the appropriate distance to that level from the neutral axis.

Since $f_b$ is bending stress that is induced in the beam by the applied loads (and the resulting moment), the steel of which the beam is composed must have sufficient strength to resist this moment without failure. (The reader is referred to Section 1-6 of this text for a discussion of *failure*). The allowable bending stres $F_b$ is specified by the ASDS based on the type of steel that is being used and other conditions that affect the strength of the beam in bending.

The stresses of Figure 4-3b exist inside the beam. The summation of the stresses acting on their appropriate areas makes up an internal force system that creates what is frequently called an *internal couple* (or internal resisting moment). For equilibrium, the internal resisting moment at any point in the beam must be equal to the external applied moment at the same point (the external applied moment may be determined from the moment diagram of Figure 4-2c).

In Figure 4-4, if the external applied moment becomes the maximum allowed, the actual bending stresses at top and bottom of the beam will be equal to the allowable bending stress $F_b$. Additional moment should not be applied, since this would cause the actual bending stress to exceed $F_b$. The internal resisting moment (or, simply, resisting moment) that exists when the bending stress is $F_b$ is termed $M_R$.

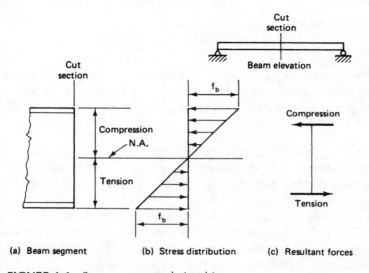

(a) Beam segment         (b) Stress distribution         (c) Resultant forces

**FIGURE 4-4**   Stress-moment relationships.

From the flexure formula, the resisting moment $M_R$ can be calculated by substituting $F_b$ for $f_b$, and $M_R$ for $M$:

$$F_b = \frac{M_R c}{I} = \frac{M_R}{S}$$

Then

$$M_R = F_b S$$

In this text $M_R$ is defined as the bending strength or allowable moment for a beam cross section. We consider this definition to be applicable for any value of unbraced length of the compression flange $L_b$ (the importance of which is discussed in the next section).

The use of the flexure formula, in any of its various forms, is basic to beam analysis and design.

## 4-3

### ALLOWABLE BENDING STRESS

In dealing with beam problems, it is necessary to have an understanding of the specified allowable bending stress $F_b$, the maximum bending stress to which a beam should be subjected. The ASDS treats this topic in Section F1.1. Neglecting later complications, the basic allowable bending stress (in both tension and compression) to be used for most rolled shapes is

$$F_b = 0.66F_y$$

where $F_y$ is the material yield stress. For a member to qualify for an allowable bending stress $F_b$ of $0.66F_y$, it must have an axis of symmetry in, and be loaded in, the plane of the web. An important condition associated with the use of this value for $F_b$ is the *lateral support* of the compression flange. The compression flange behaves somewhat like a column, and it will tend to buckle to the side, or laterally, as the stress increases if it is not restrained in some way. Varying amounts and types of lateral support may be present. In Figure 4-5a, a concrete slab encases the

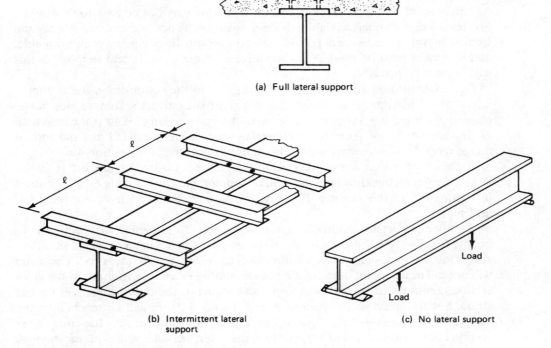

(a)  Full lateral support

(b)  Intermittent lateral
support

(c)  No lateral support

**FIGURE 4-5**  Lateral support conditions.

top flange (assumed to be in compression) and is mechanically anchored to it. The slab forms a horizontal diaphragm and effectively braces the top flange against any lateral movement. This may be termed *full lateral support*. In Figure 4-5c, *no* lateral support exists for the top flange. In Figure 4-5b, there are three points of lateral support. The *distance* between the points of lateral support (whatever it may be) *in inches* is denoted $\ell$. For convenience, we denote this distance $L_b$ when it is *in feet*. To qualify for $F_b = 0.66F_y$, the compression flange of a beam must have *adequate* lateral support such that

$$\ell \leq \frac{76b_f}{\sqrt{F_y}} \quad \text{and} \quad \frac{20,000}{(d/A_f)F_y}$$

where

$b_f$ = flange width of the beam (in.)

$F_y$ = yield stress of the steel (ksi)

$d$ = depth of the beam (in.)

$A_f$ = area of compression flange (in.$^2$)

The smaller of the two values of $\ell$ is a tabulated property for each W shape (dependent on $F_y$) and is designated $L_c$ (ft). See the ASDM, Part 2, Allowable Stress Design Selection Table.

The amount of lateral support actually available may not be easy to determine. For instance, if a concrete slab rests on a beam but is not anchored to it, only the friction between the two will provide lateral support. Its adequacy is questionable, and judgment must be used. A conservative estimate of no lateral support, in this case, would be prudent.

One must also be aware that it is the support of the *compression* flange that is important. When moments change from positive to negative, the compression flange changes from the top flange of the beam to the bottom flange. This is the case with an overhanging beam. A special detail will normally be required to brace the bottom flange where it is in compression. This is discussed further in Section 4-6.

If the compression flange of a beam has inadequate lateral support ($\ell$ is too large), the lateral buckling tendency will be counteracted by *reducing $F_b$*, as discussed in Section 4-6 of this chapter. For now, we assume beams to have adequate lateral support.

Another important condition that must be met if the beam cross section is to qualify for $F_b = 0.66F_y$ deals with the response of the beam in an overload situation. Allowable stress design assumes failure to occur when $F_y$ is first reached. The beam will not fail at this point because it has a substantial reserve of strength. If the cross section continues to strain under increased moment, the outer fibers will further strain, but the stress will remain at $F_y$ (see Figure 1-2). $F_y$ will be reached by the fibers at levels progressively closer to the neutral axis until virtually the entire cross section is stressed to $F_y$. When this occurs, the beam has achieved its *plastic moment capacity*. (This is the basis for *plastic design*.) The cross section, however, must be proportioned so that no local buckling of the flange or web occurs before the plastic moment capacity is achieved. A cross section that meets this criterion is said to be *compact*. The 1989 ASDS, Section B5.1, classifies steel sections as compact, noncompact, and slender element sections. Only compact sections qualify for $F_b = 0.66F_y$.

The test for compactness is found in the ASDS, Section B5 and Table B5.1. The governing criteria are the width/thickness ratios of the compression flange and compression web elements of the cross section. These are called, respectively, the *flange criterion* and *web criterion*. Assuming that there are no axial loads on the beam (and this assumption will be made for all beams until beam-columns are discused in Chapter 6), and using the definitions for width and thickness from the ASDS, Section B5.1, the two equations required may be simplified as follows. For a section to be considered compact,

The flange criterion is

$$\frac{b_f}{2t_f} \le \frac{65}{\sqrt{F_y}}$$

The web criterion is

$$\frac{d}{t_w} \le \frac{640}{\sqrt{F_y}}$$

where

$b_f$ = flange width of the beam (in.)

$t_f$ = flange thickness of the beam (in.)

$F_y$ = material yield stress (ksi)

$d$ = depth of the beam (in.)

$t_w$ = web thickness of the beam (in.)

*Both* the flange and the web criteria must be satisfied for a member to be considered compact.

**Example 4-1** _____

Determine whether a W18 × 76 of A36 steel ($F_y$ = 36 ksi) is compact.

**Solution:**

W18 × 76 properties are

$$b_f = 11.035 \text{ in.}$$

$$t_f = 0.680 \text{ in.}$$

$$d = 18.21 \text{ in.}$$

$$t_w = 0.425 \text{ in.}$$

Check the flange criterion:

$$\frac{b_f}{2t_f} = \frac{11.035}{2(0.68)} = 8.11$$

$$\frac{65}{\sqrt{F_y}} = \frac{65}{\sqrt{36}} = 10.8$$

$$8.11 < 10.8 \qquad\qquad\qquad \textbf{O.K.}$$

Check the web criterion:

$$\frac{d}{t_w} = \frac{18.21}{0.425} = 42.8$$

$$\frac{640}{\sqrt{F_y}} = \frac{640}{\sqrt{36}} = 106.7$$

$$42.8 < 106.7 \qquad\qquad\qquad \textbf{O.K.}$$

Therefore, the W18 × 76 is compact. The preceding could be shortened by using tabulated quantities from the ASDM, Part 1, Properties of W Shapes, or the Numerical Values furnished in Table 5 (following Appendix K) of the ASDS. The tabulated quantities are rounded slightly in some cases.

---

A faster way to determine compactness of cross section for the rolled shapes is to calculate the value of a hypothetical yield stress $F_y$ that would cause equality in each of the two criteria. For the flange criterion,

$$\frac{b_f}{2t_f} = \frac{65}{\sqrt{F_y}}$$

$$F_y = \left(\frac{65}{b_f/2t_f}\right)^2$$

For the W18 × 76,

$$F_y = \left(\frac{65}{8.11}\right)^2 = 64.2 \text{ ksi}$$

This shows that the flange criterion is satisfied provided that $F_y$ *does not exceed* 64.2 ksi. This value is termed $F_y'$ and is tabulated in the ASDM, Part 1, as a property of the W18 × 76. If the W18 × 76 is of a steel with $F_y$ in *excess* of 64.2 ksi, it is *not* compact by the flange criterion. Therefore, *for compactness by the flange criterion,* the following condition should exist:

$$F_y \leq F_y'$$

Should the calculated value of $F_y'$ be in excess of the highest available $F_y$, this is reflected by the tabulation of a dash (—) for the $F_y'$ value. For the web criterion,

$$\frac{d}{t_w} = \frac{640}{\sqrt{F_y}}$$

$$F_y = \left(\frac{640}{d/t_w}\right)^2$$

For the W18 × 76,

$$F_y = \left(\frac{640}{42.8}\right)^2 = 224 \text{ ksi}$$

This shows that the W18 × 76 is compact by the web criterion provided that $F_y$ does not exceed 224 ksi. Reference to the ASDM, Part 1, Table 1, shows that shapes with $F_y$ in excess of 65 ksi are currently not available. Therefore, the W18 × 76 is compact by the web criterion *in all steels*. This is a general rule. All rolled, W, M, and S shapes tabulated in the ASDM are compact by the web criterion (when $f_a = 0$). This does not hold true for built-up sections and plate girders, which are discussed in Chapter 5. Web noncompactness will cause $F_b$ to be reduced to $0.60F_y$ (assuming adequate lateral support).

If a shape does not satisfy the flange criterion, it is considered a noncompact shape, and $F_b$ must be reduced. The variation of $F_b$ for rolled W shapes *that have adequate lateral support* is summarized graphically in Figure 4-6. For the range of $b_f/2t_f$ between $65/\sqrt{F_y}$ and $95/\sqrt{F_y}$, the ASDS provides for a linear reduction in $F_b$ to $0.60F_y$ according to the following equation:

$$F_b = F_y\left[0.79 - 0.002\left(\frac{b_f}{2t_f}\right)\sqrt{F_y}\right] \qquad \text{ASDS Eqn. (F1-3)}$$

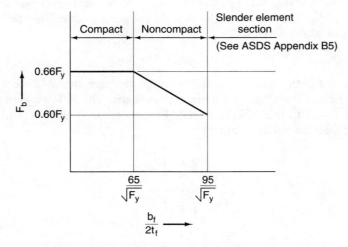

**FIGURE 4-6** $F_b$ for rolled W shapes with adequate lateral support.

This equation provides for a transition in allowable stress between the values of $0.66F_y$ and $0.60F_y$ for noncompact shapes. This assumes adequate lateral support. If $b_f/2t_f$ exceeds $95/\sqrt{F_y}$, the shape is considered to be a slender element section, and ASDS, Appendix B5, applies.

In determining $F_b$, the yield stress $F_y$ must be known. The best source for the value of $F_y$ is the ASDM, Part 1, Tables 1 and 2. For a known shape, determine the appropriate *group* from Table 2. Then, knowing the steel type, use Table 1 to determine $F_y$.

**Example 4-2** _____

Find $F_b$ for the following shapes. Assume adequate lateral support for the compression flange.

(a)   W30 × 132 of A36 steel
(b)   W12 × 65 of A242 steel

**Solution:**

(a)   All W shapes in A36 steel are compact except for the W6 × 15, since its $F_y'$ is less than 36 ksi. From the ASDM, Part 1, Table 1, all shapes in A36 steel have $F_y = 36$ ksi. Thus

$$F_b = 0.66F_y = 0.66(36)$$

where

$$F_b = 23.8 \text{ ksi (commonly rounded to 24.0 ksi)} \quad (165.5 \text{ MPa})$$

(b)   From the ASDM, Part 1, Table 2, W12 × 65 is found in *Group 2.* From Table 1, $F_y = 50$ ksi. From the ASDM properties tables for the W12 × 65, $F_y' = 43.0$ ksi. Thus

$$50.0 \text{ ksi} > 43.0 \text{ ksi}$$

Since $F_y > F_y'$, the member is not a compact shape in A242 steel. Next, we check whether the shape is a noncompact shape. The following quantities can also be found in the ASDM, Part 1, Properties of W Shapes, and Table 5 of the ASDS (Numerical Values):

$$\frac{b_f}{2t_f} = 9.9$$

$$\frac{65}{\sqrt{F_y}} = \frac{65}{\sqrt{50}} = 9.2$$

$$\frac{95}{\sqrt{F_y}} = \frac{95}{\sqrt{50}} = 13.4$$

Therefore, this is a noncompact shape, since

$$\frac{65}{\sqrt{F_y}} < \frac{b_f}{2t_f} < \frac{95}{\sqrt{F_y}}$$

Calculate $F_b$ from ASDS Equation (F1-3):

$$F_b = F_y \left[ 0.79 - 0.002 \left( \frac{b_f}{2t_f} \right) \sqrt{F_y} \right]$$

$$= 50[0.79 - 0.002(9.9) \sqrt{50}] = 32.5 \text{ ksi } (224 \text{ MPa})$$

## 4-4

## ANALYSIS OF BEAMS FOR MOMENT

The analysis problem is generally considered to be the *investigation* of a beam whose *cross section* is known. One may be concerned with checking the adequacy of a given beam, determining an allowable load, or finding the maximum existing bending stress in the beam. All these problems are related. All make use of the flexure formula and require an understanding of *allowable bending stress $F_b$*.

**Example 4-3**

A W21 × 44 beam is to span 24 ft on simple supports (as shown in Figure 4-7). Assume full lateral support and A36 steel. The load shown is a *superimposed load,* meaning that it does *not* include the weight of the beam. Determine whether the beam is adequate by

(a)  Comparing the actual bending stress with the allowable bending stress.
(b)  Comparing the actual applied moment with the resisting moment $M_R$.

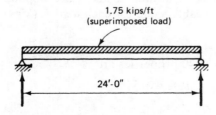

**FIGURE 4-7**  Load diagram.

**Solution:**

(a)  Determine the actual bending stress from the flexure formula

$$f_b = \frac{Mc}{I} = \frac{M}{S}$$

From the properties tables, for the W21 × 44, $S_x$ = 81.6 in.$^3$. Moment may be determined by shear and moment diagram or by formula (see the ASDM, Part 2, Beam Diagrams and Formulas, for review). The total load should include the weight of the beam:

$$
\begin{array}{ll}
1.75 \text{ kips/ft} & \text{(applied load)} \\
+0.044 & \text{(beam weight)} \\
\hline
1.794 \text{ kips/ft} & \text{(total uniform load } w\text{)}
\end{array}
$$

applied moment $M = \dfrac{wL^2}{8} = \dfrac{1.794(24)^2}{8} = 129.2$ ft-kips

$$f_b = \frac{M}{S_x} = \frac{129.2(12)}{81.6} = 19.00 \text{ ksi (131.0 MPa)}$$

With reference to the comments of Example 4-2a, $F_y' = (-)$ from the ASDM, Part 1. Therefore, since $F_y' > F_y$, the member is compact in A36 steel, and the allowable bending stress $F_b$ is 24.0 ksi. Therefore,

$$f_b < F_b \qquad\qquad \textbf{O.K.}$$

(b)  The applied moment $M$ has been determined to be 129.2 ft-kips. The resisting moment $M_R$ may be calculated from the flexure formula:

$$M_R = F_b S_x = \frac{24.0(81.6)}{12} = 163.2 \text{ ft-kips (221 kN·m)}$$

Therefore,

$$M < M_R \qquad\qquad \textbf{O.K.}$$

---

**Example 4-4**

A W18 × 40 beam spans 20 ft on a simple span, as shown in Figure 4-8. Assume A36 steel. The compression flange is supported laterally at the quarter points where equal concentrated loads $P$ are applied. Therefore, $L_b$ = 5 ft. Determine the allowable value for each load $P$ (kips).

**Solution:**

Determine $F_b$ and $M_R$. W18 × 40 properties are

$$b_f = 6.015 \text{ in.}$$

$$\frac{d}{A_f} = 5.67$$

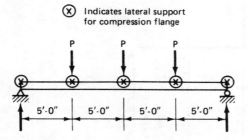

**FIGURE 4-8**  Load diagram.

$$F_y = 36 \text{ ksi}$$

$$S_x = 68.4 \text{ in.}^3$$

$$F_y' = (-)$$

Test for adequate lateral support by comparing $L_b$ with $L_c$. From the ASDM, Part 2, Allowable Stress Design Selection Table, $L_c = 6.3$ ft. $L_b = 5$ ft. Since $L_b < L_c$, this beam has adequate lateral support. $F_y'$ is $(-)$, implying that $F_y'$ is high enough so that this shape is always compact. Therefore,

$$F_b = 0.66 F_y = 24.0 \text{ ksi}$$

$$M_R = F_b S_x = \frac{24.0(68.4)}{12} = 136.8 \text{ ft-kips}$$

The resisting moment is 136.8 ft-kips. The applied moment due to the beam's own weight and the moment due to the three equal loads $P$ cannot exceed the resisting moment $M_R$. The applied moment due to beam weight is

$$M = \frac{wL^2}{8} = \frac{0.04(20)^2}{8} = 2.0 \text{ ft-kips}$$

The moment due to the concentrated loads may be determined using the shear and moment diagrams of Figure 4-9 or aids such as those found in the ASDM, Part 2, Beam Diagrams and Formulas. Thus

$$M = 10P \text{ ft-kips}$$

The resisting moment remaining to support the concentrated loads is

$$136.8 - 2.0 = 134.8 \text{ ft-kips}$$

Equating, we have

$$10P = 134.8$$

$$P = \frac{134.8}{10} = 13.48 \text{ kips}$$

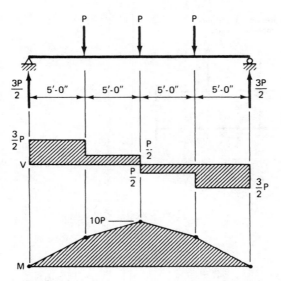

**FIGURE 4-9**  Load, shear, and moment diagrams.

Therefore, the maximum allowable value of each concentrated load $P$ on this beam is 13.48 kips (60.0 kN).

It should be noted that the weight of the beam itself in both of the foregoing examples has been a very minor part of the total load carried. This is generally true; nevertheless, it should always be considered. As one becomes more experienced, the effect of beam weight is easier to estimate. Various shortcuts are used by designers to simplify the inclusion of weight of structure in design problems. In analysis problems, where the cross section is known, inclusion of the beam weight is a simple matter.

## 4-5

## SUMMARY OF PROCEDURE: BEAM ANALYSIS FOR MOMENT ONLY

The beam analysis for moment only procedure is general and typical for the various shapes that may be used for beams, primarily wide-flange sections and, to a lesser extent, other sections. The precise method of solution will depend on the nature of the particular problem, the known conditions, and the information sought.

1.  Determine $F_y$. Use the ASDM, Part 1, Tables 1 and 2.

2.  Check the adequacy of lateral support. See the ASDS, Section F1. If lateral support is inadequate, see Section 4-6 of this chapter.

3.  Check the compactness of the cross section. Use $F'_y$ from the table of properties in the ASDM, Part 1.

4.  Using the preceding information, determine $F_b$.

5.  If the applied loads are known, the applied moment can be found. Draw shear and moment diagrams or use beam formulas from the ASDM, Part 2.

6.  If the magnitude of the applied loads is unknown, write an expression for the applied moment in terms of the unknown loads. This can then be equated to the resisting moment of the beam.

7.  The flexure formula for use in analysis is

$$f_b = \frac{Mc}{I} = \frac{M}{S}$$

$$M_R = F_b S$$

## 4-6

## INADEQUATE LATERAL SUPPORT

As the distance between points of lateral support on the compression flange ($\ell$) becomes larger, there is a tendency for the compression flange to buckle laterally. There is no upper limit for $\ell$. To guard against the buckling tendency as $\ell$ becomes larger, however, the ASDS provides that $F_b$ be reduced. This, in effect, reserves some of the beam strength to resist the lateral buckling. Figure 4-10b shows a beam that has deflected vertically with a compression flange that has buckled laterally.

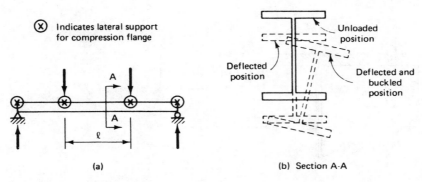

(a)                                     (b)  Section A-A

**FIGURE 4-10**   Beam deflection and lateral buckling.

The result is a twisting of the member. This is called *lateral-torsional buckling*. For simplicity, we refer to this buckling mode of the beam as *lateral buckling*. Two general resistances are available to counteract lateral buckling: torsional resistance of the cross section and lateral bending resistance of the compression flange. The total resistance to lateral buckling is the sum of the two. The ASDS conservatively considers only the *larger* of the two in the determination of a reduced $F_b$.

The ASDS, Section F1.3, establishes empirical expressions for $F_b$ for the inadequate lateral support situation. Tension and compression allowable bending stresses are treated separately. The tension $F_b$ is always $0.60F_y$. Only the compression $F_b$ is reduced. For typical rolled shapes this is of no consequence because the shapes are symmetrical and the lower $F_b$ of the two values will control. Note that the provisions of this section pertain to members having an axis of symmetry in, and loaded in, the plane of their web. They also apply to compression on extreme fibers of channels bent about their major axis.

The ASDS provides three empirical equations for the reduced compression $F_b$. The mathematical expressions that give an exact prediction of the buckling strength of beams are too complex for general use. Therefore, the ASDS equations only *approximate* this strength for purposes of determining a reasonable $F_b$. The $F_b$ that is finally used is the *larger* of the $F_b$ values determined from the applicable equations. The first two, ASDS Equations (F1-6) and (F1-7), give the $F_b$ value when the lateral bending resistance of the compression flange provides the lateral buckling resistance. The third, ASDS Equation (F1-8), gives $F_b$ when the torsional resistance of the beam section provides the primary resistance to lateral buckling. In no case should $F_b$ be greater than $0.60F_y$ for beams that have inadequate lateral support. The equations that will be applicable will depend on the value of the ratio $\ell/r_T$, where

$\ell$ = distance between points of lateral support for the compression flange (in.)

$r_T$ = radius of gyration of a section comprising the compression flange plus one-third of the compression web area taken about an axis in the plane of the web (in.), as shown in Figure 4-11

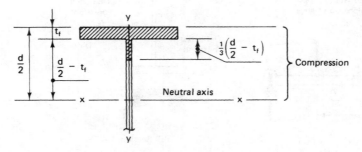

**FIGURE 4-11**   $r_T$ determination.

Here $r_T$ is a tabulated quantity for rolled shapes (see the ASDM, Part 1), and $\ell/r_T$ may be considered a slenderness ratio of the compression portion of the beam with respect to the $y$-$y$ axis. The equations for $F_b$ are as follows:

$$F_b = \left[ \frac{2}{3} - \frac{F_y(\ell/r_T)^2}{1530 \times 10^3 C_b} \right] F_y \qquad \text{ASDS Eqn. (F1-6)}$$

$$F_b = \frac{170 \times 10^3 C_b}{(\ell/r_T)^2} \qquad \text{ASDS Eqn. (F1-7)}$$

$$F_b = \frac{12 \times 10^3 C_b}{\ell d / A_f} \qquad \text{ASDS Eqn. (F1-8)}$$

where

$C_b$ = a liberalizing modifying factor whose value is between 1.0 and 2.3 that accounts for a moment gradient over the span and a decrease in the lateral buckling tendency; $C_b$ may be conservatively taken as 1.0; see ASDS, Section F1.3, for details

$d$ = depth of cross section (in.)

$A_f$ = area of compression flange (in.²)

Figure 4-12 depicts the decision-making process for the calculation of $F_b$. Note that one will use ASDS Equations (F1-6) and (F1-8) *or* ASDS Equations (F1-7) and (F1-8). The *larger* resulting $F_b$ is used. Note that Table 5 of the Numerical Values section of the ASDS provides the following numerical equivalents for A36 steel ($F_y$ = 36 ksi):

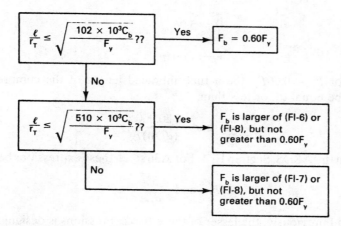

**FIGURE 4-12**   $F_b$ for beams with inadequate lateral support.

$$\sqrt{\frac{102 \times 10^3 C_b}{36}} = 53\sqrt{C_b}$$

$$\sqrt{\frac{510 \times 10^3 C_b}{36}} = 119\sqrt{C_b}$$

### Example 4-5

A W21 × 50, shown in Figure 4-13, spans 36 ft on a simple span. The compression flange is laterally supported at the third points. A36 steel is used. Determine $F_b$ for this beam.

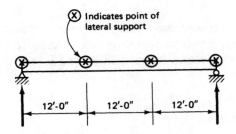

**FIGURE 4-13**   Beam diagram.

**Solution:**

The shape is compact, and $F_y = 36$ ksi. Check for adequacy of lateral support ($\ell = 144$ in. or $L_b = 12$ ft.)
For the W21 × 50,

$$b_f = 6.53 \text{ in.}$$

$$\frac{d}{A_f} = 5.96$$

$$r_T = 1.60 \text{ in.}$$

To qualify for $F_b = 0.66F_y$, the actual unbraced length of the compression flange must be equal to, or less than,

$$\frac{76b_f}{\sqrt{F_y}} \quad \text{and} \quad \frac{20,000}{(d/A_f)F_y}$$

as required in the ASDS, Section F1.1. For A36 steel, these expressions become

$$12.7b_f \quad \text{and} \quad \frac{556}{d/A_f}$$

As mentioned previously, the lesser of these two expressions is designated $L_c$ (in feet). If $L_b \leq L_c$, the beam will qualify for $F_b = 0.66F_y$.

$$L_b = 12 \text{ ft}$$

$$L_c = 6.9 \text{ ft} \qquad \text{(ASDM, Part 2)}$$

Since $L_b > L_c$, $F_b$ must be reduced. The beam compression flange is therefore said to be *inadequately braced.*

The value of $\ell/r_T$, which determines the applicable ASDS formulas for $F_b$, is compared with

$$53\sqrt{C_b} \quad \text{and} \quad 119\sqrt{C_b}$$

(from Table 5 of the Numerical Values section of the ASDS). We next compute the slenderness ratio of the compression flange:

$$\frac{\ell}{r_T} = \frac{144}{1.60} = 90.0$$

Conservatively assuming that $C_b = 1.0$, we have

$$53\sqrt{C_b} = 53$$

$$119\sqrt{C_b} = 119$$

$$53 < \frac{\ell}{r_T} < 119$$

Therefore, from Figure 4-12, the applicable $F_b$ equations are

ASDS Equation (F1-6):

$$F_b = \left[\frac{2}{3} - \frac{F_y(\ell/r_T)^2}{1530 \times 10^3 C_b}\right] F_y = \left[\frac{2}{3} - \frac{36(90)^2}{1530(10^3)(1.0)}\right] 36 = 17.1 \text{ ksi}$$

and ASDS Equation (F1-8):

$$F_b = \frac{12 \times 10^3 C_b}{\ell(d/A_f)} = \frac{12(10^3)(1.0)}{144(5.96)} = 14.0 \text{ ksi}$$

The *larger* $F_b$ is used:

$$F_b = 17.1 \text{ ksi (117.9 MPa)}$$

As indicated, the Numerical Values tables of the ASDS are useful in shortening the calculations somewhat.

---

This discussion has pertained primarily to the rolled shapes commonly used for beams and loaded for strong-axis bending. The ASDS equations for $F_b$ for beams that have inadequate lateral support are also applicable to built-up members and plate girders, provided that they have an axis of symmetry in the plane of the web.

**Example 4-6** _____

Determine the allowable superimposed uniformly distributed load that may be placed on the W21 × 50 of Example 4-5 (see Figure 4-13).

**Solution:**

$F_b$ was determined to be 17.1 ksi. Thus

$$M_R = F_b S_x$$

$$= \frac{17.1(94.5)}{12} = 134.7 \text{ ft-kips}$$

Since the beam is a simple span member with a 36-ft span,

$$M = \frac{wL^2}{8} \quad \text{and} \quad w = \frac{8M}{L^2}$$

Since, as a limit, the applied moment $M$ can equal $M_R$,

$$w = \frac{8M_R}{L^2} = \frac{8(134.7)}{36^2} = 0.831 \text{ kip/ft}$$

Subtracting the beam's own weight, the allowable superimposed load is

$$831 - 50 = 781 \text{ lb/ft} = 0.781 \text{ kip/ft (11.40 kN/m)}$$

We usually think of the lateral-torsional buckling of a beam as leading to the failure of that *member*. Additionally, however, some types of *framing systems* may suffer stability failures due to inadequate lateral support of the compression flanges of the beams. One example is the case of an overhanging beam resting on top of a steel column, as shown in Figure 4-14. In this case, lateral support of the top flange alone may be inadequate. The ASDS Commentary, Section B6, discusses this situation. In the situation shown, bracing exists only in the plane of the top flange, and the resistance of the column to lateral movement is insignificant. The bottom flange is free to buckle laterally as shown in Figure 4-14b. Unless lateral support is provided for the bottom flange by some means, such as bracing at the beam-to-column connection, lateral displacement at the top of the column accompanied by rotation of the beam about its longitudinal axis may lead to collapse of the framing.

In some cases, bracing techniques such as those shown in Figure 4-15 may be used. Additionally, when structural members do not exist on the column centerline (as may be observed in Photo 4-1), one solution is to use web stiffeners on the overhanging beam web in combination with an adequately connected bearing plate (cap plate) on top of the column. This may also be observed in Photo 4-1.

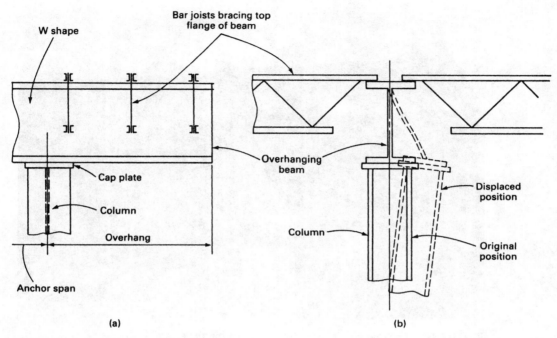

**FIGURE 4-14**   Column-supported overhanging beam.

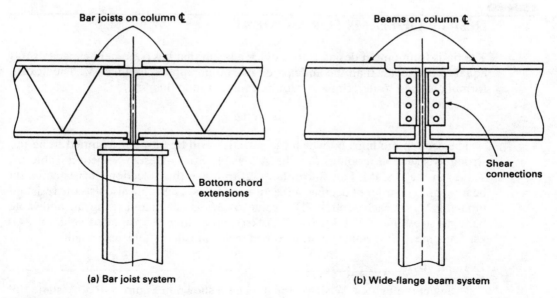

**FIGURE 4-15**   Lateral bracing for beams.

**PHOTO 4-1** Overhanging beams. Note the bearing plates and web stiffeners at the supports.

### 4-7

## DESIGN OF BEAMS FOR MOMENT

The basis for moment design is to provide a beam that has a moment capacity $(M_R)$ equal to or greater than the anticipated maximum applied moment $M$. The flexure formula is used to determine a required section modulus $S$:

$$\text{required } S = \frac{M}{F_b}$$

The section modulus on which the selection will be based is assumed to be the strong-axis section modulus $S_x$. The Allowable Stress Design Selection Table ($S_x$ Table) in the ASDM, Part 2, can be used to make this selection. It lists common beam shapes in order of decreasing section modulus. This table also lists the resisting moment $M_R$ of each section. The value of $M_R$ is calculated using an allowable bending stress $F_b$ of 23.8 ksi (or 23.76 ksi) rather than the rounded value of 24.0 ksi. This may cause some small inconsistencies in calculations and results.

### Example 4-7

Select the lightest W shape for the beam shown in Figure 4-16a. Assume full lateral support ($L_b = 0$) and A36 steel. Consider moment only.

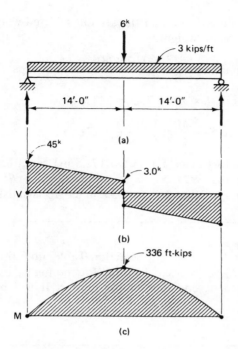

**FIGURE 4-16**   Load, shear, and moment diagrams.

## Solution:

The beam reactions are determined from the load diagram; the shear and moment diagrams are drawn. These are shown in Figure 4-16b and c. Note that no beam weight is included since the beam is unknown. An estimate of the beam weight could be made and its effect on the moment included. This is optional at this point. The beam weight must be included before the final selection of section is made, however. The maximum moment is determined to be 336 ft-kips, as shown in Figure 4-16c.

Select $F_b$, if possible. In this case the A36 steel beam has full lateral support and $F_b$ almost assuredly equals $0.66F_y$, since virtually all W shapes of A36 steel are compact. Therefore, assume that

$$F_b = 0.66F_y = 0.66(36) = 24 \text{ ksi}$$

Determine the required $S_x$:

$$\text{required } S_x = \frac{M}{F_b} = \frac{336(12)}{24} = 168.0 \text{ in.}^3$$

From the ASDM, Part 2 (Allowable Stress Design Selection Table), select a W24 × 76 with an $S_x = 176$ in.[3] It is the lightest W shape that will furnish the required section modulus. Note that the section selected weighs 76 lb/ft. Add in the effect of the beam weight:

$$\text{additional } M = \frac{wL^2}{8} = \frac{0.076(28)^2}{8} = 7.45 \text{ ft-kips}$$

$$\text{new total } M = 336 + 7.45 = 343 \text{ ft-kips}$$

$$\text{new required } S_x = \frac{M}{F_b} = \frac{343(12)}{24} = 171.5 \text{ in.}^3$$

The W24 $\times$ 76 is satisfactory since 176 in.$^3$ > 171.5 in.$^3$ required. Also, now check the assumed $F_b$; the W24 $\times$ 76 is compact, since $F_y' > F_y$, and has adequate lateral support; therefore, the assumed $F_b$ is satisfactory. **Use W24 $\times$ 76.**

In the use of the Allowable Stress Design Selection Table, note that any shape having at least the required $S_x$ will be satisfactory (for moment). If a *shallower* section is required, the choice of one is a simple matter. It will be a heavier section, however.

**Example 4-8** _____

The beam shown in Figure 4-17a is to be of A36 steel. Note the lateral support conditions. Select the lightest W shape. Consider moment only.

**Solution:**

For this design, an estimated beam weight of 40 lb/ft (0.04 kip/ft) has been added to the given uniform load. (This estimate may be based on anything from an educated guess to a rough design worked quickly on scrap paper.) The shear and moment diagrams are shown in Figure 4-17b and c. The maximum moment is 129.0 ft-kips.

Establish $F_b$. $F_y$ is 36 ksi, and a compact shape that has adequate lateral support will be assumed. Therefore, the assumed allowable bending stress is

$$F_b = 0.66F_y = 0.66(0.36) = 24 \text{ ksi}$$

from which

$$\text{required } S_x = \frac{M}{F_b} = \frac{129.0(12)}{24} = 64.5 \text{ in.}^3$$

Select a W16 $\times$ 40 with $S_x = 64.7$ in.$^3$ This is a compact shape since $F_y' > F_y$. Check the adequacy of lateral support ($\ell = 60$ in. or $L_b = 5$ ft) by comparing $L_b$ with $L_c$. From the ASDM, Part 2, $L_c = 7.4$ ft. Therefore, since $L_b < L_c$, the beam has adequate lateral support, and the assumed $F_b$ is satisfactory. The assumed beam weight is satisfactory. **Use W16 $\times$ 40.**

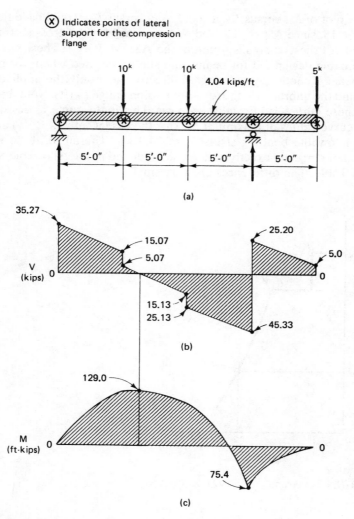

FIGURE 4-17    Load, shear, and moment diagrams.

The rapid solutions of Examples 4-7 and 4-8 have depended on the correct assumption of $F_b$. Since $F_b$ depends in part on the section to be selected, it cannot always be predetermined. Various design aids have been developed to speed the design process. In Example 4-5, $F_b$ was determined for a W21 × 50 beam of A36 steel that had inadequate lateral support. For $\ell$ of 144 in. ($L_b$ = 12 ft), $F_b$ = 17.1 ksi. For each value of $L_b$, a value for $F_b$ could be determined. A plot of $F_b$ versus $L_b$ is shown in Figure 4-18. The shape of this curve is typical for a compact cross section. A plot of $M_R$ versus $L_b$, instead of $F_b$ versus $L_b$, would have the same form as Figure 4-18, since $M_R = F_b S_x$ and $S_x$ is constant for the cross section.

A family of these curves is found in the ASDM, Part 2. These curves make up a very valuable design aid for beams and should be used whenever possible. The lightest beam section may be selected directly using only the applied moment $M$ (ft-kips) and the unbraced length of compression flange $L_b$ (ft). Note that the vertical axis for these curves is *allowable moment*. This is the same as resisting moment $M_R$. The curve values of maximum $M_R$ (for the case where $L_b \leq L_c$) are calculated using an allowable bending stress $F_b$ of 24.0 ksi. Therefore, these maximum $M$ values do not correlate exactly with those tabulated in the Allowable Stress Design Selection Table. The differences are very small.

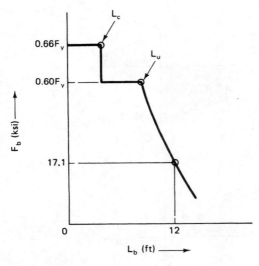

**FIGURE 4-18**   $F_b$ versus $L_b$ for a W21 × 50 (A36).

Two other terms are illustrated in Figure 4-18:

$L_c$ = Maximum unbraced length (ft) of compression flange at which the allowable bending stress may be taken at $0.66F_y$ (for compact shapes), or as determined by ASDS Equation (F1-3) or (F2-3) (when applicable). $L_c$ is the smaller value obtained from

$$\frac{76b_f}{\sqrt{F_y}} \quad \text{and} \quad \frac{20,000}{(d/A_f)F_y}$$

$L_u$ = Maximum unbraced length (ft) of the compression flange at which the allowable bending stress may be taken at $0.60F_y$.

With $F_b = 0.60F_y$ and $C_b = 1.0$, the value of $L_u$ (in feet) for *most* shapes is given as

$$\frac{20,000}{12(d/A_f)F_y}$$

from ASDS Equation (F1-8). For a few shapes, $L_u$ (in feet) is given as

$$\sqrt{\frac{102,000}{F_y}}\left(\frac{r_T}{12}\right)$$

from ASDS Equation (F1-6). $L_u$ is taken as the *larger* value obtained from these two expressions. For lengths greater than $L_c$ but not greater than $L_u$, $F_b$ may be taken as $0.60F_y$. In no case is $L_c$ taken as greater than $L_u$. $L_c$ and $L_u$ are unique for each shape and also vary with $F_y$. They are tabulated properties and may be found in the ASDM, Part 2, Allowable Stress Design Selection Table, and in the Beam Tables of Part 2, for $F_y$ of 36 ksi and 50 ksi. Their use greatly facilitates the determination of whether or not lateral support is adequate. The precise values of $L_c$ and $L_u$, however, are of no consequence when using the beam curves.

When using the curves, any shape represented by a curve that is found *above or to the right* of a particular $M_R$ and $L_b$ combination is a shape that is satisfactory. The lightest adequate shape is represented by a solid line. The dashed lines represent shapes that are also adequate but are heavier. The ASDM contains beam curves for $F_y = 36$ ksi and $F_y = 50$ ksi. These curves may be used for the range of $L_b$ from 0 up to the maximum value indicated in the curves.

**Example 4-9** _____

Select the lightest W shape for the beam shown in Figure 4-19. The depth of the beam is limited to 36 in. maximum. Consider the following cases:

(a)   A36 steel, $L_b = 4$ ft.
(b)   A36 steel, $L_b = 16$ ft.
(c)   A441 steel, $L_b = 20$ ft.

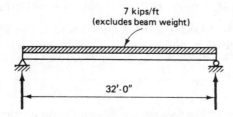

**FIGURE 4-19**   Beam load diagram.

**Solution:**

(a)   $F_y$ = 36 ksi. Determine the applied moment for a simply supported, uniformly loaded, single-span beam. Neglect the beam weight temporarily. Thus

$$M = \frac{wL^2}{8} = \frac{7.0(32)^2}{8} = 896 \text{ ft-kips}$$

From the beam curves, ASDM, Part 2, with $M$ = 896 ft-kips and unbraced length ($L_b$) = 4 ft, select a W36 × 150. Note that all lines are horizontal at the left vertical axis. For a moment of 896 ft-kips, the W36 × 150 would be satisfactory for any $L_b$ from 0 to 14.6 ft. Add the moment due to the beam weight:

$$\text{additional } M = \frac{wL^2}{8} = \frac{0.150(32)^2}{8} = 19.2 \text{ ft-kips}$$

$$\text{total } M = 896 + 19.2 = 915 \text{ ft-kips}$$

From the beam curves, $M_R$ for the W36 × 150 with $L_b$ = 4 ft is 1008 ft-kips:

$$1008 \text{ ft-kips} > 915 \text{ ft-kips} \qquad \textbf{O.K.}$$

**Use W36 × 150.**

(b)   $F_y$ = 36 ksi. $M$ = 896 ft-kips from part (a). $L_b$ = 16 ft. From the beam curves, select a W36 × 160. For $L_b$ of 16 ft, $M_R$ = 960 ft-kips for this shape. Add the moment due to the beam weight:

$$\text{additional } M = \frac{wL^2}{8} = \frac{0.16(32)^2}{8} = 20.5 \text{ ft-kips}$$

$$\text{total } M = 896 + 20.5 = 917 \text{ ft-kips} < 960 \text{ ft-kips} \qquad \textbf{O.K.}$$

**Use W36 × 160.**

(c)   *Assume* that $F_y$ = 50 ksi. $M$ = 896 ft-kips from part (a). $L_b$ = 20 ft. From the beam curves, select a W36 × 150 with $M_R$ = 960 ft-kips. Add the beam weight moment:

$$\text{additional } M = \frac{0.15(32)^2}{8} = 19.2 \text{ ft-kips}$$

$$\text{total } M = 896 + 19.2 = 915 \text{ ft-kips} < 960 \text{ ft-kips} \qquad \textbf{O.K.}$$

Check $F_y$. From the ASDM, Part 1, Tables 1 and 2, the W36 × 150 is a Group 2 shape and $F_y$ = 50 ksi.             **O.K.**

**Use W36 × 150.**

---

When the curves are not applicable (e.g., when $F_y$ equals some value other than 36 ksi or 50 ksi), $F_b$ must be assumed and subsequently verified.

**Example 4-10** _____

Rework Example 4-9. Select the lightest W shape for the beam shown in Figure 4-19. The framing system used indicates the use of an $L_b = 9.0$ ft. The steel is to be A572 Grade 60 ($F_y = 60$ ksi).

**Solution:**

*Assume* that $F_b = 0.66F_y = 0.66(60) = 40$ ksi. Using the moment calculated in Example 4-9(a), which included a 150-lb/ft beam weight,

$$\text{required } S_x = \frac{M}{F_b} = \frac{915(12)}{40} = 275 \text{ in.}^3$$

From the ASDM, Part 2 (Allowable Stress Design Selection Table), select a W30 × 108 ($S_x = 299$ in.³).

Now verify $F_b$. The section is compact since $F_y' = (—)$ (ASDM, Part 1), which indicates that it is in excess of 65 ksi. Therefore, $F_y' > F_y$. Since $L_c$ is not tabulated for a steel with $F_y = 60$ ksi, it must be computed and compared with $L_b = 9$ ft. With reference to the ASDS, Section F1.1, $L_c$ is the smaller of

$$\frac{76b_f}{\sqrt{F_y}} = \frac{76(10.475)}{\sqrt{60}} = 102.8 \text{ in.}$$

and

$$\frac{20,000}{(d/A_f)F_y} = \frac{20,000}{3.75(60)} = 88.9 \text{ in.}$$

Therefore,

$$L_c = \frac{88.9}{12} = 7.41 \text{ ft}$$

Since $L_b > L_c$, $F_b$ must be reduced to at least $0.60F_y$ and possibly lower. This will affect the required $S_x$ with a subsequent change in the section selected. Assume the new $F_b$ to be $0.60(60) = 36$ ksi:

$$\text{required } S_x = \frac{915(12)}{36} = 305 \text{ in.}^3$$

From the ASDM, Part 2, select a W30 × 116 ($S_x = 329$ in.³). For this shape, a calculation for $L_c$ (similar to that preceding) will show that $L_c = 8.3$ ft. Therefore, $L_b$ is still greater than $L_c$. Now determine $L_u$ as the larger of

$$\frac{20,000}{12(d/A_f)F_y} = \frac{20,000}{12(3.36)60} = 8.3 \text{ ft}$$

and

$$\sqrt{\frac{102{,}000}{F_y}}\left(\frac{r_T}{12}\right) = \sqrt{\frac{102{,}000}{60}}\left(\frac{2.64}{12}\right) = 9.1\ \text{ft}$$

Therefore, $L_u = 9.1$ ft and $F_b = 0.60F_y$ (as assumed), since

$$L_c < L_b < L_u$$

The beam weight included in the design moment is on the conservative side, since the assumed 150 lb/ft is greater than the actual beam weight of 116 lb/ft. No modifications of the calculations are necessary, however. **Use a W30 × 116.**

Other beam design aids are available, such as the tables of Allowable Loads on Beams also found in the ASDM, Part 2. We refer to these tables as the *Allowable Uniform Load Tables.* An explanation of how to use the tables along with examples is furnished in the manual.

The curves found in the ASDM, Part 2, may also be used for analysis problems as well as for design. If the beam size and unbraced length ($L_b$) are known, the allowable moment may be obtained from the curves.

**Example 4-11** _____

Compute the allowable superimposed uniformly distributed load that may be placed on a W36 × 150 spanning 24 ft–0 in. The beam is laterally supported at its supports only ($L_b$ = 24 ft–0 in.) and is A36 steel.

**Solution:**

Entering the curves for a W36 × 150 with an $L_b = 24$ ft, the total allowable moment = 680 ft-kips. The moment due to beam's own weight is

$$M = \frac{0.150(24)^2}{8} = 10.8\ \text{ft-kips}$$

The allowable moment for superimposed load is

$$680 - 10.8 = 669\ \text{ft-kips}$$

Since the applied moment can equal the allowable moment, as a limit,

$$M = \frac{wL^2}{8} \quad \text{or} \quad w = \frac{8M}{L^2} = \frac{8(669)}{24^2} = 9.29\ \text{kips/ft}\ (135.6\ \text{kN/m})$$

## 4-8

### SUMMARY OF PROCEDURE: BEAM DESIGN FOR MOMENT

Based on the foregoing examples, a general procedure may be established for the design of beams for moment.

1. Establish the conditions of load, span, and lateral support. This is best done with a sketch. Establish the steel type.
2. Determine the design moment. If necessary, complete shear and moment diagrams should be drawn. An estimated beam weight may be included in the applied load.
3. The beam curves should be used to select an appropriate section when possible. As an alternative, $F_b$ must be estimated and the required section modulus determined:

$$\text{required } S_x = \frac{M}{F_b}$$

   The section is then selected using the $S_x$ table.
4. After the section has been selected, recompute the design moment, including the effect of the weight of the section. Check to ensure that the section selected is still adequate.
5. Check any assumptions that may have been made concerning $F_y$ or $F_b$.
6. Be sure that the solution to the design problem is plainly stated.

## 4-9

### SHEAR IN BEAMS

Except under very special loading conditions, all beams are subjected to shear as well as moment. In the normal process of design, beams are selected on the basis of the moment to be resisted and then checked for shear. Shear rarely controls a design unless loads are very heavy (and, possibly, close to the supports) and/or spans are very short. From strength of materials, the shear stress that exists within a beam may be determined from the general shear formula

$$f_v = \frac{VQ}{Ib}$$

where

$f_v$ = shear stress on a horizontal plane located with reference to the neutral axis (ksi)

$V$ = vertical shear force at that particular section (kips)

$Q$ = statical moment of area between the plane under consideration and the outside of the section, about the neutral axis (in.³)

$I$ = moment of inertia of the section about the neutral axis (in.⁴)

$b$ = thickness of the section at the plane being considered (in.)

This formula furnishes us with the *horizontal* shear stress at a point, which, as shown in any strength of materials text, is equal in intensity to the *vertical* shear stress at the same point in a beam.

## Example 4-12

A W16 × 100 is subjected to a vertical shear of 80 kips. Determine the maximum shear stress and plot the distribution of shear stress for the entire cross section.

**Solution:**

Design dimensions for the W16 × 100 are shown in Figure 4-20a. Other properties are

$$A = 29.4 \text{ in.}^2$$

$$I_x = 1490 \text{ in.}^4$$

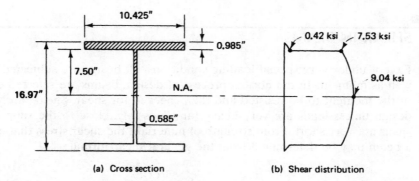

(a)  Cross section                    (b)  Shear distribution

**FIGURE 4-20**   Shear stress.

The maximum shear stress will be at the neutral axis (where $Q$ is maximum). Taking $Q$ for the crosshatched area about the neutral axis, we have

$$Q = 10.425(0.985)\left(7.50 + \frac{0.985}{2}\right) + 7.50(0.585)\left(\frac{7.50}{2}\right)$$

$$= 98.5 \text{ in.}^3$$

$$\text{maximum } f_v = \frac{VQ}{Ib} = \frac{80(98.5)}{1490(0.585)} = 9.04 \text{ ksi (62.3 MPa)}$$

If shear stress values are calculated at other levels in the cross section and the results plotted, the shear distribution will appear as in Figure 4-20b.

Note that the flanges resist very low shear stresses. Even though the areas of the flanges are large, it is the web that predominantly resists the shear in wide-flange beams. For this reason the ASDS allows the use of an *average web shear* approach for the shear stress determination:

$$f_v = \frac{V}{dt_w}$$

where

$f_v$ = computed maximum shear stress (ksi)

$V$ = vertical shear at the section considered (kips)

$d$ = depth of the beam (in.)

$t_w$ = web thickness of the beam (in.)

This method is approximate compared with the theoretically correct general shear formula and assumes that the shear is resisted by the rectangular area of the web extending the full depth of the beam. For the W16 × 100 of Example 4-12,

$$f_v = \frac{V}{dt_w} = \frac{80}{16.97(0.585)} = 8.06 \text{ ksi}$$

This is less than the shear stress of 9.04 ksi calculated by the general shear stress formula and could be considered unsafe.

Allowable shear stresses are set intentionally low to account for the fact that the computed average shear stress will be lower than the actual shear stress. In rolled beams, at locations other than end connections, the ASDS, Section F4, establishes the allowable shear stress $F_v$ as follows. For

$$\frac{h}{t_w} \leq \frac{380}{\sqrt{F_y}}$$

where

$h$ = clear distance between flanges at the section under investigation (in.)

$t_w$ = thickness of web (in.)

the allowable shear stress is based on the overall depth of the beam $d$ times the web thickness and is taken as

$$F_v = 0.40F_y \qquad \text{ASDS Eqn. (F4-1)}$$

Most rolled sections have $h/t_w \leq 380/\sqrt{F_y}$.

For $h/t_w > 380/\sqrt{F_y}$, the allowable shear stress is based on the clear distance between flanges times the web thickness and is taken as

$$F_v = \frac{F_y}{2.89}(C_v) \leq 0.40F_y \qquad \text{ASDS Eqn. (F4-2)}$$

where $C_v$ is a function of the distance between transverse stiffeners and is computed from

$$C_v = \frac{45{,}000\,k_v}{F_y(h/t_w)^2} \qquad \text{when } C_v < 0.8$$

or

$$C_v = \frac{190}{h/t_w}\sqrt{\frac{k_v}{F_y}} \qquad \text{when } C_v > 0.8$$

For a rolled section without transverse stiffeners, $k_v = 5.34$.

The localized effects of shear at connections are discussed in Chapter 7.

**Example 4-13** _____

The W18 × 50 beam shown in Figure 4-21a has been designed for moment. The uniform load includes the beam weight. Check the beam for shear. Assume A36 steel and full lateral support.

**Solution:**

The shear diagram is drawn as shown in Figure 4-21b. The maximum shear to which the beam is subjected is 78.8 kips. The W18 × 50 properties are

$$d = 17.99 \text{ in.}$$

$$t_w = 0.355 \text{ in.}$$

$$t_f = 0.570 \text{ in.}$$

Calculating the actual shear stress, we have

$$f_v = \frac{V}{dt_w} = \frac{78.8}{17.99(0.355)} = 12.34 \text{ ksi}$$

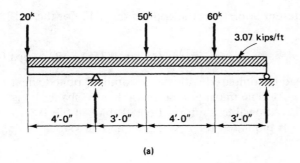

(a)

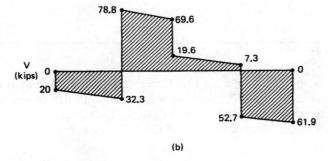

(b)

**FIGURE 4-21**   Load and shear diagrams.

Next, we calculate the allowable shear stress (ASDS, Section F4):

$$\frac{h}{t_w} = \frac{17.99 - 2(0.570)}{0.355} = 47.5$$

$$\frac{380}{\sqrt{F_y}} = \frac{380}{\sqrt{36}} = 63.3$$

$$47.5 < 63.3$$

Therefore,

$$F_v = 0.40F_y = 0.40(36) = 14.4 \text{ ksi}$$

which is rounded to 14.5 ksi in the ASDS, Numerical Values. Thus

$$f_v < F_v$$

**W18 × 50 O.K. for shear.**

The concept of *shear capacity* is also useful. The shear capacity of a beam may be determined by multiplying its web area ($d \times t_w$) by the allowable shear stress $F_v$. The ASDM calls this the *maximum web shear* and (unfortunately) designates it $V$ with no differentiation from the applied shear $V$. For the W18 × 50 of Example 4-13,

$$\text{shear capacity} = F_v dt_w = 14.5(17.99)(0.355) = 92.6 \text{ kips (412 kN)}$$

This may be readily compared with the maximum applied shear to verify that the beam is satisfactory. The maximum permissible web shear ($V$) is a tabulated quantity. Refer to the Allowable Uniform Load Tables in the ASDM, Part 2, where $V$ for each section is tabulated for $F_y = 36$ ksi and 50 ksi.

## 4-10

## DEFLECTIONS

When a beam is subjected to a load that creates bending, the beam must sag or deflect, as shown in Figure 4-22. Although a beam is safe for moment and shear, it may be unsatisfactory because it is too flexible. Therefore, the consideration of the deflection of beams is another part of the beam design process.

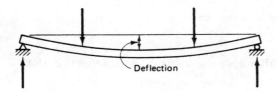

**FIGURE 4-22** Beam deflection.

Excessive deflections are to be avoided for many reasons. Among these are the effects on attached nonstructural elements such as windows and partitions, undesirable vibrations, and the proper functioning of roof drainage systems. Naturally, a visibly sagging beam tends to lessen one's confidence in both the strength of the structure and the skill of the designer.

To counteract the sag in a beam, an upward bend or *camber* may be given to the beam. This is commonly done for longer beams to cancel out the dead load deflection and, sometimes, part of the live load deflection. One production method involves cold bending of the beam by applying a point load with a hydraulic press or ram. For shorter beams, which are not intentionally cambered, the fabricator will process the beam so that any natural sweep within accepted tolerances will be placed so as to counteract expected deflections.

Normally, deflection criteria are based on some maximum limit to which the deflection of the beam must be held. This is generally in terms of some fraction of the span length. For the designer this involves a calculation of the expected deflection for the beam in question, a determination of the appropriate limit of deflection, and a comparison of the two.

The calculation of deflections is based on principles treated in most strength-of-materials texts. Various methods are available. For common beams and loadings, the ASDM, Part 2, Beam Diagrams and Formulas, contains deflection formulas. The use of some of these will be illustrated in subsequent examples.

The deflection limitations of specifications and codes are usually in the form of suggested guidelines because the strength adequacy of the beam is not at stake. Traditionally, beams that have supported plastered ceilings have been limited to maximum *live load* deflections of span/360. This is a requirement of the ASDS, Section L3.1. The span/360 deflection limit is often used for live load deflections in other situations. It is common practice, and in accordance with some codes, to limit maximum total deflection (due to live load *and* dead load) to span/240 for roofs and floors that support other than plastered ceilings.

The ASDS Commentary, Section L3.1, contains guidelines of another nature. It suggests:

1.   The *depth* of fully stressed beams and girders in floors should, if practicable, be not less than $F_y/800$ times the span.
2.   The *depth* of fully stressed roof purlins should, if practicable, be not less than $F_y/1000$ times the span, except in the case of flat roofs.

Further, it recommends that where human comfort is the criterion for limiting motion, as in the case of vibrations, the depth of a steel beam supporting large, open floor areas free of partitions or other sources of damping should be not less than $\frac{1}{20}$ of the span.

Since the moment of inertia increases with the square of the depth, the guidelines for minimum beam depth limit deflections in a general way. The ASDS Commentary, Section K2, also contains a method for checking the flexibility of roof systems when *ponding,* the retention of water on flat roofs, is a consideration.

**Example 4-14** _____

Select the lightest W shape for the beam shown in Figure 4-23. Assume full lateral support and A441 steel. Consider moment, shear, and deflection. Maximum allowable deflection for total load is to be span/360.

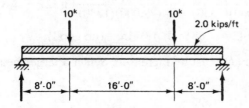

**FIGURE 4-23**   Beam load diagram.

**Solution:**

The usual procedure is to design for moment and check for the other effects. Select the section for moment:

$$M = \frac{wL^2}{8} + Pa = \frac{2.00(32)^2}{8} + 10(8) = 336 \text{ ft-kips}$$

From the beam curves, assuming that $F_y = 50$ ksi, select a W21 × 62 that has an $M_R$ of 349 ft-kips. From the ASDM, Part 1, Tables 1 and 2, $F_y = 50$ ksi. Check the additional moment due to the beam's own weight:

$$M = \frac{0.062(32)^2}{8} = 7.94 \text{ ft-kips}$$

$$\text{total } M = 336 + 7.94 = 344 \text{ ft-kips}$$

$$344 < 349 \qquad\qquad \textbf{O.K.}$$

*Check the shear:*   For this beam, the maximum shear occurs at, and is equal to, the reaction. Therefore,

$$\text{maximum shear} = \frac{2(10) + 32(2.062)}{2} = 43.0 \text{ kips}$$

From the Allowable Uniform Load Tables, Part 2, maximum permissible web shear for the W21 × 62 is 168 kips. Therefore, the shear strength is O.K.

*Check deflection* ($\Delta$):   From the ASDM properties tables, $I$ for the W21 × 62 is 1330 in.[4]:

$$\text{maximum allowable } \Delta = \frac{\text{span}}{360} = \frac{32(12)}{360} = 1.07 \text{ in.}$$

From formulas in the ASDM, Part 2, Beam Diagrams and Formulas, the actual expected deflection may be calculated. Note that the units are kips and inches.

$$\Delta = \frac{5wL^4}{384EI} + \frac{Pa(3L^2 - 4a^2)}{24EI}$$

$$= \frac{5(2.062)(32)^4(12)^3}{384(29{,}000)(1330)} + \frac{10(8)(12)^3}{24(29{,}000)(1330)}[3(32)^2 - 4(8)^2]$$

$$= 1.26 + 0.42 = 1.68 \text{ in.} > 1.07 \text{ in.} \quad \text{(no good)}$$

The moment of inertia ($I$) must be increased. Select a larger beam with an $I$ value of

$$\frac{1.68}{1.07}(1330) = 2088 \text{ in.}^4$$

From the ASDM, Part 2, Moment of Inertia Selection Tables, select W24 ×
76 ($I = 2100$ in.[4]). This shape has a higher $S_x$ and greater shear capacity than
the W21 × 62. Therefore, moment and shear are satisfactory. **Use W24 × 76.**

It is worthwhile noting in Example 4-14 that $M_R$ for an A36 steel W24 × 76 (full
lateral support) is 348 ft-kips (from the $S_x$ tables, based on $F_b$ of 23.76 ksi). Therefore,
the use of the higher-strength A441 steel is not justified, since the $M_R$ of 348 ft-
kips exceeds the total applied moment of 344 ft-kips. This is sometimes the case
where the design is governed by deflection ($I$) rather than by strength.

## 4-11

## HOLES IN BEAMS

Beams are normally found as elements of a total structural system rather than as
individual, isolated entities. They are penetrated by mechanical and electrical systems, are enveloped by nonstructural elements, and must be connected to other
structural members. They must sometimes be cut to provide clear areas. The problem of holes in beams is a common one.

Among the more evident effects of holes in beams (or, generally, any decrease
in cross-sectional area) is capacity reduction. Two such reductions may be readily
identified. Holes in beam webs reduce the shear capacity. Holes in beam flanges
reduce the moment capacity.

The ASDS is not specific concerning a recommended design procedure for beams
with web holes. The common procedure of reducing shear capacity in direct proportion to web area reduction is an oversimplification for other than small holes. Where
larger openings occur it is common practice to reinforce the beam web by welding
stiffening members around the perimeter of the hole. Some design examples for
beams with web openings are available (see References 1, 2, and 3).

Some general rules may be stated with regard to web holes. They should be
located away from areas of high shear. For uniformly loaded beams, web holes
near the center of the span will not be critical. Holes should be centered on the
neutral axis to avoid high bending stresses. The holes should be round or have
rounded corners (for rectangular holes) to avoid stress concentrations. The cutting
of web holes in the field should not be allowed without the approval of the designer.

With respect to holes in the flanges of beams, it is the moment capacity that is
affected. The cross-sectional property that governs moment capacity is moment of
inertia $I$. The *web* of a wide-flange beam contributes very little to the moment of
inertia, and the effect of web holes on moment capacity may be neglected. The
effect of flange holes, however, is to reduce the moment of inertia. The calculation
of the reduction is accomplished by subtracting from the gross moment of inertia
the quantity $Ad^2$ for each hole, where

**PHOTO 4-2**   Wide-flange beam with large web hole. Note how stiffeners have been welded around the perimeter of the hole for reinforcement.

$A$ = cross-sectional area of the hole (diameter × flange thickness) (in.²)

$d$ = distance from the neutral axis to the centroid of the hole (in.)

The neutral axis shifts very little where holes exist in only one flange and may be assumed to remain at the centroid of the gross cross section. For our discussion, no distinction will be made as to whether the flange is tension or compression, even though beams are usually controlled by the strength of the compression flange. Finally, it is the conservative practice of some designers to consider both flanges to have holes (in a symmetrical pattern) even though only one does.

It is generally agreed that flange holes for bolts do not reduce the moment capacity of beams to the extent indicated by the reduced moment of inertia as described in the preceding paragraph. The ASDS, in Section B10, states that no reduction in moment of inertia shall be made for bolt holes in either flange provided that

$$0.5F_u A_{fn} \geq 0.6F_y A_{fg}$$                    ASDS Eqn. (B10-1)

where

$A_{fn}$ = net flange area

$A_{fg}$ = gross flange area

If

$$0.5F_u A_{fn} < 0.6F_y A_{fg}$$                    ASDS Eqn. (B10-2)

then the flexural properties of the member shall be based on an *effective tension flange area* $A_{fe}$, where

$$A_{fe} = \frac{5}{6}\frac{F_u}{F_y}A_{fn} \qquad \text{ASDS Eqn. (B10-3)}$$

## Example 4-15

Using the ASDS, determine the resisting moment $M_R$ for a W18 × 71 that has two holes punched in each flange for 1-in.-diameter bolts. $F_b = 24$ ksi. Assume A36 steel. ($F_y = 36$ ksi; $F_u = 58$ ksi.)

**Solution:**

The cross section is shown in Figure 4-24. Recall from the discussion of net area computations in Section 2-2 of this text that for purposes of analysis and design, hole diameters are taken as the fastener diameter plus $\frac{1}{8}$ in. Therefore, these holes are taken as $1\frac{1}{8}$ in. in diameter. First, check whether the moment of inertia ($I$) must be reduced based on the ASDS, Section B10. For one flange,

$$A_{fg} = 7.635(0.81) = 6.18 \text{ in.}^2$$

$$A_{fn} = A_{fg} - A_{\text{holes}}$$

$$= 6.18 - 2(1.125)(0.81) = 4.36 \text{ in.}^2$$

$$0.5F_u A_{fn} = 0.5(58)(4.36) = 126.4 \text{ kips}$$

$$0.6F_y A_{fg} = 0.6(36)(6.18) = 133.5 \text{ kips}$$

Since

$$0.5F_u A_{fn} < 0.6F_y A_{fg}$$

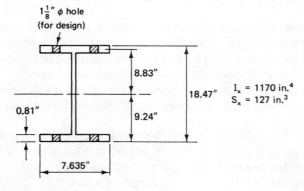

**FIGURE 4-24**   W18 × 71.

the deduction for holes must be considered, and the effective flange area is

$$A_{fe} = \frac{5}{6}\frac{F_u}{F_y}A_{fn} = \frac{5}{6}\left(\frac{58}{36}\right)(4.36) = 5.85 \text{ in.}^2$$

This is a decrease of $6.18 - 5.85 = 0.33$ in.$^2$ per flange.

$$\text{net } I_x = I_x - Ad^2$$
$$= 1170 - 2(0.33)(8.83)^2 = 1119 \text{ in.}^4$$

$$\text{net } S_x = \frac{\text{net } I_x}{c} = \frac{1119}{9.24} = 121.1 \text{ in.}^3$$

from which

$$\text{reduced } M_R = F_b S_x = \frac{24(121.1)}{12} = 242 \text{ ft-kips (328 kN·m)}$$

Calculating the percent reduction in the resisting moment, noting that $M_R$ for the gross section (from the ASDM beam curves) is 254 ft-kips, yields

$$\frac{254 - 242}{254}(100) = 4.7\%$$

The resisting moment has been reduced by 4.7%.

## 4-12

## WEB YIELDING AND WEB CRIPPLING

A beam that is subjected to concentrated loads applied normal to the flanges and symmetric to the web must be checked to ensure that a localized failure of the web does not occur.

The ASDS, Section K1, establishes requirements for beam webs under compression due to concentrated loads. When the stipulated requirements are exceeded, the webs of the beams should be reinforced or the length of bearing increased.

Two conditions are considered: *web yielding* and *web crippling*. Figure 4-25 illustrates the type of deformation failure expected. Practical and commonly used bearing lengths $N$ are usually large enough to prevent this type of failure from occurring.

With respect to web yielding, the ASDS, Section K1.3, requires that the compressive stress at the toe of the fillet, shown in Figure 4-26, not exceed $0.66F_y$. The assumption is made that the load "spreads out" so that the critical area for stress, which occurs at the toe of the fillet, has a length of $(N + 2.5k)$ or $(N + 5k)$ for end reactions and interior loads, respectively, and a width of $t_w$. The dimension $k$,

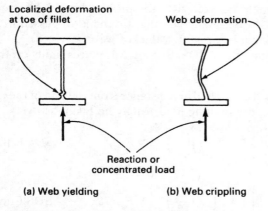

**(a) Web yielding**          **(b) Web crippling**

**FIGURE 4-25**   Web yielding/web crippling.

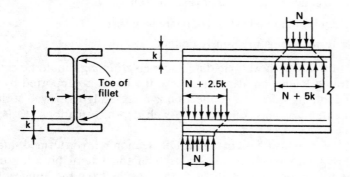

**FIGURE 4-26**   Web yielding.

which locates the toe of the fillet, is tabulated for various shapes in the ASDM, Part 1.

The controlling equations for web yielding are:

1.   For interior loads (defined as applied at a distance from the end of the member that is greater than the depth $d$ of the member),

$$\frac{R}{t_w(N + 5k)} \leq 0.66F_y \qquad \text{ASDS Eqn. (K1-2)}$$

2.   For end reactions,

$$\frac{R}{t_w(N + 2.5k)} \leq 0.66F_y \qquad \text{ASDS Eqn. (K1-3)}$$

where $R$ is the applied concentrated load and $N$ is the length of bearing.

Should the web yielding stress be excessive, the problem may be corrected by increasing the bearing length, by designing bearing stiffeners (discussed in Section 5-6 of this text), or by selecting a beam with a thicker web.

With respect to web crippling, the ASDS, Section K1, places limits on the compressive concentrated *loads*.

1.  For interior loads (defined as applied at a distance from the end of the member that is greater than $d/2$), the limiting load $R$ may be taken as

$$R = 67.5t_w^2\left[1 + 3\left(\frac{N}{d}\right)\left(\frac{t_w}{t_f}\right)^{1.5}\right]\sqrt{F_{yw}t_f/t_w} \qquad \text{ASDS Eqn. (K1-4)}$$

2.  For end reactions,

$$R = 34t_w^2\left[1 + 3\left(\frac{N}{d}\right)\left(\frac{t_w}{t_f}\right)^{1.5}\right]\sqrt{F_{yw}t_f/t_w} \qquad \text{ASDS Eqn. (K1-5)}$$

where

$R$ = maximum concentrated load or reaction (kips)

$F_{yw}$ = specified minimum yield stress of beam web (ksi)

and all other terms are as previously defined.

For unreinforced webs, both web yielding and web crippling should be checked under all concentrated loads and at points where the beam is supported by walls or pedestals or at columns when the connection is a seated type. If web stiffeners are provided and extend at least one-half the web depth, Equations (K1-4) and (K1-5) need not be checked.

A third type of failure considered in the ASDS, Section K1.5, is termed *sidesway web buckling*. This phenomenon manifests itself in the lateral buckling of the tension flange due to compression in the web that results from the application of concentrated load on the compression flange of the beam. Sidesway web buckling can be prevented by the use of lateral bracing or stiffeners at the concentrated load point. This type of failure is further discussed with regard to plate girders in Section 5-6 of this text.

**Example 4-16** _____

A W24 × 55 beam of A36 steel has an end reaction of 70 kips and is supported on a plate such that $N = 6$ in. (see Figure 4-26). Check the beam for web yielding and web crippling.

**Solution:**

For this shape,

$$t_w = 0.395 \text{ in.} \qquad k = 1\tfrac{5}{16} \text{ in.} = 1.31 \text{ in.}$$

$$t_f = 0.505 \text{ in.} \qquad d = 23.57 \text{ in.}$$

$$F_{yw} = 36 \text{ ksi}$$

Check web yielding [using Equation (K1-3) for end reactions]. The compressive stress at the toe of the fillet is

$$\frac{R}{t_w(N + 2.5k)} = \frac{70}{0.395[6 + 2.5(1.31)]} = 19.1 \text{ ksi}$$

The web yielding allowable stress is

$$0.66F_y = 0.66(36) = 23.76 \text{ ksi}$$

$$19.1 \text{ ksi} < 23.76 \text{ ksi} \qquad\qquad \textbf{O.K.}$$

Check web crippling [using Equation (K1-5) for end reactions]. The maximum compressive force is

$$R = 34t_w^2\left[1 + 3\left(\frac{N}{d}\right)\left(\frac{t_w}{t_f}\right)^{1.5}\right]\sqrt{F_{yw}t_f/t_w}$$

$$= 34(0.395)^2\left[1 + 3\left(\frac{6}{23.57}\right)\left(\frac{0.395}{0.505}\right)^{1.5}\right]\sqrt{36\left(\frac{0.505}{0.395}\right)}$$

$$= 55 \text{ kips}$$

Since 55 kips < 70 kips, the beam is inadequate with respect to web crippling. Bearing stiffeners must be provided, or the length of bearing must be increased.

---

Note that the web yielding equations [ASDS Equations (K1-2) and (K1-3)] may be expressed in different forms. To determine *allowable load* (based on allowable web yielding stress):

For end reactions,

$$R = 0.66F_{yw}(t_w)(N + 2.5k)$$

For interior loads,

$$R = 0.66F_{yw}(t_w)(N + 5k)$$

To determine *minimum length of bearing required* (based on allowable web yielding stress):

For end reactions,

$$\text{minimum } N = \frac{R}{0.66F_{yw}(t_w)} - 2.5k$$

For interior loads,

$$\text{minimum } N = \frac{R}{0.66F_{yw}(t_w)} - 5k$$

Rather than use the web yielding and web crippling equations as previously described, this checking process can be significantly simplified by using the ASDM, Part 2, Allowable Uniform Load Tables. The data furnished in the tables are summarized as follows:

1. The tabulated value $R$ (kips) is the maximum end reaction for $3\frac{1}{2}$ in. of bearing length ($N = 3\frac{1}{2}$ in.) For other values of $N$, the maximum end reaction with respect to web yielding is computed from

$$R = R_1 + NR_2$$

where $R_1$ and $R_2$ are constants tabulated for each shape in the Allowable Uniform Load Tables.

2. In a similar manner, the maximum end reaction with respect to web crippling is computed from

$$R = R_3 + NR_4$$

where $R_3$ and $R_4$ are also tabulated constants.

Using the tabular values greatly simplifies the calculations involving web yielding and web crippling.

**Example 4-17** _____

Rework Example 4-16 by using the tabulated values to determine the maximum end reaction and compare with the end reaction of 70 kips.

**Solution:**

From the Allowable Uniform Load Tables, for the W24 × 55 of A36 steel, for a bearing length $N$ of 6 in., the maximum end reaction is computed with respect to web yielding and web crippling.

*Web yielding:*

$$R = R_1 + NR_2$$
$$= 30.8 + 6(9.39)$$
$$= 87.1 \text{ kips} > 70 \text{ kips} \qquad \textbf{O.K.}$$

*Web crippling:*

$$R = R_3 + NR_4$$
$$= 36 + 6(3.17)$$
$$= 55.0 \text{ kips} < 70 \text{ kips} \qquad \textbf{N.G.}$$

The beam is not satisfactory with respect to web crippling. This checks Example 4-16.

## 4-13

### BEAM BEARING PLATES

Beams may be supported by connections to other structural members, or they may rest on concrete or masonry supports such as walls or pilasters. When the support is of some material that is weaker than steel (such as concrete), it is usually necessary to spread the load over a larger area so as not to exceed the *allowable bearing stress $F_p$*. This is achieved through the use of a *bearing plate*. The plate must be large enough so that the actual bearing pressure $f_p$ under the plate is less than $F_p$. Also, the plate must be thick enough so that the bending stress in the plate at the assumed critical section (see Figure 4-27) is less than the allowable bending stress $F_b$. An assumption is made that the pressure developed under the plate is uniformly distributed.

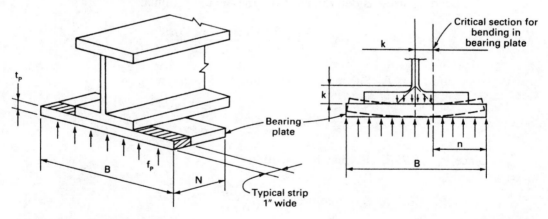

**FIGURE 4-27**   Beam bearing plate.

$F_b$, from the ASDS, Section F2, is $0.75F_y$. The allowable bearing pressure, $F_p$, for masonry or concrete may be obtained from the ASDS, Section J9, as follows:

For a plate covering the full area of concrete support,

$$F_p = 0.35f_c'$$

For a plate covering less than the full area of concrete support,

$$F_p = 0.35f_c' \sqrt{\frac{A_2}{A_1}} \leq 0.7f_c'$$

where

$f'_c$ = specified compressive strength of concrete (ksi)

$A_1$ = area of steel concentrically bearing on a concrete support (in.²)

$A_2$ = maximum area of the portion of the supporting surface that is geometrically similar to and concentric with the loaded area (in.²)

For a further explanation of bearing on concrete, see Reference 4.

The bending stress in the plate, at the critical section, may be determined with reference to Figure 4-27. The moment at the critical section for a 1-in.-wide strip of depth $t_p$ (in.), which acts like a cantilever beam, is

$$M = (\text{actual bearing pressure}) \times (\text{area}) \times (\text{moment arm})$$

$$= f_p \times (n \times 1) \times \frac{n}{2}$$

$$= \frac{f_p n^2}{2}$$

The bending stress is determined from the flexure formula:

$$f_b = \frac{Mc}{I} = \frac{(f_p n^2/2)(t_p/2)}{1(t_p^3)/12} = \frac{3 f_p n^2}{t_p^2}$$

As a limit, $f_b = F_b$. Solving for the required thickness, we have

$$\text{required } t_p = \sqrt{\frac{3 f_p n^2}{F_b}}$$

Since $F_b = 0.75 F_y$, this may be rewritten

$$\text{required } t_p = \sqrt{\frac{3 f_p n^2}{0.75 F_y}} = 2n \sqrt{\frac{f_p}{F_y}}$$

A procedure for the design of beam bearing plates is given in the ASDM, Part 2.

**Example 4-18** _____

A W16 × 50 is to be supported on a concrete wall, as shown in Figure 4-28. $f'_c$ = 3000 psi. The beam reaction is 55 kips. Design a bearing plate for the beam. Assume a 2-in. edge distance from the edge of the plate to the edge of the wall (maximum $N$ = 6 in.). All steel is A36.

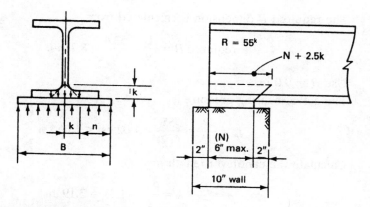

**FIGURE 4-28**   Beam bearing plate design.

**Solution:**

Use the procedure given in the ASDM, Part 2:

$$F_y = 36 \text{ ksi} \qquad R = 55 \text{ kips} \qquad k = 1\tfrac{5}{16} = 1.31 \text{ in.}$$

From the Allowable Uniform Load Tables, for the W16 × 50,

$$R_1 = 29.6 \text{ kips} \qquad R_2 = 9.03 \text{ kips/in.}$$
$$R_3 = 37.9 \text{ kips} \qquad R_4 = 3.28 \text{ kips/in.}$$

1.  Calculate the minimum bearing length $N$ based on
    (a)  Web yielding:

$$N = \frac{R - R_1}{R_2} = \frac{55 - 29.6}{9.03} = 2.81 \text{ in.}$$

    (b)  Web crippling:

$$N = \frac{R - R_3}{R_4} = \frac{55 - 37.9}{3.28} = 5.21 \text{ in.}$$

Therefore, use $N = 6$ in.

2.  Since the area of the support and the bearing area ($A_1$ and $A_2$) are unknown, conservatively assume that

$$F_p = 0.35 f'_c = 0.35(3) = 1.05 \text{ ksi}$$

3.  The required support area is

$$\text{required } A_1 = \frac{R}{F_p} = \frac{55}{1.05} = 52.4 \text{ in.}^2$$

4.   The required $B$ dimension is calculated from

$$\text{required } B = \frac{A_1}{N} = \frac{52.4}{6} = 8.73 \text{ in.}$$

Use $B = 9.0$ in.

5.   The actual bearing pressure is

$$f_p = \frac{R}{BN} = \frac{55}{6(9)} = 1.02 \text{ ksi} < 1.05 \text{ ksi} \qquad \textbf{O.K.}$$

6.   Calculate the cantilever length $n$:

$$n = \frac{B}{2} - k = \frac{9.0}{2} - 1.31 = 3.19 \text{ in.}$$

7.   Calculate the required plate thickness:

$$\text{Required } t_p = 2n \sqrt{\frac{f_p}{F_y}} = 2(3.19) \sqrt{\frac{1.02}{36}} = 1.07 \text{ in.}$$

See the ASDM, Part 1, Bars and Plates—Product Availability, for information on plate availability.

8.   **Use a bearing plate $1\frac{1}{8} \times 6 \times 0'$–9.**

---

It may be possible, if reactions are small, to support a beam in a bearing situation without the use of a bearing plate. Bearing pressure, web yielding, web crippling, and flange bending are the considerations. The critical section for flange bending is again assumed to be at a distance $k$ from the center of the section.

**Example 4-19** _____

A W24 × 76 is to be supported on a 12-in.-wide concrete wall such that there is bearing 8 in. wide. $f_c' = 3000$ psi. The beam reaction is 25 kips. Determine whether a bearing plate is required. Assume A36 steel.

**Solution:**

A diagram of the beam is shown in Figure 4-29. Beam properties and dimensions are

$$F_y = 36 \text{ ksi}$$

$$b_f = 8.99 \text{ in.}$$

$$t_f = 0.68 \text{ in.}$$

$$k = 1\frac{7}{16} = 1.44 \text{ in.}$$

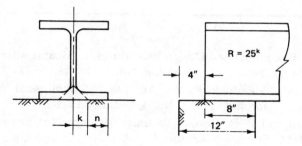

**FIGURE 4-29**   Beam without bearing plate.

1.   Check the bearing pressure:

$$f_p = \frac{25}{8.99(8)} = 0.35 \text{ ksi}$$

$$F_p = 0.35 f_c' = 0.35(3.0) = 1.05 \text{ ksi} > 0.35 \text{ ksi} \qquad \textbf{O.K.}$$

2.   Check the maximum end reaction using data from the Allowable Uniform Load Tables. Based on web yielding,

$$R = R_1 + NR_2$$

$$= 37.6 + 8(10.5) = 121.6 \text{ kips} > 25 \text{ kips} \qquad \textbf{O.K.}$$

Based on web crippling,

$$R = R_3 + NR_4$$

$$= 49.1 + 8(3.21) = 74.8 \text{ kips} > 25 \text{ kips} \qquad \textbf{O.K.}$$

3.   The bending stress in the flange may be determined using the formula developed previously for $f_b$ in the bearing plate (the flange acts exactly as the plate does).

$$n = \frac{b_f}{2} - k = \frac{8.99}{2} - 1.44 = 3.06 \text{ in.}$$

$$f_b = \frac{3 f_p n^2}{t_f^2} = \frac{3(0.35)(3.06)^2}{(0.68)^2} = 21.26 \text{ ksi}$$

$$F_b = 0.75 F_y = 0.75(36) = 27 \text{ ksi} > 21.26 \text{ ksi} \qquad \textbf{O.K.}$$

Therefore, this beam may be used on a bearing length of 8 in. without a bearing plate.

## REFERENCES

[1] J. E. Bower, "Recommended Design Procedures for Beams with Web-Openings," *AISC Engineering Journal,* Vol. 8, No. 4, October 1971.

[2] R. G. Redwood, "Simplified Plastic Analysis for Reinforced Web Holes," *AISC Engineering Journal,* Vol. 8, No. 4, October 1971.

[3] R. L. Kussman and P. B. Cooper, "Design Example for Beams with Web Openings," *AISC Engineering Journal,* Vol. 13, No. 2, 2nd Qtr., 1976.

[4] L. Spiegel and G. F. Limbrunner, *Reinforced Concrete Design,* 3rd ed. (Englewood Cliffs, NJ: Prentice Hall, Inc., 1992), p. 301.

## PROBLEMS

*Note: In the following problems and sketches, the given loads are* superimposed *loads. That is, they do not include the weights of the beams (unless noted otherwise).*

**4-1.** A floor framing plan is shown. Draw load diagrams for beams B1 and B2 and girder G1. The structural members will support a 6-in.-thick reinforced concrete floor slab. The live load is 200 psf. Assume reinforced concrete to weigh 150 pcf. Assume an additional load of 10 psf to account for the weight of the beams.

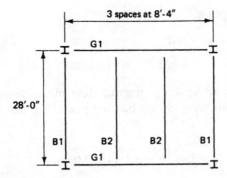

**PROBLEM 4-1**

**4-2.** The floor framing plan for a brewery process tank platform is shown. The tank legs impose loads of 40 kips at the locations shown. The floor is a 6-in. reinforced concrete slab, and the design live load is to be 200 psf. Draw load diagrams for beams B1 through B8. Assume reinforced concrete to weigh 150 pcf. Also, assume an additional load of 10 psf to account for the weight of the beams.

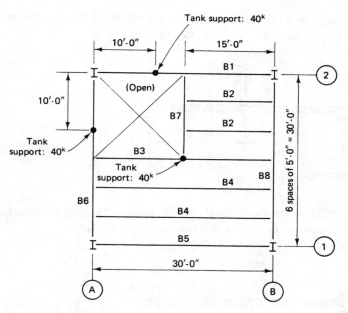

**PROBLEM 4-2**

**4-3.** A W30 × 108 simply supported beam spans 32 ft and supports a superimposed uniformly distributed load of 4 kips/ft. Assume A36 steel. Determine the maximum bending stress. Be sure to include the beam weight.

**4-4.** Calculate the maximum bending stress in a W18 × 65 beam that spans 36 ft and supports three equal concentrated loads of 12 kips each placed at the quarter points. Be sure to include the beam weight.

**4-5.** Using A588 steel, list two compact shapes and two noncompact shapes. Consider any W, M, S, or HP shape.

**4-6.** Determine $F_b$ for a W14 × 90 shape. Assume full lateral support. Use
   **(a)** A36 steel.
   **(b)** A572-Grade 42 steel.
   **(c)** A572-Grade 60 steel.

**4-7.** A W21 × 68 supports the loads shown. $F_b = 22$ ksi. Is the beam satisfactory? Consider moment only.

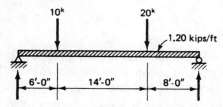

**PROBLEM 4-7**

**4-8.**  A W16 × 40 supports the loads shown. Find the maximum allowable load *P* (kips). $F_b = 24$ ksi. Consider moment only.

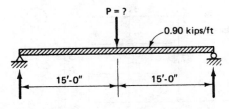

**PROBLEM 4-8**

**4-9.**  A W21 × 44 supports the loads shown. Determine whether the beam is satisfactory. $F_b = 24$ ksi. Consider moment only.

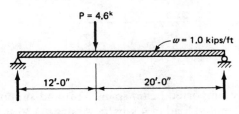

**PROBLEM 4-9**

**4-10.**  A W18 × 40 supports the loads shown. Determine the bending stress at the point of maximum moment and the bending stress at the center of the 16-ft span. Be sure to include the beam weight.

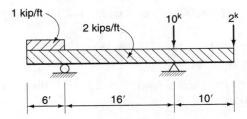

**PROBLEM 4-10**

**4-11.**  For the beam of Prob. 4-8, assume that the compression flange is laterally braced at only the supports and midspan. Assume A36 steel. Find the maximum allowable load *P* (kips) for the beam.

**4-12.**  A W21 × 44 beam supports the loads shown. Assume A36 steel. Find the maximum uniformly distributed load *w* (kips/ft) that the beam can support.

The beam's compression flange has lateral supports spaced 4 ft apart. Consider moment only.

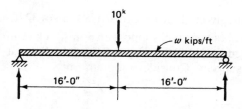

**PROBLEM 4-12**

**4-13.** A W21 × 57 simple span A36 steel beam is laterally supported at the end supports only. Use $C_b = 1.0$. Compute $F_b$ if the beam is used for the following span lengths:

**(a)** 6 ft–0 in.

**(b)** 12 ft–0 in.

**(c)** 25 ft–0 in.

**4-14.** A W21 × 62 supports a superimposed uniformly distributed load of 0.82 kips/ft and equal concentrated loads $P$ at the quarter points as shown. The compression flange is supported at the point loads and the reactions. Find the maximum allowable load $P$. Consider moment only. Use

**(a)** A36 steel.

**(b)** A588 steel.

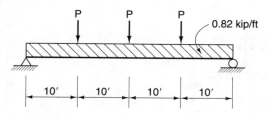

**PROBLEM 4-14**

**4-15.** Find the allowable uniformly distributed superimposed load that may be placed on a simply supported W27 × 84 beam that spans 30 ft. Lateral support exists at the ends only. The steel is A242. Use $C_b = 1.0$ and consider moment only.

**4-16.** A W12 × 30 of A36 steel is on a simple span of 22 ft. Determine $L_c$ and $L_u$. Find $F_b$ for $L_b = 14$ ft and 22 ft. Plot a curve of $F_b$ as a function of $L_b$. Assume that $C_b = 1.0$.

**4-17.** Select the lightest W shape to support a uniformly distributed load of 4 kips/ft on a simple span of 25 ft. Assume A36 steel and full lateral support. Consider moment only.

**4-18.** Rework Problem 4-17, given a load of 8 kips/ft, a span of 32 ft, and A242 steel.

**4-19.** The beams shown in plan view will support an 8-in.-thick reinforced concrete floor. The floor design live load is 175 psf. Select the lightest A36 W shape. Assume full lateral support. Consider moment only.

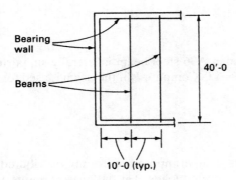

**PROBLEM 4-19**

**4-20.** Select the lightest W shape for the beam shown. Assume A36 steel and full lateral support. Consider moment only.

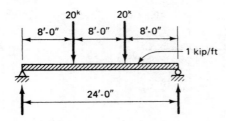

**PROBLEM 4-20**

**4-21.** Select the lightest W shape for a beam that is to span 26 ft between simple supports and will overhang one support by 8 ft. The beam supports a uniform load of 2.3 kips/ft and a concentrated load of 4 kips at the end of the overhang. Assume A242 steel and full lateral support. Consider moment only.

**4-22.** Select the lightest W shape for the beam shown. Assume A588 steel and full lateral support. Consider moment only.

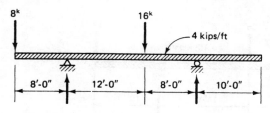

**PROBLEM 4-22**

**4-23.** A 6-in. reinforced concrete floor is supported on beams and girders with a layout for an *interior* bay, as shown. The floor is designed for a live load of 150 psf. Assume A36 steel and full lateral support for all members. Select the lightest W shapes for the beams and girders. Consider moment only.

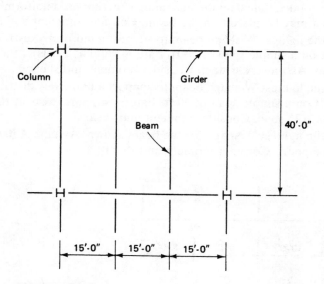

**PROBLEM 4-23**

**4-24.** Design the beams for the tank support platform of Problem 4-2. The floor slab provides lateral support for all beams except B1 and B6 in the area adjacent to the opening under the tank. Assume A36 steel. Consider moment only.

**4-25.** A simply supported beam is to support a uniform load of 2.1 kips/ft on a span of 24 ft. Lateral support exists at the reactions and at the third points. Assume A36 steel. Select the lightest W shape. Consider moment only.

**4-26.** Select the lightest W shape for the beam shown. Lateral support exists at the reactions and at the concentrated loads. Assume A36 steel. Consider moment only.

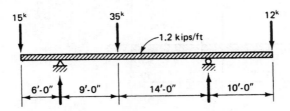

**PROBLEM 4-26**

**4-27.** A W16 × 57 beam is subjected to a shear $V$ of 78 kips. Plot the horizontal shear stress distribution for this condition. Compare with the shear stress as determined using the average web shear approach.

**4-28.** A W14 × 34 beam of A36 steel is on a simple span of 6 ft–6 in. Assume full lateral support. Calculate the maximum superimposed uniformly distributed load that may be placed on the beam. Consider moment and shear.

**4-29.** Select the lightest W-shape beam to support a uniformly distributed load of 1 kip/ft on a simple span of 30 ft. Lateral support exists at the ends and at midspan. Assume A36 steel. Consider moment and shear.

**4-30.** Select the lightest W-shape beam to support a uniformly distributed load of 5 kips/ft on a simple span of 20 ft. Lateral support exists at the ends only. Assume A36 steel. Consider moment and shear.

**4-31.** Select the lightest W shape for the beam shown. Assume A36 steel and full lateral support. Consider moment and shear.

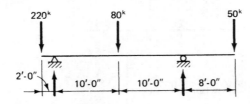

**PROBLEM 4-31**

**4-32.** Select the lightest W shape for the beam shown. Lateral support exists at points $A$ and $B$ only. Assume A36 steel. Consider moment and shear.

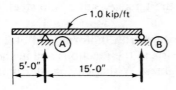

**PROBLEM 4-32**

**4-33.** Select the lightest W shape for the beam shown. Assume A36 steel and full lateral support. Consider moment and shear.

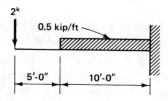

**PROBLEM 4-33**

**4-34.** Select the lightest W shape for the beam shown. Assume A36 steel. Consider moment and shear. Assume

**(a)** Full lateral support.

**(b)** Lateral support at the ends only.

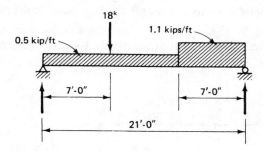

**PROBLEM 4-34**

**4-35.** A W14 × 74 of A36 steel supports a uniformly distributed load of 3 kips/ft on a simple span of 24 ft. Assume full lateral support. The maximum total allowable deflection is span/360. Determine whether the beam is adequate. Consider moment, shear, and deflection.

**4-36.** A W16 × 100 of A36 steel is on a simple span of 18 ft. Lateral support exists at the ends only. Determine the maximum superimposed uniformly distributed load that may be placed on the beam. The maximum allowable total deflection is span/360. Consider moment, shear, and deflection.

**4-37.** Rework Problem 4-36 assuming A588 steel.

**4-38.** An S12 × 50 of A36 steel is on a simple span of 30 ft. Assume full lateral support. The beam supports a uniformly distributed load of 0.25 kips/ft. Determine the maximum concentrated load (point load) that can be placed at midspan. The maximum allowable total deflection is span/360. Consider moment, shear, and deflection.

**4-39.** Select the lightest W shape for a beam that supports a uniformly distributed load of 400 lb/ft on a simple span of 22 ft. Assume A36 steel and full lateral support. The maximum allowable total deflection is span/360. Consider moment, shear, and deflection.

**4-40.** Check deflection for the beam selected in Problem 4-20 and redesign if necessary. The maximum allowable deflection for total load is span/800.

**4-41.** A set of plans that you are reviewing shows a W27 × 146 beam on a 38-ft simple span. The beam is of A36 steel and supports a superimposed uniform load of 3.4 kips/ft. Assume full lateral support. The allowable deflection is 0.85 in. Is the beam O.K.? If not, make a recommendation for a replacement beam having the *same nominal depth*. Consider moment, shear, and deflection.

**4-42.** Rework Problem 4-39 but assume that the load is 4 kips/ft, the span is 24 ft, the steel is A441, and the maximum allowable total deflection is span/1200.

**4-43.** Select the lightest W shape for a beam that supports a 20-kip concentrated load (point load) at the center of a 20-ft simple span. Assume A36 steel and that lateral support exists at the ends only. The maximum allowable total deflection is span/1000. Consider moment, shear, and deflection.

**4-44.** Select the lightest W shape for the beam shown. Assume A36 steel and full lateral support. The maximum allowable total deflection is span/240. Consider moment, shear, and deflection.

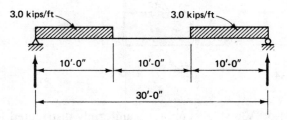

**PROBLEM 4-44**

**4-45.** Select the lightest W shape for B1 and G1, which support the floor for a typical interior bay in a small industrial building, as shown in the plan view. Use A36 steel. The floor will be a 6-in. reinforced concrete slab. Design loads are 150 psf live load and 20 psf dead load (partitions and ceilings). Assume full lateral support for the beams. Assume lateral support for the girders only where the beams are connected to them. The maximum allowable deflection for total load is span/360. Consider moment, shear, and deflection.

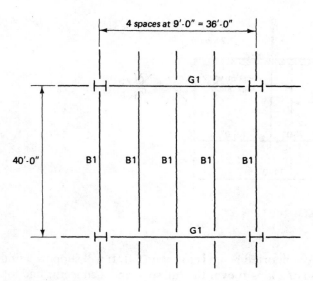

4 spaces at 9'-0" = 36'-0"

G1

B1  B1  B1  B1  B1

40'-0"

G1

**PROBLEM 4-45**

**4-46.** A W21 × 93 of A588 steel has two holes in each flange for $\frac{7}{8}$-in.-diameter bolts. Find the resisting moment $M_R$ for the beam. Assume full lateral support.

**4-47.** Select the lightest W shape for a beam that will support a uniformly distributed load of 2.5 kips/ft on an 18-ft simple span. Use A36 steel and assume that the beam has full lateral support. The beam is a spandrel beam supporting an exterior wall. Therefore, it requires a bottom flange detail that uses two $\frac{3}{4}$-in.-diameter bolts in any given plane through the bottom flange. Consider moment only.

**4-48.** The steel for a building is being erected. You are the engineer's representative on the site. A question has arisen and a quick answer is necessary. A simple span beam W16 × 45, A36 steel, is subjected to a maximum moment of 121.5 ft-kips. The beam is laterally supported ($L_b = 0$). Would you permit the drilling of two holes in the tension flange (for $\frac{7}{8}$-in.-diameter bolts)? Show all computations to support your answer. Assume that the given maximum moment includes the beam's own weight. Consider moment only.

**4-49.** Select the lightest W shape for the beam shown on the next page. Assume A36 steel and full lateral support. There will be two holes in each flange for $\frac{7}{8}$-in.-diameter bolts located 9 ft–0 in. from the left support.

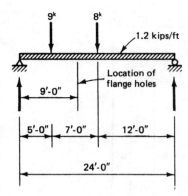

**PROBLEM 4-49**

**4-50.** A simply supported beam is to span 15 ft. It will support a uniformly distributed load of 2 kips/ft over the full span and a concentrated load (point load) of 60 kips at midspan. Assume A36 steel and full lateral support.

   **(a)** Select the lightest W shape. Consider moment and shear.

   **(b)** Determine the minimum bearing length required at the supports and under the concentrated load.

**4-51.** An A36 steel W14 × 53 beam is on a simple span of 11 ft. The beam supports a total uniformly distributed load of 112 kips (includes the beam weight). Determine the minimum required bearing length $N$ at the reactions.

**4-52.** Rework Problem 4-32 assuming a uniformly distributed load of 12 kips/ft (includes an estimated beam weight), A36 steel, and full lateral support. Select a beam with a maximum nominal depth of 16 in. Determine the minimum bearing length required at each support.

**4-53.** Design a beam bearing plate for a W18 × 50 beam supported by a masonry wall. $F_p = 0.25$ ksi. The beam reaction is 49 kips and the length of bearing $N$ is limited to 10 in. Use A36 steel.

**4-54.** Design a beam bearing plate for a W30 × 116 beam supported by a concrete wall. $F_p = 0.75$ ksi. The beam reaction is 105 kips and the length of bearing $N$ is limited to 6 in. Use A36 steel.

**4-55.** A W12 × 65 beam is to be supported on a brick (and mortar) pier. The beam reaction is 4 kips. $F_p = 300$ psi. Determine whether a bearing plate is required. The beam is A36 steel and has a bearing length of $3\frac{1}{2}$ in.

**4-56.** A W18 × 106 beam is supported on a concrete wall. The reaction is 40 kips. The maximum bearing length $N$ is 8 in. Use $F_p = 750$ psi and A36 steel. Design a beam bearing plate, if required.

**4-57.** A beam is supported on a bearing plate on a concrete wall as shown. Find the maximum allowable reaction. Assume A36 steel. Concrete $f'_c = 3000$ psi. Use the ASDS for $F_p$.

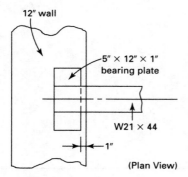

12" wall

5" × 12" × 1"
bearing plate

W21 × 44

1"

(Plan View)

**PROBLEM 4-57**

**4-58.** A wide-flange beam is to support the loads shown. Assume full lateral support and A36 steel.

**(a)** Select an economical wide-flange shape. Consider moment and shear.

**(b)** Design bearing plates for the beam supports. Assume maximum $N$ of 6 in. and $F_p = 1050$ psi.

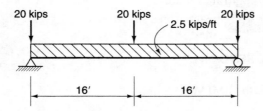

20 kips      20 kips          20 kips

2.5 kips/ft

16'          16'

**PROBLEM 4-58**

# CHAPTER 5

# Special Beams

## 5-1

### LINTELS

In the construction of walls, it is necessary to provide beams over openings, such as doors or windows. These beams, commonly called *lintels,* are required to support the weight of the wall and any other loads above the opening. Lintels are usually found only in building construction. Their loadings and design considerations deserve special attention. Structural steel lintels that support masonry walls will be treated here. Lintels are not limited to this type of material or loading, however.

Lintels may be composed of various structural steel shapes, as shown in Figure 5-1. The type used depends on the wall to be supported and the span between

162

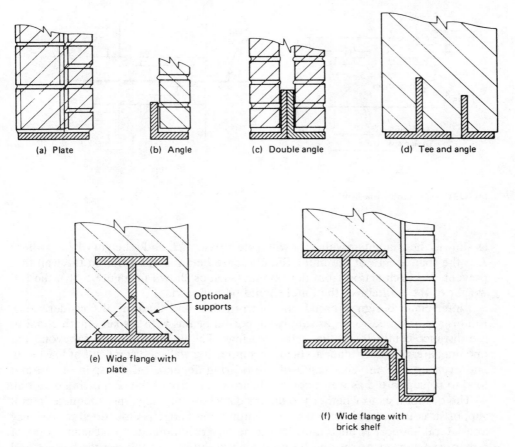

(a)  Plate          (b)  Angle          (c)  Double angle          (d)  Tee and angle

(e)  Wide flange with
     plate

Optional
supports

(f)  Wide flange with
     brick shelf

**FIGURE 5-1**   Lintels.

supports. The simple plate lintel shown in Figure 5-1a may be satisfactory for only the very shortest of spans, whereas the wide-flange lintels shown in Figure 5-1e and f are appropriate for the longer spans.

The load that a lintel supports should be carefully considered. If the wall is sufficiently continuous (both horizontally and vertically), it will probably support itself by an arching action that develops above the lintel, leaving only an approximately triangular section of the wall to be supported, as shown in Figure 5-2a. One rule of thumb suggests that the wall should be continuous above the opening a distance at least equal to the span of the lintel. Substantial piers or supports must also be furnished *adjacent* to the opening and must be capable of supporting the horizontal and vertical forces transmitted by the arch action. This action is usually assumed to develop only after the mortar has set and attained adequate strength. If the assumption is made that the load area is bounded by two sloping 45° lines,

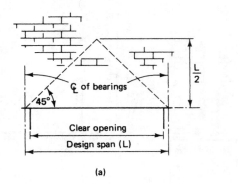

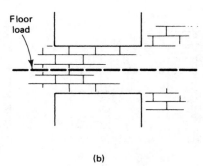

**FIGURE 5-2** Lintel loadings.

as shown, the triangular wall area will have a base of $L$ and a height of $L/2$, where $L$ is the span length of the lintel. Should there be an interruption in the wall that prevents the arch action from developing (such as shown in Figure 5-2b), the full weight of the wall above the lintel should be included.

Lintel beam design is based on both practical and theoretical considerations. Lintels are assumed to act as simply supported beams with a span length equal to the distance center to center of the bearings. The ends of the lintel, beyond the opening, are generally made to bear on supporting walls a distance of at least 4 in. and even up to 8 in. The length of the bearing depends on the span length and load to be supported as well as on the bearing capacity of the supporting material.

The question as to whether the lintel's compression flange has adequate lateral support is controversial since the top compression flange is embedded in masonry and lateral support is uncertain. It is generally recommended that some form of temporary support or shoring be used for the lintel until the mortar has set and the arching action has developed above the opening. Where lateral buckling is of concern due to long spans and heavy loads, a special investigation should be made and a reduced $F_b$ considered.

It is usually desirable to have lintels fully or partially buried in walls and hidden from view from at least one side. If two shapes are used, as in Figure 5-1c and d, they need not be mechanically fastened together. Individual loose lintel members are easier and more convenient to place than heavier, built-up sections. The selection of the two members should be based on approximately equal deflections of the individual parts so that unsightly cracking does not develop in the walls.

When selecting angles or tees to support brick walls, the preferred length of the horizontal outstanding leg or half-flange is $3\frac{1}{2}$ in. so that the steel edge does not project beyond the edge of the bricks while still providing sufficient width to support the bricks. In addition, the horizontal leg thickness should not exceed $\frac{3}{8}$ in. if possible, so that it can be buried in a mortar joint that is usually less than $\frac{1}{2}$ in. in thickness.

Shear is generally neglected in the design of steel lintels. It may be checked in the same way that normal beams are checked by using the average web shear approach, however, as discussed in Chapter 4 of this text. Average web shear is calculated from

$$f_v = \frac{V}{dt_w}$$

where

$d$ = full depth of vertical leg

$t_w$ = thickness of vertical leg

$f_v$ and $V$ are as previously defined

The allowable shear stress is as stipulated in the ASDS, Section F4.

**Example 5-1**

Design a lintel to span a clear opening of 8 ft–0 in. The masonry wall to be supported is 8 in. thick and weighs 130 lb/ft³ (see Figure 5-2a for reference). Use an inverted structural tee. Assume A36 steel, a 6-in. bearing on each side of the opening, and an allowable deflection of span/240. Assume that arching action can develop.

**Solution:**

The lintel is loaded as shown in Figure 5-3. The span length may be taken as the distance center to center of supports: 8 ft–6 in. Assume the weight of lintel to be 20 lb/ft. The maximum moment due to the triangular load (see the ASDM, Part 2, Beam Diagrams and Formulas, Case 3) and the moment due to the lintel weight are added:

$$\text{maximum } M = \frac{WL}{6} + \frac{wL^2}{8}$$

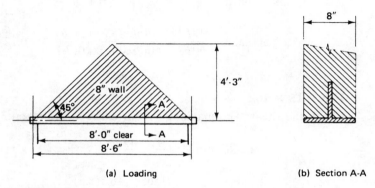

(a)  Loading                                          (b)  Section A-A

**FIGURE 5-3**    Lintel design.

1.  The total weight of the triangular load is

$$W = \frac{8}{12}(8.5)\left(\frac{4.25}{2}\right)(130) = 1565 \text{ lb}$$

2.  The maximum bending moment is

$$M = \frac{1565(8.5)}{6} + \frac{20(8.5)^2}{8} = 2400 \text{ ft-lb}$$

3.  Determine the required section modulus. An inverted tee used as a beam places the stem (or web) in compression. The rather thin stem may be subject to localized buckling. Tees used in this way are governed by the provisions of the ASDS, Section B5 and Appendix B, which provide that

$$F_b \leq 0.60 F_y Q_s$$

where $Q_s$ is a reduction factor based on the width–thickness ratio of the stem. $Q_s$ is tabulated as a property for the structural tees.

Assume that $F_b = 0.60 F_y = 22$ ksi. This must be verified later. Calculate the required section modulus:

$$\text{required } S_x = \frac{M}{F_b} = \frac{2400(12)}{22,000} = 1.31 \text{ in.}^3$$

4.  Select an appropriate WT. The tee selected should have a nominal flange width no larger than 8 in. Therefore, try a WT5 × 16.5. There is no value tabulated for $Q_s$, which means that the tee complies with the ASDS, Section B5. Therefore, the assumed $F_b$ is satisfactory. The assumed lintel weight is slightly conservative.

5.  Check the deflection. The deflection due to lintel weight is very small and is therefore neglected:

$$\text{allowable } \Delta = \frac{\text{span}}{240} = \frac{8.5(12)}{240} = 0.43 \text{ in.}$$

$$\text{actual } \Delta = \frac{WL^3}{60EI} = \frac{1565(8.5)^3(1728)}{60(29,000,000)(7.71)} = 0.12 \text{ in.}$$

$$0.12 \text{ in.} < 0.43 \text{ in.} \qquad \textbf{O.K.}$$

**Use WT5 × 16.5.**

6.  Check the shear. The maximum shear exists at the reaction:

$$\text{maximum } V = \frac{W}{2} + \frac{wL}{2}$$

$$= \frac{1565}{2} + \frac{16.5(8.5)}{2} = 853 \text{ lb}$$

For the WT5 × 16.5, $d$ = 4.865 in. and $t_w$ = 0.290 in. Thus

$$f_v = \frac{V}{dt_w} = \frac{853}{4.865(0.290)} = 605 \text{ psi} = 0.605 \text{ ksi}$$

Determine the allowable shear stress:

$$\frac{h}{t_w} = \frac{4.865 - 0.435}{0.290} = 15.3$$

$$\frac{380}{\sqrt{F_y}} = \frac{380}{\sqrt{36}} = 63.3$$

$$15.3 < 63.3$$

Therefore,

$$F_v = 0.40F_y = 0.40(36) = 14.4 \text{ ksi}$$

Since 14.4 ksi > 0.605 ksi, the WT is satisfactory for shear.

## Example 5-2

Design a lintel to carry a 12-in.-thick brick wall over a clear opening of 12 ft–6 in. Use A36 steel. The maximum allowable deflection is span/240. Use a structural tee (WT) and an angle (L), as shown in Figure 5-4b. The weight of the wall is 120 lb/ft³. The bearing length at the supports is 8 in.

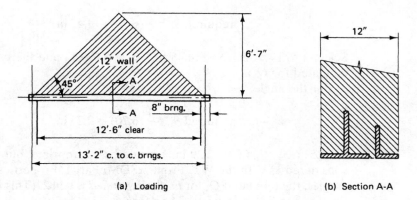

(a) Loading                                   (b) Section A-A

**FIGURE 5-4**   Lintel design.

## Solution:

1. Assume a triangular load due to wall arching action and refer to Figure 5-4 for load diagram. Assume 40 lb/ft for lintel weight. The total tri-

angular load on the lintel is $W$. Thus

$$W = \frac{1}{2}(13.17)(6.58)\left(\frac{12}{12}\right)(120) = 5200 \, \text{lb}$$

2. The maximum bending moment is

$$M = \frac{WL}{6} + \frac{wL^2}{8} = \frac{5200(13.17)}{6} + \frac{40(13.17)^2}{8} = 12{,}280 \, \text{ft-lb}$$

3. Determine the required section modulus. Assume that $F_b = 0.60F_y = 22$ ksi or 22,000 psi:

$$\text{required } S_x = \frac{M}{F_b} = \frac{12{,}280(12)}{22{,}000} = 6.70 \, \text{in.}^3$$

4. Figure 5-4b shows that the tee will support two-thirds of the wall. Therefore, the tee should supply two-thirds of the $S_x$. $F_b$ for the tee, however, may be affected by a $Q_s$ factor. $Q_s$ may be thought of as effectively reducing the $S_x$ of a shape by the same ratio that it reduces the $F_b$:

$$\text{reduced } S_x = (\text{original } S_x)Q_s$$

In the selection process, the structural tee should have a flange width of 8 in., or slightly less, and the horizontal leg of the angle should not exceed 4 in. This will make the sum of the two dimensions slightly less than the wall thickness of 12 in.

For the tee,

$$\text{required } S_x = \frac{2}{3}(6.70) = 4.47 \, \text{in.}^3$$

Select a WT7 × 24 ($S_x = 4.48$ in.$^3$, $b_f = 8.03$ in., and there is no value tabulated for $Q_s$).

For the angle,

$$\text{required } S_x = \frac{1}{3}(6.70) = 2.23 \, \text{in.}^3$$

An L5 × $3\frac{1}{2}$ × $\frac{3}{8}$ ($S_x = 2.29$ in.$^3$) would be appropriate, but $Q_s$ must be considered as with the WT. From ASDM, Part 1, Properties of Double Angles, the tabulated $Q_s$ for *angles separated* is 0.982. (This is conservative for bending members.) Therefore,

$$\text{reduced } S_x = 0.982(2.29) = 2.25 \, \text{in.}^3 > 2.23 \, \text{in.}^3 \qquad \textbf{O.K.}$$

The total lintel weight is

$$24 + 10.4 = 34.4 \, \text{lb/ft}$$

Therefore, the assumed weight of 40 lb/ft is slightly conservative.

5. Check the deflection for the selected steel shapes. The shapes should be selected so that the deflections are approximately equal. If one of the shapes is overly stiff, relative to the other, it will support a disproportionate amount of the load and will be overstressed. Since the WT supports two-thirds of the wall load, for equal deflections its moment of inertia $I$ should be twice the $I$ of the angle. For the WT, $I_{WT} = 24.9$ in.[4] Therefore, $I$ for the angle should be approximately $24.9/2 = 12.5$ in.[4] For the selected angle, $I$ is 7.78 in.[4] Therefore, select another angle with $I$ of approximately 12.5 in.[4], a minimum $S_x$ of 2.23 in.[3], and a horizontal leg of $3\frac{1}{2}$ in.

An L6 × $3\frac{1}{2}$ × $\frac{3}{8}$ will be selected. The weight of the lintel will therefore be

$$24 + 11.7 = 35.7 \text{ lb/ft}$$

For the angle, $I_x = 12.9$ in.[4], $S_x = 3.24$ in.[3], and $Q_s = 0.911$:

$$\text{reduced } S_x = 0.911(3.24) = 2.95 \text{ in.}^3 > 2.23 \text{ in.}^3 \qquad \textbf{O.K.}$$

The total $I_x$ is

$$24.9 + 12.9 = 37.8 \text{ in.}^4$$

Neglecting deflection due to the weight of the lintel,

$$\Delta_{\text{actual}} = \frac{WL^3}{60EI} = \frac{5200(13.17)^3(1728)}{60(29,000,000)(37.8)} = 0.31 \text{ in.}$$

$$\Delta_{\text{allowable}} = \frac{\text{span}}{240} = \frac{13.17(12)}{240} = 0.66 \text{ in.}$$

$$0.31 \text{ in.} < 0.66 \text{ in.} \qquad \textbf{O.K.}$$

Use a WT7 × 24 and an L6 × $3\frac{1}{2}$ × $\frac{3}{8}$. Shapes are to be loose, and the 6-in. leg is to be vertical.

6. Check the shear. The area stressed in shear is assumed to be the web of the WT and the vertical leg of the angle. The maximum shear occurs at the reaction:

$$\text{maximum } V = \frac{W}{2} + \frac{wL}{2}$$

$$= \frac{5200}{2} + \frac{35.7(13.17)}{2} = 2835 \text{ lb}$$

$$f_v = \frac{V}{\text{area stressed in shear}}$$

$$= \frac{2835}{6.895(0.340) + 6\left(\dfrac{3}{8}\right)} = 617 \text{ psi} = 0.617 \text{ ksi}$$

Check both elements to determine the allowable shear stress. For the WT7 × 24,

$$\frac{h}{t_w} = \frac{6.895 - 0.595}{0.340} = 18.5$$

$$\frac{380}{\sqrt{F_y}} = \frac{380}{\sqrt{36}} = 63.3$$

$$18.5 < 63.3$$

For the L6 × $3\frac{1}{2}$ × $\frac{3}{8}$,

$$\frac{h}{t_w} = \frac{6.00 - 0.375}{0.375} = 15.0$$

$$15.0 < 63.3$$

Therefore, the allowable shear stress is

$$F_v = 0.40F_y = 14.4 \text{ ksi}$$

Since 14.4 ksi > 0.617 ksi, the lintel is satisfactory for shear.

---

The lintel of Example 5-2 could also be designed to be made up from three individual angles. In Figure 5-4b, a double angle would replace the WT shape. (Refer also to Figure 5-1c and d.) The selection of the angles would proceed as follows:

$$\text{required } S_x = 6.70 \text{ in.}^3 \text{ (no change)}$$

Try three angles L5 × $3\frac{1}{2}$ × $\frac{3}{8}$ (with the 5-in. leg vertical):

$$S_x = 3(2.29) = 6.87 \text{ in.}^3$$

The foregoing assumes that $Q_s = 1.0$, which would be the case for angles in contact back-to-back. However, from the ASDM, Part 1, Properties of Double Angles, $Q_s = 0.982$ for angles separated. Therefore, assuming that this $Q_s$ applies to all three angles,

$$S_x = 6.87(0.982) = 6.75 \text{ in.}^3$$

$$6.75 \text{ in.}^3 > 6.70 \text{ in.}^3 \qquad \textbf{O.K.}$$

Check the deflection:

$$\Delta_{\text{allowable}} = 0.66 \text{ in. (no change)}$$

$$\Delta_{\text{actual}} = \frac{WL^3}{60EI} = \frac{5200(13.17)^3(1728)}{60(29,000,000)(3 \times 7.78)} = 0.51 \text{ in.}$$

$$0.51 \text{ in.} < 0.66 \text{ in.} \qquad \textbf{O.K.}$$

The three angles are identical, support equal loads, and therefore deflect equally. The weight for the three-angle lintel is

$$3(10.4) = 31.2 \text{ lb/ft}$$

It is slightly lighter than the WT–angle combination (35.7 lb/ft) of Example 5-2. Additionally, the three-angle lintel would be easier to handle and place.

## 5-2

## FLITCH BEAMS

Beams that are composed of more than one type of material are called *composite beams.* One type of composite beam, called a *flitch beam,* which has been used for many years, is made up of wood sections reinforced with structural steel shapes or plates. One resulting cross section is shown in Figure 5-5a. Flitch beams are sometimes found in wood-framed structures where long clear openings are desired (such as over a single-door two-car garage in residential construction). They have also been found to be useful in rehabilitation projects where existing wood members must be strengthened. This is accomplished by bolting steel plates or channels on the outside of the existing wood beams.

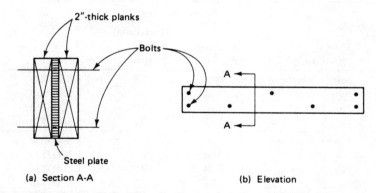

**(a)  Section A-A**                                     **(b)  Elevation**

**FIGURE 5-5**    Flitch beam.

The design of a flitch beam with a cross section such as shown in Figure 5-5 involves the selection of both wood and steel components. The planks and the steel plate are bolted together so that they *deflect* together. Since the deflection and the curvature of the planks and the plate are the same, the pattern of the *bending strains* will be identical over the cross section. The strains will be maximum at the top and bottom of the flitch beam and will vary linearly to zero at the neutral axis.

From strength of materials, for elastic composite members that are equally strained, it can be shown that the induced stresses are proportional to the modulus of elasticity values:

$$\frac{f_s}{f_w} = \frac{E_s}{E_w} = n$$

from which

$$f_s = n f_w$$

where

$f_s$ = bending stress in the steel

$f_w$ = bending stress in the wood

$E_s$ = modulus of elasticity for steel (29,000,000 psi)

$E_w$ = modulus of elasticity for wood

$n$ = modular ratio

The relationship between the stress in the steel and the stress in the wood is determined once $n$ is calculated. The value for $E_s$ is taken to be 29,000,000 psi. The value for $E_w$ varies with the species of wood. Table 5-1 lists some design values for

**TABLE 5-1   Design Values for Lumber**

| Species or group | Grade | Allowable stress Bending $F_{bw}$ (psi) | Horizontal shear, $F_{vw}$ (psi) | $E_w$ (psi) | $n$ |
|---|---|---|---|---|---|
| Douglas fir– larch | Select Structural | 1800 | 95 | 1,800,000 | 16.1 |
| | No. 1 | 1500 | 95 | 1,800,000 | 16.1 |
| | No. 2 | 1250 | 95 | 1,700,000 | 17.1 |
| | No. 3 | 725 | 95 | 1,500,000 | 19.3 |
| Hem– Fir | Select Structural | 1400 | 75 | 1,500,000 | 19.3 |
| | No. 1 | 1200 | 75 | 1,500,000 | 19.3 |
| | No. 2 | 1000 | 75 | 1,400,000 | 20.7 |
| | No. 3 | 575 | 75 | 1,200,000 | 24.2 |
| Eastern hemlock | Select Structural | 1550 | 85 | 1,200,000 | 24.2 |
| | No. 1 | 1300 | 85 | 1,200,000 | 24.2 |
| | No. 2 | 1050 | 85 | 1,100,000 | 26.4 |
| | No. 3 | 625 | 85 | 1,000,000 | 29.0 |
| Eastern spruce | Select Structural | 1200 | 70 | 1,500,000 | 19.3 |
| | No. 1 | 1000 | 70 | 1,500,000 | 19.3 |
| | No. 2 | 825 | 70 | 1,400,000 | 20.7 |
| | No. 3 | 475 | 70 | 1,200,000 | 24.2 |

commonly used structural lumber. This table is simplified and is primarily intended as a resource to accompany the examples and problems of this text. Those who require more detailed information should obtain Reference 1. Both wood and steel have allowable bending stresses that should not be exceeded. The allowable bending stress in the wood is denoted $F_{bw}$. Since the ratio between the induced maximum bending stresses is fixed, the bending stress in one material may be at the allowable whereas the bending stress in the other material will usually be at less than the allowable. Once the bending stresses in the materials (whether at allowable or less) are known, analysis or design may proceed. In general, A36 steel would be used with a maximum allowable bending stress of $0.6F_y$.

The ratio of *loads* supported by each of the two materials is the same as ratio of the respective resisting moments. Therefore, the ratio of the induced *shears* in each material is the same as the ratio of the respective resisting moments. The *shear capacity* of each material must be greater than the respective induced shear. Recalling that for *rectangular sections*,

$$\text{maximum } f_v = \frac{3V}{2A}$$

and that shear capacity may be expressed as

$$V_{\text{capacity}} = \frac{2}{3}F_v A = \frac{F_v A}{1.5}$$

the foregoing may be expressed as governing equations for shear:

$$V_{\text{capacity(wood)}} = \frac{F_{vw}A_w}{1.5} \geq \text{shear in wood}$$

$$V_{\text{capacity(steel)}} = \frac{F_{vs}A_s}{1.5} \geq \text{shear in steel}$$

where

$f_v$ = shear stress

$V$ = applied vertical shear force

$A$ = rectangular area of wood ($A_w$) or steel ($A_s$)

$F_v$ = allowable shear stress for wood ($F_{vw}$) or steel ($F_{vs}$)

Table 5-2 contains properties of dressed lumber (S4S) that will be needed for analysis and design problems. For properties of other size sections, see Reference 1, standard strength of materials textbooks, or determine the properties by calculation.

■■■  **TABLE 5-2**   **Properties of Dressed Lumber (S4S)**

| Nominal size (in.) | Dressed size (S4S) (in.) | $S_x^a$ (in.$^3$) | Weight per foot[b] (lb) |
|---|---|---|---|
| 2 × 6 | $1\frac{1}{2} \times 5\frac{1}{2}$ | 7.56 | 2.0 |
| 2 × 8 | $1\frac{1}{2} \times 7\frac{1}{4}$ | 13.14 | 2.6 |
| 2 × 10 | $1\frac{1}{2} \times 9\frac{1}{4}$ | 21.39 | 3.4 |
| 2 × 12 | $1\frac{1}{2} \times 11\frac{1}{4}$ | 31.64 | 4.1 |

[a]Centroidal axis x-x is parallel to the short dimension.
[b]Based on 35 lb/ft$^3$.

**Example 5-3** _____

Determine the adequacy (with respect to moment and shear) of a flitch beam composed of two S4S 2 × 12 planks reinforced with a $\frac{1}{2}$-in. steel plate as shown in Figure 5-6. Assume that continuous lateral support is furnished by a floor system supported by the beam. Use A36 steel with $F_b = 22$ ksi. Planks are No. 2 Eastern hemlock. The beam is on a simple span of 16 ft and supports a uniformly distributed load of 0.7 kip/ft, which includes the weight of the beam.

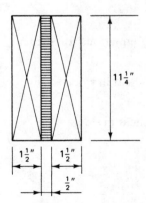

**FIGURE 5-6**   Flitch beam analysis.

**Solution:**

From Table 5-1, $E_w = 1,100,000$ psi:

$$n = \frac{E_s}{E_w} = \frac{29,000,000}{1,100,000} = 26.4$$

Assume that the wood is stressed to its allowable bending stress (see Table 5-1):

$$f_w = F_{bw} = 1050 \text{ psi}$$

$$f_s = nf_w = 26.4(1050) = 27{,}720 \text{ psi}$$

The allowable bending stress for the steel, $F_b$, however, is 22 ksi or 22,000 psi. Therefore, the maximum bending stress in the steel must be limited to 22,000 psi, and the maximum bending stress in the wood may be calculated as

$$f_w = \frac{f_s}{n} = \frac{22{,}000}{26.4} = 833 \text{ psi}$$

The two bending stresses that exist without either stress being excessive have now been established. The resisting moment for the wood $M_{Rw}$ (at a stress less than the allowable stress) is

$$M_{Rw} = f_w S_x \qquad \text{(use } S_x \text{ from Table 5-2)}$$

$$= \frac{0.833(2)(31.64)}{12} = 4.39 \text{ ft-kips}$$

For the steel plate,

$$S_x = \frac{I}{c} = \frac{bh^2}{6} = \frac{0.5(11.25)^2}{6} = 10.55 \text{ in.}^3$$

$$M_{Rs} = F_b S_x = \frac{22(10.55)}{12} = 19.34 \text{ ft-kips}$$

For the flitch beam,

$$M_R = M_{Rw} + M_{Rs} = 4.39 + 19.34 = 23.73 \text{ ft-kips}$$

Therefore, the total resisting moment is comprised as follows:

$$M_{Rw}: \quad \frac{4.39}{23.73}(100) = 18.5\%$$

$$M_{Rs}: \quad \frac{19.34}{23.73}(100) = 81.5\%$$

The flitch beam supports a uniformly distributed load of 0.7 kip/ft (which includes the weight of the beam) on a span of 16 ft. The maximum applied moment may be calculated as

$$M = \frac{wL^2}{8} = \frac{0.70(16)^2}{8} = 22.4 \text{ ft-kips}$$

Since 22.4 ft-kips < 23.7 ft-kips, the beam is satisfactory with respect to moment.

The maximum applied shear may be calculated as

$$V = \frac{wL}{2} = \frac{0.70(16)}{2} = 5.6 \text{ kips}$$

This shear is divided between the two materials in the same ratio as is moment; therefore, for wood,

$$V_w = 0.185(5.6) = 1.04 \text{ kips}$$

and for steel,

$$V_s = 0.815(5.6) = 4.56 \text{ kips}$$

The allowable shear stress $F_{vs}$ for the steel is 14,500 psi since

$$\frac{h}{t_w} = \frac{11.25}{0.5} = 22.5 < \frac{380}{\sqrt{F_y}} = 63.3$$

The allowable shear stress for wood is $F_{vw} = 85$ psi. The shear capacities may be calculated as

$$V_{\text{capacity(wood)}} = \frac{F_{vw}A_w}{1.5} = \frac{85(2)(1.5)(11.25)}{1.5} = 1913 \text{ lb} = 1.91 \text{ kips}$$

$$V_{\text{capacity(steel)}} = \frac{F_{vs}A_s}{1.5} = \frac{14,500(0.5)(11.25)}{1.5} = 54,400 \text{ lb} = 54.4 \text{ kips}$$

Since 1.04 kips < 1.91 kips and 4.56 kips < 54.4 kips, both materials are satisfactory for shear.

## Example 5-4

A flitch beam is to support a uniform load of 600 lb/ft (this includes an estimated flitch beam weight of 30 lb/ft) on a simple span of 14 ft. Assume that $2 \times 12$ S4S planks of No. 1 Hem–Fir will be used. Use A36 steel with $F_b$ of 22 ksi. Design the flitch beam. Assume continuous lateral support.

**Solution:**

Determine the applied moment:

$$M = \frac{wL^2}{8} = \frac{600(14)^2}{8} = 14,700 \text{ ft-lb}$$

Determine if a flitch beam is required. From Table 5-1, $F_{bw} = 1200$ psi:

$$\text{required } S_x = \frac{M}{F_{bw}} = \frac{14,700(12)}{1200} = 147 \text{ in.}^3$$

This would require five $2 \times 12$ planks. Therefore, use a flitch beam composed of two $2 \times 12$s and a steel plate. The plate will be $11\frac{1}{4}$ in. deep to match the depth of the planks. The thickness of the plate must be determined. From Table 5-1, $n = 19.3$. If the steel is at the allowable bending stress ($F_b = 22$ ksi),

$$f_w = \frac{f_s}{n} = \frac{F_b}{n} = \frac{22}{19.3} = 1.140 \text{ ksi} = 1140 \text{ psi}$$

For this wood, from Table 5-1, $F_{bw} = 1200$ psi. Therefore,

$$f_s = F_b = 22 \text{ ksi}$$

$$f_w = 1140 \text{ psi} < F_{bw} \qquad\qquad \textbf{O.K.}$$

The resisting moment of the wood is

$$M_{Rw} = f_w S_x = \frac{1.140(2)(31.64)}{12} = 6.01 \text{ ft-kips}$$

The moment to be resisted by the steel plate is

$$\text{required } M_{Rs} = 14.70 - 6.01 = 8.69 \text{ ft-kips}$$

The actual resisting moment of the steel plate is

$$M_{Rs} = F_b S_x = F_b\left(\frac{th^2}{6}\right)$$

Equating the required and the actual $M_{Rs}$ values and solving for the required thickness $t$ gives

$$\text{required } t = \frac{6M_{Rs}}{F_b h^2} = \frac{6(8.69)(12 \text{ in./ft})}{22(11.25)^2} = 0.22 \text{ in.}$$

Try a plate $\frac{1}{4}$ in. $\times$ $11\frac{1}{4}$ in. and check the adequacy of the flitch beam for shear. The actual resisting moment of the steel plate is

$$M_{Rs} = F_b S_x = \frac{22(0.25)(11.25)^2}{6(12 \text{ in./ft})} = 9.67 \text{ ft-kips}$$

Therefore, the total resisting moment is comprised as

$$M_{Rw}: \quad \frac{6.01}{6.01 + 9.67}(100) = 38.3\%$$

$$M_{Rs}: \quad \frac{9.67}{6.01 + 9.67}(100) = 61.7\%$$

The maximum applied shear is

$$V = \frac{wL}{2} = \frac{600(14)}{2} = 4200 \text{ lb}$$

For the steel, the shear stress is calculated as

$$f_{vs} = \frac{3V_s}{2A_s} = \frac{3(0.617)(4200)}{2(0.25)(11.25)} = 1382 \text{ psi}$$

and for the wood,

$$f_{vw} = \frac{3V_w}{2A_w} = \frac{3(0.383)(4200)}{2(2)(1.5)(11.25)} = 71.5 \text{ psi}$$

The allowable shear stress $F_{vs}$ for the steel is 14,500 psi since

$$\frac{h}{t_w} = \frac{11.25}{0.25} = 45.0 < \frac{380}{\sqrt{F_y}} = 63.3$$

The allowable shear stress $F_{vw}$ for the wood is 75 psi. Therefore, both are satisfactory for shear. Use two 2 × 12 S4S planks of No. 1 Hem–Fir and a steel plate $\frac{1}{4}$ in. × $11\frac{1}{4}$ in.

The flitch beam should be bolted together to ensure that the individual parts act as a unit. Bolts $\frac{3}{4}$ in. in diameter staggered on 2 ft–0 in. centers are sufficient for applications of the type discussed. Flitch beams composed of heavier plates and larger timbers require further investigation.

## 5-3

### COVER-PLATED BEAMS

Sometimes, available rolled shapes will have inadequate bending strength and will not satisfy the requirements for a given beam. Also, depth restrictions may require the use of shallower rolled shapes that do not possess the required bending strength. Although the W40 is the deepest shape rolled in the United States, rolled shapes of 44 in. nominal depth in standard sections are also available from nondomestic producers. These large rolled wide-flange beams may offer considerable cost savings when compared with welded plate girders or cover-plated beams. They are appropriate for use in some situations. If the available hot-rolled shapes are inadequate, the other traditional methods may be used to solve the problem.

First, it may be possible to double-up two shapes, that is, place them side by side to develop the required strength. In some structures that contain repetitive beams, it may be possible to decrease the lateral spacing and thereby decrease the load carried by each beam. Second, a *plate girder* may be devised by welding together plates of the required sizes and often in the shape typical of wide-flange sections. The design of plate girders is discussed in Section 5-6 of this text. Third, an appropriate W shape can be strengthened by adding *cover plates* to its flanges, as shown in Figure 5-7. In the past, the attachment of the cover plates has been by

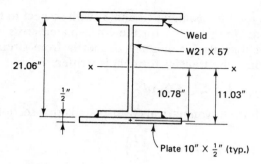

**FIGURE 5-7**   Cover-plated beam analysis.

riveting, and many of these cover-plated beams may still be observed. Cover-plated beams (and plate girders) are now fabricated predominantly by welding, however.

The analysis of a cover-plated beam involves the determination of the moment capacity, $M_R = F_b I/c = F_b S$.* The moment of inertia (and section modulus) are readily determined using familiar methods from strength of materials. The allowable bending stress for members in which the cover plate is wider than the flange must be determined in accordance with the requirements of the ASDS, Section F1. In addition, cover plates possibly may not extend for the full length of the beam. Therefore, the strength of the section without cover plates must be checked against the applied moment that it must resist.

**Example 5-5** _____

Find the resisting moment of the A36 steel cross section shown in Figure 5-7. The cover plates are attached with continuous welds as shown. Assume full lateral support for the compression flange.

**Solution:**

Properties of the W21 × 57 are as follows:

$$d = 21.06 \text{ in.}$$

$$b_f = 6.56 \text{ in.}$$

$$I_W = 1170 \text{ in.}^4$$

$$S_W = 111 \text{ in.}^3$$

---

\* In this section, for the discussion of cover-plated beams, all bending calculations are with respect to the strong $x$-$x$ axis of the cross sections. The $x$ subscript for $I$ and $S$ is omitted.

Note that a $W$ subscript is used to distinguish $I$ and $S$ (with respect to the $x$-$x$ axis) for the W shape alone. Find $I$ and $S$ for the cover-plated cross section using the familiar transfer formula for moment of inertia from strength of materials. In its general form, the transfer formula is written

$$I = \Sigma I_c + \Sigma Ad^2$$

Rewriting this expression for the symmetrical cover-plated cross section gives

$$I = I_W + 2(A)\left(\frac{d}{2} + \frac{t}{2}\right)^2$$

where

$I$ = moment of inertia of the cover-plated cross section with respect to the $x$-$x$ axis

$A$ = area of one cover plate

$t$ = thickness of one cover plate

$I_W$ and $d$ are as described previously

Therefore,

$$I = 1170 + 2(10)(0.5)\left(\frac{21.06}{2} + \frac{0.5}{2}\right)^2 = 2332 \text{ in.}^4$$

$$S = \frac{I}{c} = \frac{2332}{11.03} = 211.4 \text{ in.}^3$$

Note that the moments of inertia of the plates about their own centroidal axes parallel to the $x$-$x$ axis are very small and have been neglected. We will next determine $F_b$. The ASDS, Section B5, covers compactness for built-up members. Since the W21 × 57 is itself compact, only the width–thickness ratio of the compression flange must be checked. Note that according to the ASDS, Section B5, Table B5.1, the width–thickness ratio may not exceed $65/\sqrt{F_y}$ and that according to the ASDS, Section B5.1, the width of the projecting element is taken from the free edge to the weld. Therefore,

$$\frac{\text{width}}{\text{thickness}} = \frac{(10 - 6.56)/2}{0.5} = 3.44$$

$$\frac{65}{\sqrt{F_y}} = \frac{65}{\sqrt{36}} = 10.8$$

$$3.44 < 10.8 \qquad\qquad \textbf{O.K.}$$

Also in the ASDS, Section B5 and Table B5.1, the width–thickness ratio of flange cover plates *between* lines of welds may not exceed $190/\sqrt{F_y}$. The flange cover plate is considered to be a *stiffened element*.

$$\frac{\text{width}}{\text{thickness}} = \frac{6.56}{0.5} = 13.12$$

$$190/\sqrt{F_y} = 190/\sqrt{36} = 31.7$$

$$13.12 < 31.7 \qquad\qquad \textbf{O.K.}$$

Therefore, $F_b = 0.66F_y = 24.0$ ksi. Calculate the resisting moment for the built-up section:

$$M_R = F_b S = 24.0(211.4) = 5074 \text{ in.-kips}$$

$$= \frac{5074}{12} = 422.8 \text{ ft-kips } (573 \text{ kN} \cdot \text{m})$$

The *design* of symmetrical cover-plated beams involves the proportioning of the required plates, which, when added to the chosen rolled shape, will result in moment capacity sufficient to enable the beam to resist the applied moment. In Figure 5-8, the rolled shape has been strengthened with two plates each having area $A$ and thickness $t$. The areas $A$ are to be chosen so that $M_R \geq M$, where $M$ is the applied moment:

$$M_R = F_b S$$

$$\text{required } S = \frac{M}{F_b}$$

As before, let $I_W$ and $S_W$ be properties of the W shape (about the $x$-$x$ axis) and $S$ and $I$ be properties of the cover-plated section (also about the $x$-$x$ axis). The

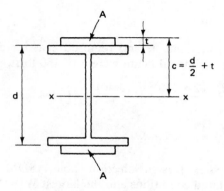

**FIGURE 5-8**  Cover-plated beam cross section.

quantities $d$, $t$, and $c$ are as shown in Figure 5-8. Thus

$$I = I_W + 2A\left(\frac{d}{2} + \frac{t}{2}\right)^2$$

$$S = \frac{I}{c} = \frac{I_W + 2A[(d/2) + (t/2)]^2}{(d/2) + t}$$

Since $t$ is usually unknown, an approximate expression for $S$ may be written by assuming $t$ to be small compared with $d$. Neglecting $t$, we have

$$\text{approximate furnished } S = \frac{I_W + 2A(d/2)^2}{d/2} = \frac{I_W}{d/2} + Ad = S_W + Ad$$

If the required $S$ is equated to the foregoing, we have

$$\text{required } S = S_W + Ad$$

and (*approximately*)

$$\text{required } A = \frac{\text{required } S - S_W}{d}$$

The area determined by the foregoing expression will be on the low side. An analysis check must be made following selection of the plates.

### Example 5-6

A simply supported beam is to support a uniform load of 2.4 kips/ft and a 10-kip concentrated load at midspan. The span is 40 ft, A36 steel, with full lateral support for the compression flange. Maximum overall beam depth is not to exceed 20 in. This is a rush job. Steel is to be obtained locally, and the local steel supplier currently has nothing in stock larger than a W18 × 119 wide-flange section. Design a symmetrical beam cross section for maximum applied moment.

**Solution:**

Compute the maximum applied moment $M$ and required $S$ assuming that $F_b = 0.66F_y = 24$ ksi. Include an estimated beam weight of 150 lb/ft.

$$M = \frac{wL^2}{8} + \frac{PL}{4} = \frac{2.55(40)^2}{8} + \frac{10(40)}{4} = 610 \text{ ft-kips}$$

$$\text{required } S = \frac{M}{F_b} = \frac{610(12)}{24} = 305 \text{ in.}^3$$

With reference to the Allowable Stress Design Selection Table, ASDM, Part 2, a W21 will be too deep for the stated conditions and the largest W18 (W18 × 119) has insufficient moment capacity ($M_R = 457$ ft-kips, $S_W = 231$ in.$^3$).

Therefore, design a cover-plated beam. Use the W18 × 119 and select appropriate cover plates.

Properties of the W18 × 119 are

$$d = 18.97 \text{ in.}$$

$$I_W = 2190 \text{ in.}^4$$

$$S_W = 231 \text{ in.}^3$$

$$b_f = 11.265 \text{ in.}$$

The *approximate* required area $A$ for each cover plate is

$$A = \frac{\text{required } S - S_w}{d} = \frac{305 - 231}{18.97} = 3.90 \text{ in.}^2$$

As noted previously, because of simplifying assumptions, this calculated required area will be on the low side. Therefore, try two cover plates that are 9 in. × $\frac{1}{2}$ in. The cover-plated section is shown in Figure 5-9. Compute the section modulus $S$ that is furnished by the cover-plated section:

$$I = I_W + 2(A)\left(\frac{d}{2} + \frac{t}{2}\right)^2$$

$$= 2190 + 2(9)(0.50)(9.74)^2 = 3044 \text{ in.}^4$$

$$\text{furnished } S = \frac{I}{c} = \frac{3044}{9.99} = 304.7 \text{ in.}^3$$

The required section modulus is 305 in.³. This is close enough to be considered satisfactory. For a check of beam weight (150 lb/ft was assumed), use a unit weight of steel of 490 lb/ft³ or see the Table for Weight of Rectangular Sections, ASDM, Part 1:

$$\text{beam weight} = 119 + \frac{2(9)(0.5)(12)}{1728}(490) = 150 \text{ lb/ft}$$

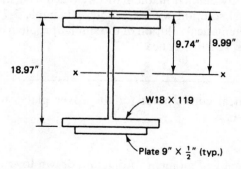

**FIGURE 5-9**  Cover-plated beam analysis.

Check the width/thickness ratio of the plate as a stiffened element of the compression flange:

$$\frac{\text{width}}{\text{thickness}} = \frac{9.0}{0.5} = 18.0$$

$$190/\sqrt{F_y} = 190/\sqrt{36} = 31.7$$

$$18.0 < 31.7 \qquad\qquad \textbf{O.K.}$$

The cover-plated section qualifies for $F_b = 0.66F_y$ as assumed. Therefore, use a W18 × 119 with top and bottom cover plates 9 × $\frac{1}{2}$.

---

In Example 5-6, equal-area plates were chosen. In some situations this will not be the case. For example, where a concrete slab acts in conjunction with the compression flange of a composite plate girder, the compression flange of the plate girder will normally be smaller than the plate used on the tension side. The design of such an unsymmetrical section is necessarily a trial-and-error procedure. The previous approach for symmetrical sections does not apply. For a discussion of the required welded connection between the cover plate and the flange, see Section 5-6 of this chapter.

Cover plates may extend the full length of the beam. This is not necessary, however, and they may be discontinued in areas where the applied moment is low enough so that the resisting moment of the wide-flange section is sufficient. The point where the resisting moment of the wide-flange section is equal to the applied moment is called the *theoretical cutoff point*. The ASDS, Section B10, requires that partial-length cover plates be extended a definite length $a'$ beyond the theoretical cutoff point. As a maximum, $a'$ is to be taken as two times the plate width. This is the length required when there is no weld across the end of the plate, but continuous welds along both edges of the cover plate in the length $a'$. For the determination of $a'$, reference should be made to the ASDS, Section B10.

The determination of the theoretical cutoff point for the cover plates involves the superposition of the applied moment $M$ diagram and the moment capacity $M_R$ of the wide-flange section. The solution may be either graphical or mathematical. If the moment diagram is easily defined mathematically (such as a straight line or a parabola), the mathematical solution will be simpler. Example 5-7 illustrates a graphical solution. For a mathematical solution of the similar problem of bar cutoffs in reinforced concrete beams, see Reference 2.

### Example 5-7

Determine the theoretical cutoff point for the cover plates for the beam designed in Example 5-6.

### Solution:

Figure 5-10 shows the applied moment $M$ diagram drawn to scale. Moments were computed at the eighth points for purposes of drawing the diagram. $M_R$

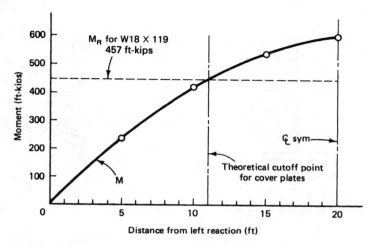

**FIGURE 5-10**   Determination of theoretical cutoff point for cover plates.

for the W18 × 119, 457 ft-kips, is superimposed. The point at which the $M$ diagram crosses the $M_R$ line is the theoretical cutoff point. This is seen to be 11 ft from the left support. Symmetry exists. The required total length of cover plate would be 18 ft plus any required extensions at the ends of the plates.

## 5-4

## UNSYMMETRICAL BENDING

Thus far, the beams considered have been loaded so that bending occurs about the strong axis (the $x$-$x$ axis). That is, the loads have been applied in the *plane* of the weak axis. This is the normal situation for gravity loads on wide-flange beams that are oriented with their webs in the vertical plane. The $x$-$x$ and $y$-$y$ axes of beam cross sections are also called the *principal axes*. Occasionally, beams are subjected to loads that are not in the plane of the weak axis, or a beam may have to support two or more systems of loads that are applied simultaneously but in different directions. When this occurs, the beam is said to be subjected to *unsymmetrical bending,* or bending about two axes. Unsymmetrical bending may further be defined as bending about any other than one of the principal axes. Some examples are shown in Figure 5-11. The beam of Figure 5-11a supports a load that passes through the centroid of the cross section. The roof purlin of Figure 5-11b, supported on the sloping top chord of a roof truss, must support roof loads, the wind load, and its own weight, applied as shown. Note that the wind load is assumed to be applied

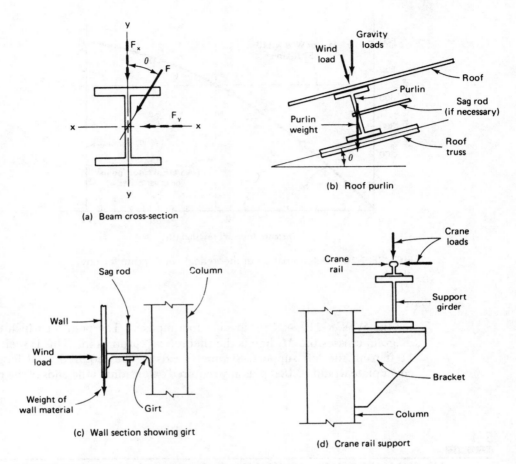

**FIGURE 5-11**   Unsymmetrical bending.

perpendicular to the roof surface. The girt of Figure 5-11c must support vertical and horizontal loads, as shown, in addition to its own weight. The support girder for the crane rail, shown in Figure 5-11d, must support vertical loads as well as the lateral thrust due to the moving crane.

In the case where the load passes through the centroid of the cross section (Figure 5-11a), it may be broken into its components that are parallel and perpendicular to the principal (*x-x* and *y-y*) axes. Thus (where $\theta$ is the angle between the applied force and the *y-y* axis)

$$F_x = F(\cos \theta)$$
$$F_y = F(\sin \theta)$$

and moments about the principal axes $M_x$ and $M_y$ may be found. Note that the subscripts are such that, for example, $F_x$ is the force that creates bending ($M_x$) about the $x$-$x$ axis. The stresses may be calculated separately for bending about each axis and added algebraically. Thus

$$f_b = \frac{M_x}{S_x} \pm \frac{M_y}{S_y}$$

The $\pm$ sign indicates that the stresses may be additive (i.e., both tension or both compression) or they may be of opposite sign and be subtractive.

In most cases the applied load will not be positioned so that its line of action passes through the centroid of the cross section. It is more common for the load to be applied at the top flange. When this occurs, the top flange must resist most of the lateral force component, as shown in Figure 5-11b and d. Actually, this situation results in twisting of the beam. It is commonly assumed that the top flange acts alone in resisting the lateral force component, however. The formula for bending stress in the top flange then becomes

$$f_b = \frac{M_x}{S_x} \pm \frac{M_y}{S_{y\text{(top flange)}}}$$

For typical wide-flange shapes, the section modulus of the top flange about the $y$-$y$ axis ($S_{y\text{(top flange)}}$) is approximately equal to $S_y/2$. Hence the formula is commonly written

$$f_b = \frac{M_x}{S_x} \pm \frac{M_y}{S_y/2}$$

Since the *allowable* bending stresses with respect to the $x$-$x$ and $y$-$y$ axes are different, the ASDS, Section H1, utilizes an interaction formula of stress ratios for the purpose of establishing a design criterion. For low axial stress ($f_a/F_a \leq 0.15$), the applicable formula is

$$\frac{f_a}{F_a} + \frac{f_{bx}}{F_{bx}} + \frac{f_{by}}{F_{by}} \leq 1.0 \qquad\qquad \text{ASDS Eqn. (H1-3)}$$

where $f_a$ is the computed axial stress, $F_a$ is the allowable axial stress permitted in the absence of bending, $f_b$ and $F_b$ are as previously defined, and the $x$ and $y$ subscripts refer to the axis about which bending takes place. For the beams that will be considered here, axial stress will be zero; therefore, the preceding formula reduces to

$$\frac{f_{bx}}{F_{bx}} + \frac{f_{by}}{F_{by}} \leq 1.0$$

To utilize the interaction formula for the design of an unsymmetrically loaded beam, a trial-and-error process must be used. The beam size must be estimated, checked by the interaction formula, and revised if necessary. Some design aids can be used, and computer programs are available to help in the selection process.

**Example 5-8** _____

Check the adequacy of a W10 × 22 that carries a uniform (gravity) load of 0.50 kip/ft on a simple span of 15 ft. The beam is placed on a slope of 4 : 12 as shown in Figure 5-12. Use A36 steel and assume that the load passes through the centroid of the section. Further, assume allowable bending stresses of $F_{bx} = 24$ ksi and $F_{by} = 27$ ksi.

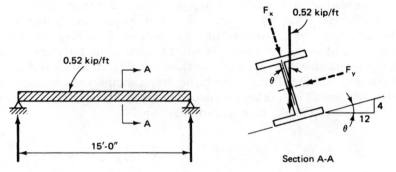

**FIGURE 5-12**   Purlin analysis.

**Solution:**

The beam weight has been included in the uniform load of 0.52 kip/ft shown in Figure 5-12.

Properties of the W10 × 22 are

$$S_x = 23.2 \text{ in.}^3$$

$$S_y = 3.97 \text{ in.}^3$$

Resolve the load into its components, $F_x$ and $F_y$:

$$\theta = \tan^{-1}\left(\frac{4}{12}\right) = 18.4°$$

$$F_x = 0.52(\cos \theta) = 0.49 \text{ kip/ft}$$

$$F_y = 0.52(\sin \theta) = 0.16 \text{ kip/ft}$$

Calculate the component moments:

$$M_x = \frac{w_x L^2}{8} = \frac{0.49(15)^2}{8} = 13.8 \text{ ft-kips}$$

$$M_y = \frac{w_y L^2}{8} = \frac{0.16(15)^2}{8} = 4.5 \text{ ft-kips}$$

Calculate the actual bending stresses:

$$f_{bx} = \frac{M_x}{S_x} = \frac{13.8(12)}{23.2} = 7.1 \text{ ksi}$$

$$f_{by} = \frac{M_y}{S_y} = \frac{4.5(12)}{3.97} = 13.6 \text{ ksi}$$

Check ASDS Equation (H1-3) (with $f_a = 0$):

$$\frac{f_{bx}}{F_{bx}} + \frac{f_{by}}{F_{by}} \le 1.0$$

$$\frac{7.1}{24} + \frac{13.6}{27} = 0.8$$

Therefore, the W10 × 22 is satisfactory because 0.8 < 1.0.

For *design* purposes, a reasonable approximation for a required beam size may be obtained by modifying the expression

$$f_b = \frac{M_x}{S_x} \pm \frac{M_y}{S_y/2}$$

Assuming that it is desired to select a section on the basis of $S_x$, we set $f_b = F_b$ (proper subscripts to be included shortly) and solve for the required $S_x$:

$$F_b = \frac{M_x}{S_x} \pm \frac{2M_y}{S_y} = \frac{1}{S_x}\left[ M_x \pm 2M_y\left(\frac{S_x}{S_y}\right) \right]$$

$$\text{required } S_x = \frac{M_x}{F_b} \pm \frac{2M_y}{F_b}\left(\frac{S_x}{S_y}\right)$$

Ignoring the minus sign and introducing the proper subscripts, we have

$$\text{required } S_x = \frac{M_x}{F_{bx}} + \frac{2M_y}{F_{by}}\left(\frac{S_x}{S_y}\right)$$

Knowing $M_x$, $M_y$, $F_{bx}$, and $F_{by}$ and estimating a ratio of $S_x/S_y$ (with the aid of Table 5-3), one can compute an approximate required section modulus ($S_x$). A member may then be selected and a check made using the ASDS interaction equation.

As a guide in selecting which numerical value of $S_x/S_y$ to use, the smaller number in each W group applies to shapes with relatively wide flanges and square profiles.

The following design example uses an overly simplified crane loading. Longitudinal forces are neglected, as is the fact that there would be two (or more) wheel loadings on the rail. See the ASDS, Section A4, for brief discussion on minimum crane runway loads. Also, Reference 3 contains an excellent discussion of some of the aspects of crane runway design.

■■■ **TABLE 5-3**   $S_x/S_y$ **Ratios for Shapes**

| Shape | Nominal depth, $d$ (in.) | Approximate range of $S_x/S_y$[a] |
|-------|--------------------------|-----------------------------------|
| W     | 4–5<br>6<br>8            | 3<br>3–5<br>3–7                   |
|       | 10–14                    | Over 50 lb/ft: 2.5–5<br>Under 50 lb/ft: 3.5–11 |
|       | 16–18<br>21–24<br>27–36<br>40–44 | 5–11<br>6–13<br>7–11<br>8–13 |
| S     | 6–8<br>10–18<br>20–24    | $d$ (depth)<br>$0.75d$<br>$0.6d$  |

[a]Ratio decreases as weight increases for the same nominal depth.

## Example 5-9 _____

Select a wide-flange shape to be used as a bridge crane runway girder. The girder is on a simple span of 20 ft. Assume that the crane wheel imparts a vertical load of 16 kips and a lateral load of 1.6 kips, applied at the top flange of the girder. A standard 85-lb/yd rail will be used. Refer to Figure 5-13, and use the ASDS and A36 steel. (Neglect shear and deflection.)

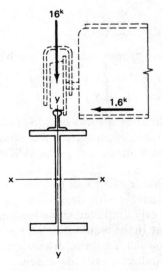

**FIGURE 5-13**   Bridge crane runway girder.

**Solution:**

Use the previously developed formula for approximate required $S_x$ to select a member. Then check ASDS Equation (H1-3). Assume a girder weight of 60 lb/ft. Note that rail weights are given in pounds per yard. The weight per foot is therefore

$$\frac{85}{3} = 28.3 \text{ lb/ft}$$

from which the total uniform load (girder and rail) is

$$60 + 28.3 = 88.3 \text{ lb/ft}$$

The applied moments with respect to the $x$-$x$ and $y$-$y$ axes are calculated as

$$M_x = \frac{wL^2}{8} + \frac{PL}{4} = \frac{0.0883(20)^2}{8} + \frac{16(20)}{4} = 84.4 \text{ ft-kips}$$

$$M_y = \frac{PL}{4} = \frac{1.6(20)}{4} = 8.0 \text{ ft-kips}$$

$F_{by}$, from the ASDS, Section F2-1, is

$$0.75F_y = 0.75(36) = 27.0 \text{ ksi}$$

Assuming that the top flange is not laterally braced between end supports, we note that $F_{bx}$ will probably be reduced below $0.66F_y$ or $0.60F_y$. Having no other guideline, and subject to later change, assume that $F_{bx} = 0.60F_y = 22.0$ ksi. For a span of 20 ft, a beam depth in the range 10 to 14 in. would be a reasonable minimum. Therefore, from Table 5-3, pick an estimated $S_x/S_y$ ratio of 4.0:

$$\text{required } S_x = \frac{M_x}{F_{bx}} + \frac{2M_y}{F_{by}}\left(\frac{S_x}{S_y}\right)$$

$$= \frac{84.4(12)}{22} + \frac{2(8)(12)}{27}(4.0) = 74.5 \text{ in.}^3$$

Try a W14 × 53:

$$S_x = 77.8 \text{ in.}^3$$

$$S_y = 14.3 \text{ in.}^3$$

$$F_{by} = 27.0 \text{ ksi}$$

Using the beam curves of the ASDM, Part 2, with $L_b = 20$ ft, the allowable moment $M_R$ may be obtained from the W14 × 53 curve ($M_R = 124$ ft-kips). Since $M_R = F_{bx}S_x$,

$$F_{bx} = \frac{M_R}{S_x} = \frac{124(12)}{77.8} = 19.1 \text{ ksi}$$

Also, the actual bending stress is calculated from

$$f_{bx} = \frac{M_x}{S_x} = \frac{84.4(12)}{77.8} = 13.0 \text{ ksi}$$

Assuming that the top flange resists the lateral load,

$$f_{by} = \frac{M_y}{S_y/2} = \frac{8.0(12)}{14.3/2} = 13.4 \text{ ksi}$$

Check ASDS Equation (H1-3) (with $f_a = 0$):

$$\frac{f_{bx}}{F_{bx}} + \frac{f_{by}}{F_{by}} = \frac{13.0}{19.1} + \frac{13.4}{27} = 1.18 > 1.0 \qquad\qquad \textbf{N.G.}$$

A different section must be selected. Try a W12 × 58:

$$S_x = 78.0 \text{ in.}^3$$

$$S_y = 21.4 \text{ in.}^3$$

$$F_{bx} = \frac{143(12)}{78.0} = 22.0 \text{ ksi}$$

$$f_{bx} = \frac{84.4(12)}{78.0} = 12.98 \text{ ksi}$$

$$F_{by} = 27.0 \text{ ksi}$$

$$f_{by} = \frac{8.0(12)}{21.4/2} = 8.97 \text{ ksi}$$

Check ASDS Equation (H1-3):

$$\frac{f_{bx}}{F_{bx}} + \frac{f_{by}}{F_{by}} = \frac{12.98}{22.0} + \frac{8.97}{27} = 0.92 < 1.0 \qquad\qquad \textbf{O.K.}$$

**Use a W12 × 58.**

As shown in Figure 5-11b and as stated previously, a common example of unsymmetrical bending (biaxial bending) is the purlin supported on the sloping top chord of roof trusses. The force parallel to the roof surface must be carried in transverse bending of the purlin (bending with respect to the y-y axis). Since the section modulus in this direction is small for most sections, it is usually a more economical design to brace the purlins with sag rods that serve as intermediate supports for the loading parallel to the roof surface. Sag rods are usually placed at the third points or the middle of the purlin span. Hence the purlins act as three-span or two-span continuous beams with respect to bending about the y-y axis and as simple beams, spanning from truss to truss, with respect to bending about the x-x axis. This in effect reduces the bending moment with respect to the y-y axis and would lend itself to the use of a lighter purlin section.

## 5-5

## COMPOSITE BENDING MEMBERS

Bending members composed of two distinct elements that act as one are called *composite bending members.* An example of a common type of composite bending member is the combination of a structural steel beam and a cast-in-place reinforced concrete slab, as shown in Figure 5-14. The connection between the steel beam and the concrete slab is accomplished through the use of a mechanical device termed a *shear connector* that in effect makes the beam and slab act as a unit in resisting the induced shears and bending moments.

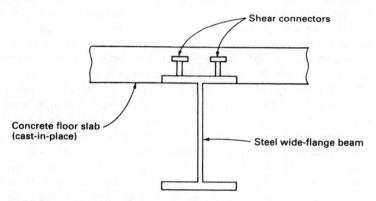

**FIGURE 5-14**   Typical composite cross section.

A significant advantage of composite construction is the use of the structural steel beams as supports for the slab forms. Referring to Figure 5-14, the interconnection between the two elements must be designed so that as the total composite unit deflects, there is no relative movement (slip) between the cast-in-place concrete slab and the top of the steel beam. The connection can, therefore, be seen to be primarily a horizontal *shear* connection between the two elements. Steel studs are the most common mechanical shear connector and are applied by welding to the flange as shown in Figure 5-14. Also see Photo 1-2.

In the absence of shear connectors between the slab and steel beam, the resulting system is said to be *noncomposite.* The amount of shear transfer due to friction is unreliable and therefore is neglected. The supporting beam alone is then assumed to carry the total vertical loads from the slab. In some cases, if a limited number of shear connectors is installed that will transfer a limited but known amount of shear, the system is termed *partially composite.*

Bridges having reinforced concrete decks supported on steel girders are commonly designed for composite action. In buildings, composite construction is most efficient with heavy loadings, relatively long spans, and beams that are spaced as far apart as practical with repetitive bay framing. Composite floor systems are stiffer and stronger than similar noncomposite floor systems with beams of the same size. In practice, this means that lighter and shallower beams may be used in composite systems. It also means that deflections of the steel section, if it alone must carry construction loads, may be excessive. Excessive dead load deflections as well as high dead load stresses can be prevented through the use of temporary shoring. This is a method of temporarily supporting the beam during the placing of the concrete slab.

If shoring is not used, the beam itself must support all loads until the slab has developed its design strength, at which time the composite section will be available to resist all further loads. If temporary shoring is used because of strength and/or deflection considerations, it will be used only until the slab has developed sufficient strength; at such a time the full composite section will be available to resist both dead and live loads. Tests to destruction have shown that there is no difference in the ultimate strength of similar members, whether constructed shored or unshored.

Figure 5-15 shows stress patterns in a composite beam constructed *without* temporary shoring. The steel beam alone supports the construction loads (formwork, fresh concrete, construction live loads) and any other loads that may be put in place before the slab has gained sufficient strength. Figure 5-15a shows a typical stress diagram with the neutral axis at the center of the steel beam. Once the slab has gained strength, the *composite section* will resist further loads (primarily live loads) because of the mechanical interconnection provided between the beam and slab by the shear connectors. The stress pattern induced by these loads is shown in Figure 5-15b. Note that the neutral axis is much higher in the full composite section than it is for the steel beam alone. The load-carrying capacity of the composite beam is greater than that of the steel beam acting alone. Note also that the action of the full composite section is used only in resisting the loads placed *after* the slab has gained strength.

The stress pattern in a composite beam constructed *with* shoring is depicted in Figure 5-16. The stress diagram is for a *transformed section,* which is discussed

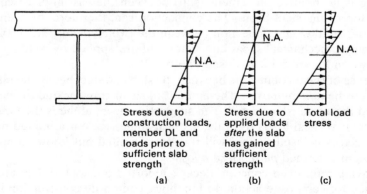

Stress due to construction loads, member DL and loads prior to sufficient slab strength

(a)

Stress due to applied loads *after* the slab has gained sufficient strength

(b)

Total load stress

(c)

**FIGURE 5-15**   Composite beam stresses without shoring.

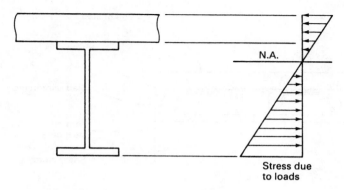

**FIGURE 5-16**  Composite beam stresses with shoring.

shortly. The amount of shoring used governs the amount of stress induced in the steel beam during construction. Sufficient shoring can be placed so that the construction load stresses are minor. Thus the beam remains virtually unstressed until after the slab has hardened and attained sufficient strength. After the temporary shores are removed, the composite section resists both dead and live loads. Over a long period of time, creep in the concrete slab due to sustained load will cause the steel beam to carry an additional portion of the sustained load while the full composite section resists the transient live loads.

It is generally assumed that for composite bending members, adequate lateral support for the compression flange of the steel beam is provided by the concrete slab after hardening. During construction, however, lateral support must be provided or considered in the design process. Properly constructed concrete forms or adequately attached steel deck will usually provide the necessary lateral support.

Composite bending members usually incorporate a solid concrete slab. There are several types of composite systems using *stay-in-place* formed steel deck, however. A typical composite system is composed of structural steel beams, formed steel deck with a maximum rib height of 3 in., and a concrete slab with a minimum thickness above the steel deck of 2 in. (see Figure 5-17). The composite action is developed through the use of stud shear connectors (with a minimum length of the rib height plus $1\frac{1}{2}$ in.) that are welded through the steel deck to the beam flange. This type of system offers quick and straightforward construction that results in lower cost.

The ASDM provides general notes, design examples, and composite beam selection tables for a wide variety of steel beams. The tables constitute a valuable design aid and are applicable to both solid slab and formed steel deck floors. They are only applicable for the case of full composite action for the designated beam size. There is no restriction on concrete strength. Design examples may be reviewed in Part 2 in the ASDM.

The design assumptions are furnished in the ASDS, Chapter 1, and may be summarized as follows:

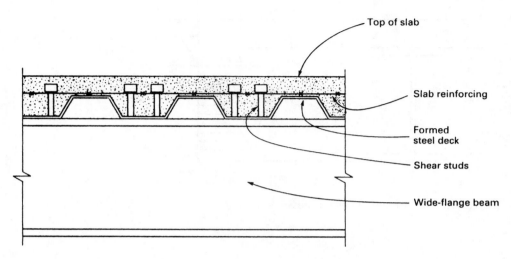

**FIGURE 5-17**   Composite system using formed steel deck.

1. The maximum steel stress is $0.66F_y$, where $F_y$ is the specified minimum yield stress of the steel beam.
2. The maximum concrete stress is $0.45f'_c$, where $f'_c$ is the ultimate compressive strength of concrete. Concrete tensile stresses are neglected.
3. The section properties of the composite section are computed in accordance with the elastic theory (ASD).
4. The compression area of the concrete on the compression side of the neutral axis is to be treated as an equivalent area of steel by dividing the concrete area by the modular ratio $n$, where $n$ is defined as $E_s/E_c$.

The following example will lend an understanding of the behavioral aspects of a simple composite steel-concrete beam.

**Example 5-10** _____

Compute the resisting moment $M_R$ for the fully composite steel-concrete beam shown in Figure 5-18. Assume shoring so that the full capacity of the composite section is available to resist the applied moment. The steel beam is a W16 × 40, and the slab is 5 in. thick. The steel is A36 ($F_y = 36$ ksi), and the concrete $f'_c = 3000$ psi. Use a modular ratio $n = 9$.

**Solution:**

To determine the resisting moment, the concrete slab will be transformed into an equivalent steel area. The neutral axis location will be determined, the moment of inertia will be calculated, the governing allowable stress will be determined, and the resisting moment will be found.

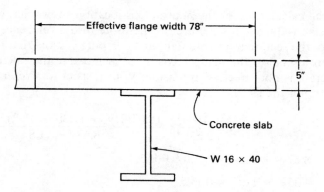

**FIGURE 5-18** Cross section for Example 5-10.

With a modular ratio $n = 9$, the width of the equivalent steel area may be computed as

$$\frac{\text{slab width}}{n} = \frac{78}{9} = 8.67 \text{ in.}$$

The transformed steel section is shown in Figure 5-19a. Properties of the W16 × 40 are

$$A = 11.8 \text{ in.}^2$$
$$d = 16.01 \text{ in.}$$
$$I_x = 518 \text{ in.}^4$$

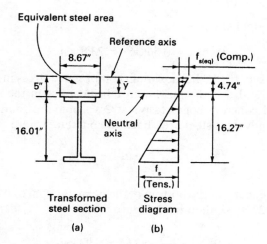

**FIGURE 5-19** Sketch for Example 5-10.

1.  A trial calculation for the neutral axis location shows that it will fall in the slab (equivalent steel area). Therefore, use a reference axis at the top of the section (concrete slab) and compute $\bar{y}$. Note that the concrete in tension below the neutral axis is not included in the calculation since concrete is not effective in tension. With respect to a reference axis at the top of the slab,

$$\bar{y} = \frac{\Sigma(A_y)}{\Sigma A} = \frac{(8.67/2)\bar{y}^2 + 11.8[(16.01/2) + 5.0]}{8.67\bar{y} + 11.8}$$

$$8.67\bar{y}^2 + 11.8\bar{y} = 4.33\bar{y}^2 + 153.5$$

$$4.33\bar{y}^2 + 11.8\bar{y} = 153.5$$

$$\bar{y}^2 + 2.73\bar{y} = 35.45$$

Solving this quadratic equation yields

$$\bar{y} = 4.74 \text{ in.}$$

2.  Compute the moment of inertia of the effective transformed area:

$$I = \Sigma I_c + \Sigma A d^2$$

$$= \frac{1}{3}(8.67)(4.74)^3 + 518 + 11.8\left(\frac{16.01}{2} + 0.26\right)^2$$

$$= 1632 \text{ in.}^4$$

3.  Determine the governing allowable stress. The allowable concrete stress is

$$0.45 f_c' = 0.45(3000) = 1.35 \text{ ksi}$$

The allowable steel stress is

$$0.66 F_y = 0.66(36) = 23.8 \text{ ksi}$$

In Figure 5-19b, $f_{s(eq)}$ is the maximum compressive stress in the equivalent steel area and $f_s$ is the maximum tensile stress in the steel beam. The computed concrete compressive stress $f_c$ may be found by transforming the equivalent steel area back to concrete, where

$$f_c = \frac{f_{s(eq)}}{n}$$

If we assume that $f_s$ at the bottom of the beam is equal to the allowable steel stress of 23.8 ksi, then from Figure 5-19b:

$$f_{s(eq)} = \frac{4.74}{16.27}(23.8) = 6.93 \text{ ksi}$$

from which

$$f_c = \frac{6.93}{9} = 0.770 \text{ ksi}$$

Since 0.770 ksi < 1.35 ksi (the concrete allowable stress), it can be seen that the steel will reach its allowable stress first. Figure 5-19b represents the stress diagram at the maximum resisting moment with $f_s$ = 23.8 ksi and $f_{s(eq)}$ = 6.93 ksi.

4.   Next, compute the resisting moment $M_R$ using the flexure formula ($f = Mc/I$) rearranged to yield resisting moment:

$$M_R = \frac{fI}{c}$$

where $f$ is the allowable steel stress. Substituting,

$$M_R = \frac{23.8(1632)}{16.27(12)} = 199 \text{ ft-kips } (270 \text{ kN·m})$$

## 5-6

## WELDED PLATE GIRDERS

### Introduction

Plate girders are built-up bending members designed and fabricated to fulfill requirements that exceed those of usual rolled sections. The most common form of plate girder currently being designed consists of two flange plates welded to a relatively thin web plate.

It is sometimes economical to change the width, thickness, or both of the *flange* plates somewhere along the span length. The *thickness* of the *web* plate is generally constant. The *depth* of the web plate may be constant or it may be increased in areas of higher moment such as at supports of continuous or overhanging beams. Variable-depth plate girders are normally used only for long-span structures (see Photo 5-1). For reasonable proportions and economical design, the depth-to-span ratio of a girder should range anywhere from $\frac{1}{8}$ to $\frac{1}{12}$. The basic elements of a typical welded plate girder are shown in Figure 5-20.

Since plate girders are built-up from individual plates, economy is sometimes achieved by using steels of varying yield stress. Stronger steels are used in areas of higher stress, and weaker steels are used where stresses are lower. This results in what is termed a *hybrid member*. As seen shortly, it necessitates design expressions that specify yield strength of a particular *element* of the cross section.

Bearing stiffeners are used at reactions and concentrated load points to transfer the concentrated loads to the full depth of the web. Transverse intermediate stiffen-

**PHOTO 5-1**   Plate girder highway grade separation bridge near
Rochester, New York. Girders are simple spans and carry a curved upper
roadway. Note variable-thickness bottom flange and intermediate
transverse stiffeners. (Courtesy of the New York State Department of
Transportation.)

ers are utilized at various spacings along the span length and serve to increase the
web buckling strength, thereby increasing web resistance to shear and moment
combinations. For deeper web plates, particularly in areas of high moments, longitudinal web stiffeners may also be required. Overall economy is sometimes realized
by using a web plate of such thickness that stiffeners are not required.

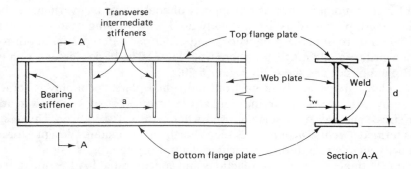

**FIGURE 5-20**   Welded plate girder.

The following four subsections review the AISC ASD design criteria for the preliminary selection of web plates and flange plates, transverse intermediate stiffeners, bearing stiffeners, and the connection of the plate girder elements.

## Preliminary Selection of Plate Girder Webs and Flanges

The ASDS, Section B10, states that, in general, plate girders should be proportioned by the moment-of-inertia method. This approach requires the selection of a suitable trial cross section that would then be checked by the moment-of-inertia method.

As mentioned previously, the total girder depth should generally range from $\frac{1}{8}$ to $\frac{1}{12}$ of the span length, depending on load and span requirements. Therefore, the web depth may be estimated to be from 2 to 4 in. less than the assumed girder total depth. The web thickness may then be selected based on permissible depth–thickness ratios as established in the ASDS. These ratios are based on buckling considerations. The web must have sufficient thickness to resist buckling tendencies that are created by girder curvature under load. As a girder deflects, a vertical compression is induced in the web due to the components of the flange stresses, the result of which constitutes a squeezing action. The buckling strength of the web must be capable of resisting this squeezing action. This is the basis for the ASDS criteria (Section G1) that the ratio of the clear distance between flanges to the web thickness must not exceed

$$\frac{h}{t_w} > \frac{14,000}{\sqrt{F_{yf}(F_{yf} + 16.5)}} \qquad \text{ASDS Eqn. (G1-1)}$$

where

$h$ = clear distance between flanges

$t_w$ = web thickness

$F_{yf}$ = specified minimum yield stress of the flange (ksi)

It is allowed for this ratio to be exceeded if transverse intermediate stiffeners are provided with a spacing not in excess of 1.5 times the distance between flanges. The maximum permissible $h/t_w$ ratio then becomes

$$\text{maximum } \frac{h}{t_w} = \frac{2000}{\sqrt{F_{yf}}} \qquad \text{ASDS Eqn. (G1-2)}$$

Resulting values for the preceding two expressions, as functions of various $F_{yf}$ values, are shown in Table 5-4.

In addition, web buckling considerations may require a reduction of the allowable bending stress in the compression flange. According to the ASDS, Section G2, when the web depth–thickness ratio exceeds $760/\sqrt{F_b}$, the maximum bending stress

TABLE 5-4   Maximum $h/t_w$ Ratios

| $h/t_w$ | $F_{yf}$ | | | |
|---|---|---|---|---|
| | 36 ksi | 42 ksi | 46 ksi | 50 ksi |
| $\dfrac{14{,}000}{\sqrt{F_{yf}(F_{yf}+16.5)}}$ | 322 | 282 | 261 | 243 |
| $\dfrac{2000}{\sqrt{F_{yf}}}$ | 333 | 309 | 295 | 283 |

in the compression flange must be reduced to a value that may be computed from ASDS Equation (G2-1).

Plate girder webs that depend on *tension field action* (to be defined shortly), as discussed in ASDS, Section G3, must be proportioned so that the web bending tensile stress due to moment in the plane of the web does not exceed $0.60F_y$ or

$$\left(0.825 - 0.375\frac{f_v}{F_v}\right)F_y \qquad \text{ASDS Eqn. (G5-1)}$$

where

$f_v$ = computed average web shear stress (total shear divided by web area)(ksi)

$F_v$ = allowable web shear stress according to ASDS Equation (G3-1)(ksi)

This expression in effect constitutes an allowable bending stress reduction due to the interaction of concurrent bending and shear stress (ASDS, Section G5).

After preliminary web dimensions are selected, the required flange area may be determined using an approximate approach as follows. With reference to Figure 5-21, the moment of inertia of the total section with respect to axis x-x is

$$I_x = I_{x(\text{web})} + I_{x(\text{flanges})}$$

Neglecting the moment of inertia of the flange areas about their own centroidal axes and assuming that $h \simeq (d - t_f)$, an approximate gross moment of inertia may be expressed as

$$I_x = \frac{t_w h^3}{12} + 2A_f\left(\frac{h}{2}\right)^2$$

Expressing this in terms of the section modulus ($S$) and also assuming that $h \simeq d$:

$$S_x = \frac{t_w h^3/12}{h/2} + \frac{2A_f(h/2)^2}{h/2}$$

$$= \frac{t_w h^2}{6} + A_f h$$

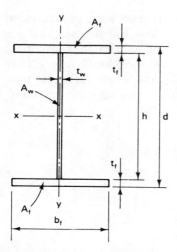

**FIGURE 5-21**   Girder nomenclature.

The required $S_x = M/F_b$; therefore,

$$\frac{M}{F_b} = \frac{t_w h^2}{6} + A_f h$$

and

$$\text{required } A_f = \frac{M}{F_b h} - \frac{t_w h}{6}$$

where

$A_f$ = area of one girder flange

$h$ = depth of the girder web

$F_b$ = allowable bending stress for the compression flange

$M$ = maximum bending moment with respect to $x$-$x$ axis

   The first portion of this expression $M/(F_b h)$ represents the required flange area necessary to resist the bending moment $M$, assuming no contribution by the girder web. Since the web does furnish some bending moment resistance, however, the second term $(t_w h/6)$ is included.

   Based on the computed required flange area, actual proportions of the flange can be determined taking into account additional ASDS criteria.

   To prevent a localized buckling of the compression flange, the ASDS, Section B5 and Table B5.1, places an upper limit on the width–thickness ratio of the flange. This upper limit for the flange of an I-shaped plate girder is that tabulated for a noncompact shape in Table B5.1. The noncompact classification is used since it is not likely that girder dimensions will be such that a compact section is produced. Therefore, in general, the maximum allowable bending stress is taken as $0.60F_y$. For the flange to be considered fully effective (not subject to further allowable stress reductions), the width–thickness ratio of the compression flange may not exceed $95.0\sqrt{F_y/k_c}$. This may be expressed mathematically as

$$\frac{b}{t} \le \frac{95}{\sqrt{F_y/k_c}}$$

where

$b$ = half the full nominal flange width $(b_f/2)$

$t$ = the flange thickness $(t_f)$

$F_y$ = the specified minimum yield stress (ksi) (for a hybrid girder, use the yield strength of the flange $F_{yf}$ instead of $F_y$)

$k_c$ = a compressive element restraint coefficient

If $h/t > 70$, $k_c$ is determined from

$$k_c = \frac{4.05}{(h/t)^{0.46}}$$

Otherwise, $k_c$ is taken as 1.0. $h$ is defined as the clear distance between flanges.

Values of $95.0\sqrt{F_y/k_c}$ are tabulated in Table 5 of the Numerical Values section of the ASDS for the case where $k_c = 1.0$. For A36 steel, the maximum flange plate width (for a fully effective flange) is determined from

$$\frac{b}{t} = \frac{b_f}{2t_f} = 15.8$$

from which

$$\text{maximum } b_f = 2(15.8t_f) = 31.6t_f$$

As mentioned previously, a reduction of the allowable bending stress in the compression flange due to web buckling will be necessary if the web depth–thickness ratio exceeds $760/\sqrt{F_b}$. When this occurs, the allowable flange stress may not exceed

$$F_b' \le F_b\left[1.0 - 0.0005\frac{A_w}{A_f}\left(\frac{h}{t_w} - \frac{760}{\sqrt{F_b}}\right)\right]R_e \qquad \text{ASDS Eqn. (G2-1) Modified}$$

where

$F_b$ = applicable bending stress as established by the ASDS, Chapter F (ksi)

$A_w$ = area of web (in.$^2$)

$A_f$ = area of compression flange (in.$^2$)

$F_b'$ = allowable bending stress in compression flange of plate girders as reduced because of large web depth–thickness ratio (ksi)

$R_e$ = a hybrid girder factor that is taken as 1.0 for nonhybrid girders

The term within the brackets [ ] is a plate girder bending strength reduction factor and is designated $R_{PG}$ in ASDS, Section G2.

After completing the preliminary selection of the girder web and flanges, the actual moment of inertia and section modulus must be calculated. The actual bending stress should then be calculated and compared with the allowable bending stress. Due consideration must be given to a laterally unsupported compression flange.

The flange plates whose sizes are determined based on the *maximum* bending moment may extend the full length of the girder. That is not necessary, however, and they may be reduced in size when the applied moment has decreased appreciably. Changes in flange plates are best achieved by changing plate thickness, width, or both, with the ends of the two flange plates being joined by a full-penetration groove butt weld. Any such reduction in plate size should be made only if the saving in the cost of the flange material more than offsets the added expense of making the butt welds at the transition locations.

The determination of the theoretical transition points for the flange plates is similar to the determination of the theoretical cutoff points for the cover plates of cover-plated beams. This was discussed in Section 5-3.

## Transverse Intermediate Stiffeners

Transverse intermediate stiffeners primarily serve the purpose of stiffening the deep, thin girder webs against buckling. The ASDS, however, permits the girder web to go into the postbuckling range since research has shown that after a *stiffened* thin web panel buckles in shear, it can still continue to resist increasing load. When this occurs, the buckled web is subject to a diagonal tension and the intermediate stiffeners to a compressive force. This behavior is termed *tension field action,* and the design of the stiffeners must consider the added compressive force.

No intermediate stiffeners are required, and tension field action is not considered, if the ratio $h/t_w$ for the web is less than 260 (as well as being less than the limit stipulated in Table 5-4) and the maximum web shear stress $f_v$ is less than that permitted by ASDS Equation (F4-2), where

$$\text{maximum } f_v = \frac{V_{\max}}{h t_w}$$

and the allowable shear stress is

$$F_v = \frac{F_y}{2.89}(C_v) \leq 0.40F_y \qquad\qquad \text{ASDS Eqn. (F4-2)}$$

where

$$C_v = \frac{45,000k}{F_y(h/t_w)^2} \qquad \text{when } C_v < 0.8$$

$$= \frac{190}{h/t_w}\sqrt{\frac{k}{F_y}} \qquad \text{when } C_v > 0.8$$

$$k = 4.0 + \frac{5.34}{(a/h)^2} \qquad \text{when } a/h < 1.0$$

$$= 5.34 + \frac{4.00}{(a/h)^2} \qquad \text{when } a/h > 1.0$$

where

$t_w$ = web thickness

$a$ = clear distance between intermediate stiffeners

$h$ = clear distance between flanges

The allowable shear stress $F_v$, based on ASDS Equation (F4-2), may also be obtained from the ASDM, Part 2, Tables 1-36 and 1-50, for 36 ksi yield stress steel and 50 ksi yield stress steel, respectively. These values are based on tension field action *not* occurring. *With* tension field action included, for girders other than hybrid girders (and assuming that proper intermediate stiffeners are provided), the allowable shear stress $F_v$ may be obtained from ASDS Equation (G3-1), or the ASDM, Part 2, Tables 2-36 and 2-50, for 36 ksi yield stress steel and 50 ksi yield stress steel, respectively.

The spacing of intermediate stiffeners, where stiffeners are required, must be such that the actual web shear stress does not exceed the value of $F_v$ given by ASDS Equations (F4-2) or (G3-1) as applicable. The ratio $a/h$ (sometimes called the *aspect ratio*) must not exceed the value given by ASDS Equation (F5-1):

$$\frac{a}{h} \leq \left(\frac{260}{h/t_w}\right)^2 \qquad\qquad \text{ASDS Eqn. (F5-1)}$$

with a maximum spacing of three times the girder web depth $h$.

When intermediate stiffeners are required, the design procedure is to locate the first intermediate stiffener relative to the end bearing stiffener at the girder support. This must be based on the use of ASDS Equation (F4-2) or Tables 1-36 and 1-50 of the ASDM, Part 2, since this panel must be designed without any benefit of tension field action (ASDS, Section G4).

The spacing for the remaining intermediate stiffeners must then be computed and may be based on the conventional design method. ASDS Equation (F4-2) or

Tables 1-36 and 1-50 may be used to determine the allowable design shear stress $F_v$ or, if designing on the basis of tension field action, ASDS Equation (G3-1) or Tables 2-36 and 2-50 of the ASDM, Part 2, may be used. Note that the use of the tables, in combination with a maximum shear stress diagram along the girder, assists in a rapid selection of stiffener spacing.

The size of the stiffener is then determined. Generally, for welded plate girders, the stiffeners are plates welded alternately on each side of the web.

Whenever stiffeners are required, they must satisfy minimum moment-of-inertia requirements, whether tension field action is counted upon or not. To provide adequate lateral support for the web, the ASDS, Section G4, requires that all intermediate stiffeners (whether a pair or single) have a moment of inertia $I_{st}$ with reference to an axis in the plane of the web, as follows:

$$I_{st} \geq \left(\frac{h}{50}\right)^4 \qquad \text{ASDS Eqn. (G4-1)}$$

The stiffeners must also satisfy a minimum cross-sectional area requirement as provided by ASDS Equation (G4-2). The gross area (in.$^2$) of intermediate stiffeners, spaced as required for ASDS Equation (G3-1), must not be less than

$$A_{st} = \frac{1 - C_v}{2} \left[ \frac{a}{h} - \frac{(a/h)^2}{\sqrt{1 + (a/h)^2}} \right] YDht_w \qquad \text{ASDS Eqn. (G4-2)}$$

where

$C_v$, $a$, $h$, and $t_w$ are as previously defined

$Y$ = ratio of yield stress of web steel to yield stress of stiffener steel

$D$ = 1.0 for stiffeners furnished in pairs

= 1.8 for single-angle stiffeners

= 2.4 for single-plate stiffeners

When stiffeners are furnished in pairs, the area determined is *total* area. This area requirement is for the additional purpose of supplying adequate compression capacity for the intermediate stiffener during tension field action. Hence it should *only* be used when the design is based on tension field action.

The required $A_{st}$ may also be obtained in most cases from the ASDM, Part 2, Table 2-36 or 2-50, using the italicized tabulated values. This gross area requirement may be reduced by the ratio $f_v/F_v$ when $f_v < F_v$ in a panel (ASDS, Section G4).

In addition, the ASDS, Section B5, Table B5.1, states that the ratio of width to thickness for plate girder stiffeners must not exceed $95/\sqrt{F_y}$.

Generally, intermediate stiffeners are stopped short of the girder tension flange. A minimum length of stiffener, based on ASDS requirements for the attaching weld (ASDS, Section G4), may be taken as

minimum length = web depth − 6(web thickness) − web to flange weld size

## Bearing Stiffeners

Bearing stiffeners are generally placed in pairs at unframed ends on the webs of plate girders and where required at points of concentrated loads. In addition to transferring reactions or concentrated loads to the web, bearing stiffeners prevent localized web yielding as well as a more general web crippling and sidesway web buckling. The first two of these considerations were discussed in Chapter 4 of this text. The determination of whether bearing stiffeners are required under concentrated load or at reactions makes use of the same criteria: ASDS Equations (K1-2) and (K1-3) for web yielding and (K1-4) and (K1-5) for web crippling. The $k$ dimension, for use in the web yielding equations, (K1-2) and (K1-3), is taken as the distance from the junction of the fillet weld and girder web to the outside face of the flange. (See Figure 5-22.) ASDS Equations (K1-4) and (K1-5) need not be checked if stiffeners are provided and extend at least $h/2$.

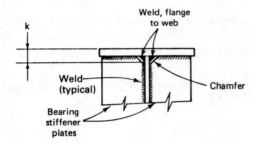

**FIGURE 5-22**   Bearing stiffeners.

For the third consideration, sidesway web buckling, the following applies where flanges are not restrained against relative horizontal movement. Bearing stiffeners are required when the compressive load exceeds the following values.

If the loaded flange *is* restrained against rotation and the value of

$$\frac{d_c/t_w}{L/b_f} < 2.3$$

then

$$R = \frac{6800 t_w^3}{h}\left[1 + 0.4\left(\frac{d_c/t_w}{L/b_f}\right)^3\right] \qquad \text{ASDS Eqn. (K1-6)}$$

Note for this case that the flanges are assumed to remain parallel to each other as the web distorts.

If the loaded flange *is not* restrained against rotation and the value of

$$\frac{d_c/t_w}{L/b_f} < 1.7$$

then

$$R = \frac{6800 t_w^3}{h} \left[ 0.4 \left( \frac{d_c/t_w}{L/b_f} \right)^3 \right] \qquad \text{ASDS Eqn. (K1-7)}$$

where

$R$ = maximum reaction or concentrated load (kips)

$L$ = largest laterally unbraced length along either flange at the point of load (in.)

$b_f$ = flange width (in.)

$d_c = d - 2k$ = web depth clear of the fillet welds (in.)

Equations (K1-6) and (K1-7) need not be checked if

$$\frac{d_c/t_w}{L/b_f}$$

exceeds 2.3 or 1.7, respectively, or when the web is subject to a uniformly distributed load.

If a plate girder is connected to columns at its ends by plates and/or angles, end bearing stiffeners are usually unnecessary.

Bearing stiffeners should have close contact against the flanges and should extend approximately to the edges of the flanges, as shown in Figure 5-22.

Although not required by the ASDS, it is recommended that all bearing stiffeners be full depth and be designed as columns, assuming that the column section will be composed of the pair of stiffeners and a centrally located strip of the web whose width is equal to not more than 12 times its thickness when the stiffeners are located at the end of the web or 25 times its thickness when the stiffeners are located at an interior load (see Figure 5-23). The effective column length shall be taken as not less than three-fourths the length of the stiffeners in computing the slenderness ratio $\ell/r$ (ASDS, Section K1.8). The stiffeners must also be checked for local bearing pressure. Only that portion of the stiffener outside the flange to web welds shall

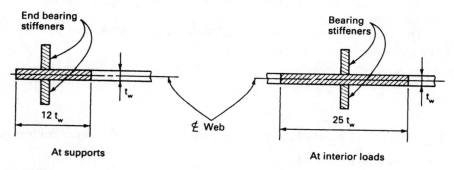

**FIGURE 5-23**   Bearing stiffeners as columns.

be considered effective in bearing, and the bearing stress shall not exceed the allowable value of $0.90F_y$ (ASDS, Section J8).

## Connection of Girder Elements

*Connection of intermediate stiffeners to web:*   ASDS, Section G4, estimates the total shear ($f_{vs}$) in kips per linear inch that must be transferred between the intermediate stiffeners and web due to tension field action. This expression furnishes a minimum value

$$f_{vs} = h \sqrt{\left(\frac{F_y}{340}\right)^3} \qquad \text{ASDS Eqn. (G4-3)}$$

If the actual web shear based on

$$f_v = \frac{V}{ht_w}$$

is less than the allowable shear based on ASDS Equation (G3-1), however, the shear to be transferred ($f_{vs}$) may be reduced in direct proportion.

Generally, this connection is made with intermittent fillet welds where the clear distance between welds cannot exceed 16 times the web thickness or 10 in.

*Connection of bearing stiffener to web:*   Since bearing stiffeners are load-carrying elements, the weld connection is generally a continuous fillet weld on both sides of each stiffener plate. The weld is designed to transmit the total reaction or concentrated load into the web.

*Connection of flange plate to web:*   The connections of flange plate to web are designed to resist the total horizontal shear resulting from the bending forces on the girder. In addition, the welds must be proportioned to transmit to the web any loads applied directly to the flange, unless provision is made to transmit such loads by direct bearing such as through bearing stiffeners.

This weld may be designed as an intermittent fillet weld; it is the contention of the authors, however, that for the same reasons that the weld of the bearing stiffeners to the web should be continuous, the weld of the flange to the web should also be continuous.

The total horizontal shear force $v_h$ (kips per linear inch) may be obtained from the expression (see any strength-of-materials text)

$$v_h = \frac{VQ_f}{I}$$

where

$Q_f$ = statical moment of the flange area with respect to the girder neutral axis (in.$^3$)

$V$ = maximum shear (kips)

$I$ = moment of inertia of total girder section (in.$^4$)

The load applied directly to the flange (where no bearing stiffeners exist) may be considered a vertical shear force per inch and added vectorially to the horizontal shear to determine the resultant shear between the web and the flange.

Using $v_v = w/12$ (where $w$ is distributed load in kips per foot or pounds per foot) as the shear per linear inch applied directly to the flange,

$$v_r = \sqrt{v_h^2 + v_v^2}$$

The connection must be capable of transferring the shear force $v_r$.

The ASDM, Part 2, provides general notes, design examples, and welded plate girder property tables for a wide range of sections with nominal depths from 45 to 92 in. The tables serve as a guide for selecting welded plate girders of economic proportions. Because of the coverage of plate girder design in the ASDM, encompassing four different design examples, further design example treatments are not included in this text.

## REFERENCES

[1]  American Forest and Paper Association, *National Design Specification for Wood Construction,* with supplements, 1995.

[2]  L. Spiegel and G. F. Limbrunner, *Reinforced Concrete Design, 3rd Ed.* (Englewood Cliffs, NJ: Prentice-Hall, Inc., 1992), pp. 151–155.

[3]  D. T. Ricker, "Tips for Avoiding Crane Runway Problems," *Engineering Journal, American Institute of Steel Construction,* Vol. 19, No. 4, 1982, pp. 181–205.

## PROBLEMS

**5-1.**  Design the lightest double-angle lintel to support an 8-in. wall. The weight of wall is 120 lb/ft³. Use A588 steel. The angles are in contact. The maximum allowable deflection = span/240. Assume that arching action can develop in the wall.

    **(a)** Clear opening is 6 ft–8 in. with a 4-in. bearing at each end of the lintel.

    **(b)** Clear opening is 8 ft–0 in. with a 6-in. bearing at each end of the lintel.

**5-2.**  Design the lightest single-angle lintel to support a 4-in. masonry partition wall over a clear opening of 5 ft–4 in. The weight of wall is 120 lb/ft³. Use A36 steel. Assume a 4-in. bearing at each end of the lintel. The maximum allowable deflection = span/240. Assume that arching action can develop in the wall.

**5-3.**   Design the lightest lintel to support a 12-in. brick wall over a clear opening of 12 ft–0 in. The weight of wall is 140 lb/ft³. Use A36 steel. Assume an 8-in. bearing at each end of the lintel. The maximum allowable deflection = span/240. Select a lintel composed of a WT and an L. Assume that arching action can develop in the wall.

**5-4.**   Design the lightest structural tee (WT) lintel to support an 8-in. wall over a clear opening of 12 ft–4 in. The weight of wall is 125 lb/ft³. Use A36 steel. Assume an 8-in. bearing at each end of the lintel. The maximum allowable deflection = span/240. Assume that arching action can develop in the wall.

**5-5.**   A flitch beam is composed of two 2 × 12 (S4S) planks of No. 1 Eastern hemlock and an A36 steel plate $11\frac{1}{4} \times \frac{3}{8}$ in. Use $F_b = 22,000$ psi.

   **(a)** Find the resisting moment $M_R$ (ft-kips) for the flitch beam.

   **(b)** Find the allowable uniformly distributed load (lb/ft) for the flitch beam on a design span length of 14 ft–0 in.

   **(c)** Check the shear.

**5-6.**   A 6 × 10 in. beam of Douglas fir is reinforced by the addition of two A36 C10 × 15.3 channels bolted to the sides of the beam, as shown. Assume that $F_b = 22,000$ psi for the steel. Use nominal dimensions for the wood beam. $F_{bw} = 1600$ psi for this thickness, and $E_w = 1,600,000$ psi.

   **(a)** Find the resisting moment for the wood beam alone.

   **(b)** Find the resisting moment for the reinforced beam.

   **(c)** Find the percent increase in resisting moment due to the addition of the two channels to the 6 × 10 in. beam.

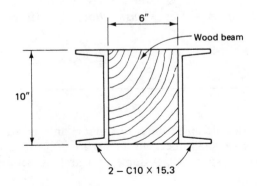

**PROBLEM 5-6**

**5-7.**   A flitch beam is made up of two wood beams and a steel plate, as shown. The steel plate is A36. If $W$ equals the total uniformly distributed load supported by the beam, what portion of the load will be supported by the steel and what portion by the wood? Use nominal dimensions for the wood. The materials are so connected that they will act together as a unit. $F_{bw} =$

1050 psi and $E_w = 1,400,000$ psi. Assume continuous lateral support. Neglect shear.

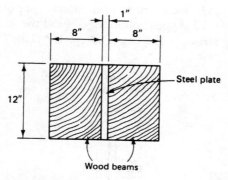

**PROBLEM 5-7**

**5-8.** Design a flitch beam having the cross section shown for a maximum bending moment of 96 ft-kips. The beam is limited to a maximum depth of 12 in. The steel is to be A36 with an allowable bending stress of $F_b = 22,000$ psi. Available steel plates are 1 in. thick. The wood is to have an allowable bending stress of $F_{bw} = 1300$ psi and a modulus of elasticity of $E_w = 1,300,000$ psi. Use nominal dimensions for the wood. Consider moment only.

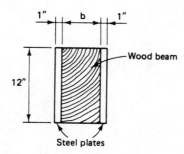

**PROBLEM 5-8**

**5-9.** A flitch beam over a double garage door is to support roof trusses that superimpose a load (assumed to be uniformly distributed) of 900 lb/ft on the beam. The design span length is 15 ft–8 in. Design the flitch beam using two S4S wood sections of No. 1 Hem–Fir and an A36 steel plate. Assume continuous lateral support. Consider moment only.

  **(a)** Use 2 × 12s.

  **(b)** Use 2 × 10s.

**5-10.** An S12 × 31.8 has a section modulus with respect to its $x$-$x$ axis of 36.4 in.³. What thickness of plate 6 in. wide must be welded to each flange to double the value of the section modulus?

**5-11.** The beam shown is of A36 steel and has full lateral support. The beam is composed of two W10 × 22 shapes bolted together at the flanges. Compute the maximum bending stress. Is the beam adequate? (Use $F_b$ = 24 ksi.)

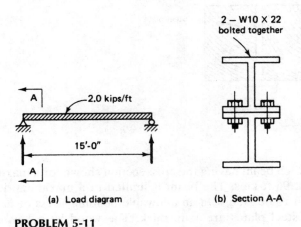

(a)  Load diagram                                (b)  Section A-A

**PROBLEM 5-11**

**5-12.** Rework Problem 5-11 assuming a cross section composed of a W16 × 26 and a C9 × 15 welded together as shown.

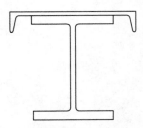

**PROBLEM 5-12**

**5-13.** An A36 W12 × 35 simply supported beam must support a uniformly distributed load of 4.0 kips/ft over a span length of 20 ft. Determine the size of cover plate that should be welded to each flange if the compression flange has full lateral support. Consider moment only.

**5-14.** Compute the resisting moment with respect to the $x$-$x$ axis for the A36 steel member shown. The cover plates are attached with continuous fillet welds, and the compression flange has full lateral support.

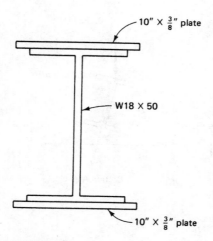

$10'' \times \frac{3}{8}''$ plate

W18 × 50

$10'' \times \frac{3}{8}''$ plate

**PROBLEM 5-14**

**5-15.** An A36 W24 × 76 is used as a 30-ft simple span beam. The beam compression flange has full lateral support. Consider moment only. Compute the superimposed uniformly distributed load that the beam can support if

(a) A 6 × ¾ plate is added to the top flange and a 12 × ⅜ plate is added on the bottom flange.

(b) A 12 × ⅜ plate is added on the bottom flange only.

**5-16.** A beam has been designed to span 40 ft on simple supports. The beam is an A36 W30 × 116. Due to unforeseen circumstances, the design load must be changed to 3.8 kips/ft uniform load. It is desired to use the same beam strengthened with cover plates on each flange. Determine the cover plate size and required length. Assume full lateral support. Consider moment only.

**5-17.** A W16 × 31 beam of A36 steel is subjected to a strong axis moment $M_x$ of 38 ft-kips and a weak axis moment $M_y$ of 7.0 ft-kips. Assume full lateral support. Is the beam O.K.?

**5-18.** A W14 × 61 beam of A588 steel supports a superimposed uniformly distributed gravity load of 2.90 kips/ft on a simple span of 20 ft. Assume full lateral support. Determine the maximum weak axis moment $M_y$ (ft-kips) that this beam could safely support.

**5-19.** Check the adequacy of a W8 × 35 that supports a uniform load of 0.325 kip/ft on a simple span of 18 ft, as shown. Use allowable bending stresses of $F_{bx} = 24$ ksi and $F_{by} = 27$ ksi.

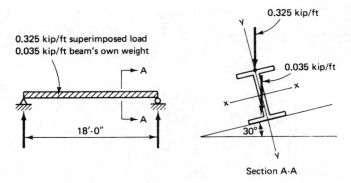

0.325 kip/ft superimposed load
0.035 kip/ft beam's own weight

18'-0"

0.325 kip/ft

0.035 kip/ft

30°

Section A-A

**PROBLEM 5-19**

**5-20.** Select a W10 section to serve as a purlin that is simply supported on a 20-ft span between roof trusses. The roof is assumed to support a superimposed dead load of 15 psf and a snow load of 35 psf. The slope of the roof truss top chord is 1 vertically and 2 horizontally, and the purlins are spaced 10 ft–0 in. on centers. Use A36 steel and assume that the purlin compression flange has full lateral support. Consider moment only.

**5-21.** Calculate the resisting moment $M_R$ for the cross section shown. The steel beam is a W21 × 68 of A36 steel. Concrete $f'_c = 3000$ psi. Assume full shoring and full composite action. Use a modular ratio $n = 9$.

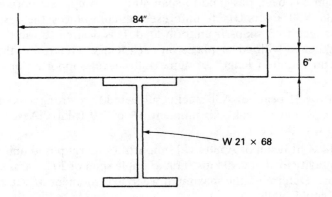

84"

6"

W 21 × 68

**PROBLEM 5-21**

**5-22.** A composite beam, having the cross section shown, is on a simple span of 20 ft. The beam supports a uniformly distributed load of 2.78 kips/ft, which includes the weight of the steel beam and the concrete slab. The beam is a W12 × 40 of A36 steel, and the concrete $f'_c$ is 3500 psi. Compute the stresses in the concrete and the steel and determine if the section is adequate. Use a modular ratio $n$ of 9. Assume full composite action and full shoring.

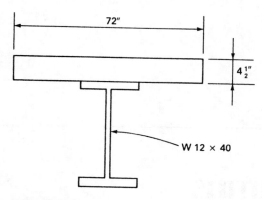

**PROBLEM 5-22**

**5-23.** Select the web and flange plates for a preliminary design of a welded plate girder of 48 in. total depth. The girder is on a simple span of 48 ft, frames into supporting columns, is laterally supported for its full span length, and supports a total loading of 7.0 kips/ft (which includes an assumed girder weight). Use A36 steel. No intermediate stiffeners are desired.

**5-24.** Rework Problem 5-23 assuming a span length of 58 ft and A572 Grade 50 steel. ($F_y = 50$ ksi.)

# CHAPTER 6

# Beam-Columns

## 6-1

## INTRODUCTION

It is generally accepted that axially (concentrically) loaded compression members are nonexistent in actual structures and that all compression members are subjected to some amount of bending moment. The bending moment may be induced by an eccentric load, as shown in Figure 6-1a. The interior column of Figure 6-1b, shown with a concentric load, will not be concentrically loaded if the live loads are not symmetrical. Bending moments may be induced in columns through continuous frame action. Since columns may be subjected to varying amounts of axial load and bending moment, two extremes may exist. If the *bending moment* approaches zero, as a limit, the member is theoretically subjected to an axial load only. The analysis and design of such a member are the same as for an axially loaded compression member as treated in Chapter 3. If the eccentricity, *e*, becomes large (and the

218

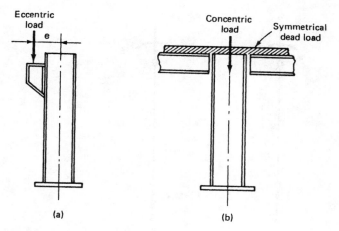

**FIGURE 6-1**   Column loadings.

axial load approaches zero), the member is theoretically subjected to a bending moment only, and the analysis and design are the same as for a beam (bending member), as treated in Chapter 4. A structural member that is subjected to varying amounts of both axial compression and bending moment is commonly termed a *beam-column*.

The actual stresses induced in a beam-column by axial compression and bending moment are not directly additive since the combination of the two generates a secondary moment that cannot be ignored. This secondary moment results from a lateral deflection initially caused by the bending moment, as shown in Figure 6-2a. The product of this deflection and the axial load ($P \times \Delta$, sometimes called the *P-delta moment*) causes further bending and creates secondary stresses that normally are not considered in individual beam or column analysis and design.

Neglecting the secondary moment for now, an approximate expression for the combined stresses for a short beam-column subjected to an axial load and bending moment with respect to one axis only may be expressed as

$$f_{\max} = \frac{P}{A} \pm \frac{Mc}{I}$$

where

$f_{\max}$ = computed maximum stress

$P$ = axial load

$A$ = gross cross-sectional area

$M$ = applied moment

$c$ = distance from the neutral axis to the extreme outside of the cross section

$I$ = moment of inertia of the cross section about the bending neutral axis

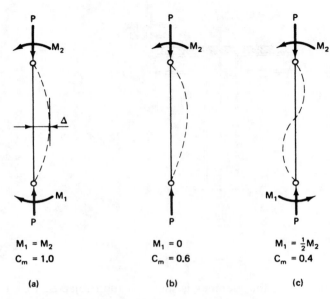

$M_1 = M_2$
$C_m = 1.0$

(a)

$M_1 = 0$
$C_m = 0.6$

(b)

$M_1 = \frac{1}{2}M_2$
$C_m = 0.4$

(c)

**FIGURE 6-2**   Values of $C_m$.

If bending occurs with respect to *both* axes, the expression becomes

$$f_{max} = \frac{P}{A} \pm \left(\frac{M_x c}{I_x}\right) \pm \left(\frac{M_y c}{I_y}\right)$$

Utilizing this expression, it is a simple problem to compute approximate actual combined stresses for beam-columns. The results are of little significance, however, since the allowable bending stress and the allowable compressive stress have always been appreciably different and an allowable combined stress has never been established by code.

In an effort to simplify the combined stress problem, the previous expression may be rewritten as

$$f_{max} = f_a + f_{bx} + f_{by}$$

with the negative signs neglected. Dividing both sides by $f_{max}$, we have

$$1 = \frac{f_a}{f_{max}} + \frac{f_{bx}}{f_{max}} + \frac{f_{by}}{f_{max}}$$

This may be further modified by substituting the applicable allowable stresses in place of the $f_{max}$ terms:

$$1 = \frac{f_a}{F_a} + \frac{f_{bx}}{F_{bx}} + \frac{f_{by}}{F_{by}}$$

An equation of this type is commonly called an *interaction equation*. With this arrangement, if any two of the computed stresses become zero, the correct allowable stress is approached either as an axially loaded column or as a beam subjected to bending about either axis.

## 6-2

## ANALYSIS OF BEAM-COLUMNS (ASDS)

The expression developed in Section 6-1 is the basis for ASDS Equation (H1-3). It may be rewritten as

$$\frac{f_a}{F_a} + \frac{f_{bx}}{F_{bx}} + \frac{f_{by}}{F_{by}} \leq 1.0 \qquad \text{ASDS Eqn. (H1-3)}$$

where

$f_a$ = actual axial compressive stress

$f_b$ = actual maximum compressive bending stress

$F_a$ = allowable axial compressive stress for axial force alone

$F_b$ = allowable compressive bending stress for bending moment alone

It applies to members subjected to both axial compression and bending stresses when $f_a/F_a$ is less than or equal to 0.15.

When $f_a/F_a > 0.15$, the secondary moment due to the member deflection may be of a significant magnitude. The effect of this moment may be approximated by multiplying $f_{bx}$ and $f_{by}$ by an *amplification factor,*

$$\frac{1}{1 - (f_a/F_e')} \qquad \text{ASDS Eqn. (C-H1-1)}$$

where $f_a$ is as previously defined and $F_e'$ is the Euler stress divided by a factor of safety of 23/12 and is expressed as follows:

$$F_e' = \frac{12\pi^2 E}{23(K\ell_b/r_b)^2}$$

In this expression, $K$ is the effective length factor in the plane of bending, $\ell_b$ is the actual unbraced length in the plane of bending, and $r_b$ is the corresponding radius of gyration. Values of $F_e'$ may be obtained from Table 8 in the Numerical Values section of the ASDS or through the use of the properties tabulated at the bottom of the column load tables in the ASDM, Part 3.

Under some combinations of loading, it was found that this amplification factor overestimated the effect of the secondary moment. To compensate for this condition, the amplification factor was modified by a reduction factor $C_m$.

With the introduction of the two factors, ASDS Equation (H1-3) was modified for the case when $f_a/F_a > 0.15$ and expressed as follows:

$$\frac{f_a}{F_a} + \frac{C_{mx}f_{bx}}{(1 - f_a/F'_{ex})F_{bx}} + \frac{C_{my}f_{by}}{(1 - f_a/F'_{ey})F_{by}} \leq 1.0 \qquad \text{ASDS Eqn. (H1-1)}$$

where all terms are as previously defined and $C_m$ is a coefficient defined as follows:

$$C_m = 0.6 - 0.4\left(\frac{M_1}{M_2}\right)$$

in which $M_1/M_2$ is the ratio of the smaller end moment to the larger end moment. $M_1/M_2$ is taken as positive if the moments tend to cause reverse curvature and negative if they tend to cause single curvature. Examples of $C_m$ values are shown in Figure 6-2.

If the column bending moment is a result of a lateral load placed between column support points, $C_m$ may be conservatively taken as unity. If frame *sidesway* is not prevented by adequate bracing or other means, $C_m$ should not be taken as less than 0.85, since in this case the column ends move out of alignment, causing an additional secondary moment from the axial load.

The question as to whether adequate bracing exists to prevent sidesway is difficult to answer and is usually a judgment factor. Sidesway itself may be described as a kind of deformation whereby one end of a member moves laterally with respect to the other. A simple example is a column fixed at one end and entirely free at the other (cantilever column or flagpole). Such a column will buckle, as shown in Figure 6-3. The upper end will move laterally with respect to the lower end.

ASDS Equation (H1-1) applies where stability of a member is a problem and the critical buckling stresses are assumed to occur away from the points of bracing. To guard against overstressing at one end of a member where no buckling action is present, stresses are limited by a modified interaction expression,

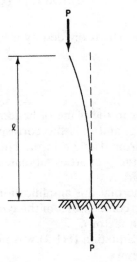

**FIGURE 6-3**   Sidesway for a flagpole-type column.

$$\frac{f_a}{0.60F_y} + \frac{f_{bx}}{F_{bx}} + \frac{f_{by}}{F_{by}} \le 1.0 \qquad \text{ASDS Eqn. (H1-2)}$$

If only one axis of bending is involved in a problem, one of the terms will equal zero with the remaining formula still applicable.

In determining $F_b$ for the interaction equations, the compactness of the beam-column must be established. As discussed in Section 4-3, the web compactness of a *beam* is based on $f_a = 0$. This applies to a beam subjected to bending only, with no axial load. With a beam-column ($f_a$ is not zero), the web compactness must be checked using equations from ASDS, Table B5.1, for "Webs in combined flexural and axial compression." This is simplified in the ASDM through the use of $F_y'''$, which is defined as the theoretical maximum yield stress (ksi) based on the depth–thickness ratio of the web below which a particular shape may be considered compact for any condition of combined bending and axial stresses. $F_y'''$ is tabulated in the properties table of Part 1. It is determined in the same way that $F_y'$ is determined, as discussed in Section 4-3. If $F_y''' \ge F_y$, the member is compact based on the web criterion. If $F_y''' < F_y$, the member is not compact based on the web criterion and $F_b$ cannot exceed $0.60F_y$.

In summary, to establish whether a beam-column is satisfactory, the following applies:

1.   When $f_a/F_a \le 0.15$, use ASDS Equation (H1-3).
2.   When $f_a/F_a > 0.15$, use ASDS Equations (H1-1) and (H1-2). *Both* equations must be satisfied.

**Example 6-1** _____

An A36 steel W6 × 25 column is subjected to an eccentric load of 32 kips, as shown in Figure 6-4. The column has an unbraced length of 15 ft and may be assumed to have pinned ends. Bracing prevents sidesway. Determine whether the column is adequate.

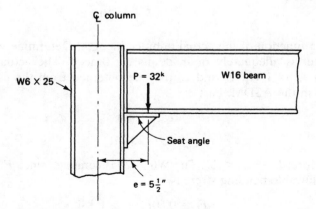

**FIGURE 6-4**   Beam-column connection.

**Solution:**

For the W6 × 25,

$$A = 7.34 \text{ in.}^2$$

$$S_x = 16.7 \text{ in.}^3$$

$$r_y = 1.52 \text{ in.}$$

$$r_x = 2.70 \text{ in.}$$

$$K = 1 \text{ (pinned ends)}$$

$$F_y''' = -$$

1.   Replacing the eccentric load with a concentric load and a couple (moment),

$$P = 32 \text{ kips}$$

$$M_x = Pe = 32(5.5) = 176 \text{ in.-kips}$$

2.   Calculating actual axial compressive stress,

$$f_a = \frac{P}{A} = \frac{32}{7.34} = 4.36 \text{ ksi}$$

3.   $F_a$ is a function of $K\ell/r_y$. Thus

$$\frac{K\ell}{r_y} = \frac{1(15)(12)}{1.52} = 118.4$$

Rounding to 118, from the ASDM, Part 3, Table C-36, we obtain

$$F_a = 10.57 \text{ ksi}$$

4.   Calculating actual maximum compressive bending stress, we have

$$f_{bx} = \frac{M_x}{S_x} = \frac{176}{16.7} = 10.5 \text{ ksi}$$

5.   $F_{bx}$ is a function of the actual unbraced length. Determine whether the member is adequately or inadequately braced. The actual unbraced length $L_b = 15$ ft. $L_c$ and $L_u$ may be obtained from the column load tables in the ASDM, Part 3:

$$L_c = 6.4 \text{ ft}$$

$$L_u = 20.0 \text{ ft}$$

Therefore, $L_c < L_b < L_u$. The W6 × 25 is compact, since $F_y''' > F_y$, and the allowable bending stress is

$$F_b = 0.60F_y = 21.6 \text{ ksi}$$

6. $f_a/F_a = 4.36/10.57 = 0.41 > 0.15$. Therefore, use ASDS Equations (H1-1) and (H1-2).

7. Calculate $C_m$ and $F'_{ex}$. Since $M_1 = 0$ and $M_2 = 176$ in.-kips, and sidesway is prevented

$$C_m = 0.6 - 0.4\left(\frac{M_1}{M_2}\right)$$

$$= 0.6$$

$F'_{ex}$ is a function of $K\ell_b/r_b$, which in this case is $K\ell/r_x$:

$$\frac{K\ell}{r_x} = \frac{1(15)(12)}{2.70} = 66.7$$

Rounding to 67, from Table 8 in the Numerical Values section of the ASDS we obtain

$$F'_{ex} = 33.27 \text{ ksi}$$

8. Checking ASDS Equation (H1-1), we have

$$\frac{f_a}{F_a} + \frac{C_{mx} f_{bx}}{(1 - f_a/F'_{ex}) F_{bx}} \leq 1.0$$

$$\frac{4.36}{10.57} + \frac{0.6(10.5)}{(1 - 4.36/33.27)(21.6)} \leq 1.0$$

$$0.41 + 0.34 = 0.75 < 1.0 \qquad \textbf{O.K.}$$

9. Checking ASDS Equation (H1-2), we have

$$\frac{f_a}{0.6F_y} + \frac{f_{bx}}{F_{bx}} \leq 1.0$$

$$\frac{4.36}{21.6} + \frac{10.5}{21.6} \leq 1.0$$

$$0.20 + 0.49 = 0.69 < 1.0 \qquad \textbf{O.K.}$$

The beam-column is adequate.

---

## Example 6-2

An A572 ($F_y = 50$ ksi) W12 $\times$ 136 column supports beams framing into it, as shown in Figure 6-5. The connections are moment connections. The column supports an axial load of 600 kips, which includes the beam reactions at its

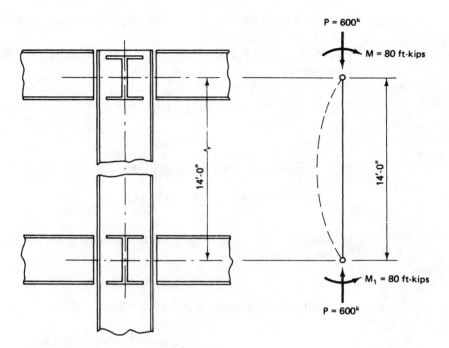

**FIGURE 6-5**    Beam-column.

top. Due to unbalanced floor loading, moments of 80 ft-kips each are applied in opposite directions at the top and bottom of columns as shown. Sidesway is prevented by a bracing system. $K_y = 1.0$ and $K_x$ is estimated to be 0.9. Determine whether the member is adequate.

**Solution:**

For the W12 × 136,

$$A = 39.9 \text{ in.}^2$$
$$S_x = 186 \text{ in.}^3$$
$$r_y = 3.16 \text{ in.}$$
$$r_x = 5.58 \text{ in.}$$
$$K_x = 0.9$$
$$K_y = 1.0$$
$$F_y''' = —$$

1.  $P = 600$ kips, $M = 80$ ft-kips.

2.  $f_a = \dfrac{P}{A} = \dfrac{600}{39.9} = 15.0$ ksi.

3.  $F_a$ is a function of the largest slenderness ratio:

$$\frac{K_y \ell}{r_y} = \frac{1(14)(12)}{3.16} = 53.2$$

$$\frac{K_x \ell}{r_x} = \frac{0.9(14)(12)}{5.58} = 27.1$$

Rounding to 53, from the ASDM, Part 3, Table C-50, we obtain

$$F_a = 23.88 \text{ ksi}$$

4.  $f_{bx} = \dfrac{M_x}{S_x} = \dfrac{80(12)}{186} = 5.16$ ksi.

5.  $F_{bx}$ is a function of $L_b$, $L_c$, and $L_u$.

$$L_b = 14 \text{ ft} \qquad L_c = 11.1 \text{ ft} \qquad L_u = 38.3 \text{ ft}$$

Since $L_c < L_b < L_u$, and since W12 × 136 is compact ($F_y''' > F_y$), the allowable bending stress is

$$F_b = 0.60F_y = 30 \text{ ksi}$$

6.  $f_a/F_a = 15.0/23.88 = 0.63 > 0.15$. Therefore, use ASDS Equations (H1-1) and (H1-2).

7.  Calculate $C_m$ and $F'_{ex}$. Since $M_1 = M_2 = 80$ ft-kips and causes single curvature, the ratio $M_1/M_2$ is negative (see the ASDS, Section H1). Thus

$$C_m = 0.6 - 0.4\left(\frac{M_1}{M_2}\right)$$

$$= 0.6 - 0.4(-1)$$

$$= 1.0$$

$F'_{ex}$ is a function of $K\ell_b/r_b$, which in this case is $K_x\ell/r_x$. Note that this is with respect to the $x$-$x$ axis. Using the properties tabulated in the ASDM, Part 3, the column load table for the W12 × 136, we have

$$\frac{F'_{ex}(K_x L_x)^2}{10^2} = 323$$

from which

$$F'_{ex} = \frac{323(10)^2}{[0.9(14)]^2} = 203 \text{ ksi}$$

Note that $F'_{ex}$ may also be obtained using $K_x \ell / r_x$ and Table 8 of the Numerical Values section of the ASDS.

8.    Checking ASDS Equation (H1-1) gives us

$$\frac{f_a}{F_a} + \frac{C_{mx} f_{bx}}{(1 - f_a / F'_{ex}) F_{bx}} \leq 1.0$$

$$\frac{15.0}{23.88} + \frac{1.0(5.16)}{(1 - 15.0/203)(30.0)} \leq 1.0$$

$$0.63 + 0.19 = 0.82 < 1.0 \qquad\qquad \textbf{O.K.}$$

9.    Checking ASDS Equation (H1-2) yields

$$\frac{f_a}{0.6 F_y} + \frac{f_{bx}}{F_{bx}} \leq 1.0$$

$$\frac{15.0}{30.0} + \frac{5.16}{30.0} < 1.0$$

$$0.5 + 0.17 = 0.67 < 1.0 \qquad\qquad \textbf{O.K.}$$

The beam-column is adequate.

## 6-3

## DESIGN OF BEAM-COLUMNS (ASDS)

The use of the interaction formulas furnishes a convenient means of beam-column analysis. These may also be used for beam-column design. A trial section must first be selected, however. After the selection is made, the problem becomes one of analysis. In essence, the design process is one of trial and error, since no simple design procedure exists whereby a most economical member can be selected in one quick step.

The ASDM furnishes a method of design whereby a trial section may be obtained using an equivalent axial load in conjunction with the ASDM axial load table of

Part 3. The ASDM also furnishes modified interaction equations that may be used in place of the previously discussed interaction equations. As the modified equations do not simplify the analysis, they will not be introduced in this text. The reader is referred to Part 3 of the ASDM.

Using the ASDM approach to determine a trial section, the reader is referred to the ASDM, Part 3, Table B. The equivalent axial load, for design purposes, is designated $P_{eff}$:

$$P_{eff} = P_0 + M_x m + M_y m U$$

where

$P_0$ = actual axial load (kips)

$M_x$ = bending moment about the strong axis (ft-kips)

$M_y$ = bending moment about the weak axis (ft-kips)

$m$ = factor taken from the ASDM, Part 3, Table B

$U$ = factor taken from the ASDM, Part 3, column load tables

The procedure for selection of a trial section is as follows:

1.  With the known value of $KL$ (in feet), select a value of $m$ from the first approximation section of Table B and assume that $U = 3$.
2.  Solve for $P_{eff}$.
3.  From the column load table in Part 3 of the ASDM, select a trial section to support $P_{eff}$.
4.  Based on the section selected, obtain a *subsequent approximate value of m* from Table B and a $U$ value from the column load table. Solve for $P_{eff}$ again.
5.  Select another section (if necessary) and continue the process until the values of $m$ and $U$ stabilize.

Using this trial section, the beam-column may then be analyzed in the manner discussed previously using the ASDM interaction equations.

## Example 6-3

Using A36 steel and the ASDS, select a wide-flange column for the conditions shown in Figure 6-6. The column has pinned ends and sidesway is prevented. Bending occurs with respect to the strong ($x$-$x$) axis.

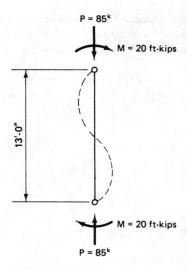

**FIGURE 6-6**    Beam-column.

**Solution:**

From the ASDM, Part 3, Table B, with $KL = 13$ ft, select a value of $m = 2.25$ from the first approximation portion. Since $M_y = 0$, the expression for the effective axial load becomes

$$P_{eff} = P_0 + M_x m$$

$$= 85 + 20(2.25)$$

$$= 130 \text{ kips}$$

From the column load table of the ASDM, Part 3, select a W8 $\times$ 31 (allowable load $P_a$ is 143 kips). From Table B again, select a value of $m = 2.85$ from the Subsequent Approximations portion of the table.

$$P_{eff} = 85 + 20(2.85)$$

$$= 142 \text{ kips}$$

Since 142 kips $<$ 143 kips, the W8 $\times$ 31 remains as the trial section and should be checked using the interaction equations. The checking procedure is identical to that in Examples 6-1 and 6-2. ASDS Equation (H1-1) results in a value of 0.77, and Equation (H1-2) results in a value of 0.82. The W8 $\times$ 31 is therefore satisfactory. The reader may wish to verify the results of this analysis.

## Example 6-4

Using A36 steel and the ASDS, select a wide-flange column for the conditions shown in Figure 6-7. Architectural requirements indicate the use of a W8, if possible. The column is pinned at both ends. Bending occurs with respect to both axes. Sidesway is prevented in both directions.

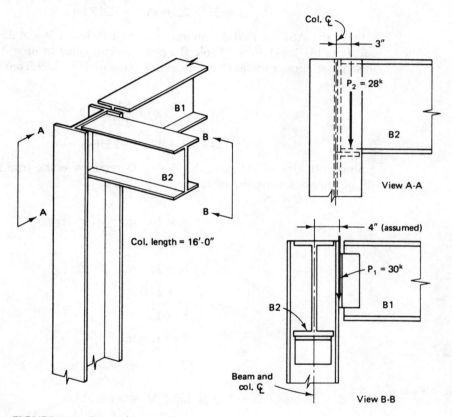

**FIGURE 6-7**   Corner beam-column.

## Solution:

Replace the eccentric loads with concentric loads and couples (moments). A 4-in. eccentricity is assumed for strong-axis bending:

$$P = 58 \text{ kips}$$

$$M_x = 30(4) = 120 \text{ in.-kips}$$

$$M_y = 28(3) = 84 \text{ in.-kips}$$

From the ASDM, Part 3, Table B, with $KL = 16$ ft, select a value of $m = 2.2$ from the first approximation portion. Let $U = 3$; therefore,

$$P_{eff} = P_0 + M_x m + M_y mU$$

$$= 58 + \frac{120}{12}(2.2) + \frac{84}{12}(3)(2.2)$$

$$= 58 + 22 + 46.2 = 126.2 \text{ kips}$$

From the ASDM, Part 3, column load table, select a W8 × 35 (allowable load $P_a$ is 141 kips). From Table B again, select a value of $m = 2.6$ from the subsequent approximation portion and a value of $U = 2.59$ from the column load table.

$$P_{eff} = 58 + \frac{120}{12}(2.6) + \frac{84}{12}(2.6)(2.59)$$

$$= 58 + 26 + 47 = 131 \text{ kips}$$

Since 131 kips < 141 kips, the W8 × 35 remains as the trial section and will be checked using the interaction equations.

For the W8 × 35,

$$A = 10.3 \text{ in.}^2$$

$$S_x = 31.2 \text{ in.}^3$$

$$r_x = 3.51 \text{ in.}$$

$$r_y = 2.03 \text{ in.}$$

$$K = 1$$

$$S_y = 10.6 \text{ in.}^3$$

$$F_y''' = —$$

1.  $P = 58$ kips, $M_x = 120$ in.-kips, $M_y = 84$ in.-kips.

2.  $f_a = \dfrac{P}{A} = \dfrac{58}{10.3} = 5.63$ ksi.

3.  $F_a$ is a function of $K\ell/r_y$:

$$\frac{K\ell}{r_y} = \frac{1(16)(12)}{2.03} = 94.6$$

Rounding to 95, from the ASDM, Part 3, Table C-36, we obtain

$$F_a = 13.60 \text{ ksi}$$

4.  $f_{bx} = \dfrac{M_x}{S_x} = \dfrac{120}{31.2} = 3.85$ ksi; $f_{by} = \dfrac{M_y}{S_y} = \dfrac{84}{10.6} = 7.92$ ksi.

5.   $F_{bx}$ is a function of $L_b$, $L_c$, and $L_u$:

$$L_b = 16 \text{ ft} \qquad L_c = 8.5 \text{ ft} \qquad L_u = 22.6 \text{ ft}$$

Since $L_c < L_b < L_u$, and since the W8 × 35 is compact ($F_y''' > F_y$), the allowable bending stress $F_{bx}$ is

$$F_{bx} = 0.60F_y = 21.6 \text{ ksi}$$

From the ASDS, Section F2:

$$F_{by} = 0.75F_y = 27.0 \text{ ksi}$$

6.   $f_a/F_a = 5.63/13.60 = 0.41 > 0.15$. Therefore, use ASDS Equations (H1-1) and (H1-2).

7.   Calculate $C_m$ and $F_e'$. Sidesway is prevented in both directions; therefore,

$$C_m = 0.6 - 0.4 \left(\frac{M_1}{M_2}\right) \geq 0.4$$

Since $M_1 = 0$,

$$C_{mx} = C_{my} = 0.6$$

Since bending occurs with respect to both axes, $F_e'$ values must be obtained with respect to each axis. Using properties for the W8 × 35 in the column load table,

$$\frac{F_{ex}'(K_x L_x)^2}{10^2} = 128$$

$$\frac{F_{ey}'(K_y L_y)^2}{10^2} = 42.7$$

from which

$$F_{ex}' = \frac{128(10)^2}{16^2} = 50.0 \text{ ksi}$$

$$F_{ey}' = \frac{42.7(10)^2}{16^2} = 16.7 \text{ ksi}$$

8.   Checking ASDS Equation (H1-1) gives us

$$\frac{f_a}{F_a} + \frac{C_{mx}f_{bx}}{(1 - f_a/F_{ex}')F_{bx}} + \frac{C_{my}f_{by}}{(1 - f_a/F_{ey}')F_{by}} \leq 1.0$$

$$\frac{5.63}{13.60} + \frac{0.60(3.85)}{(1 - 5.63/50.0)(21.6)} + \frac{0.60(7.92)}{(1 - 5.63/16.7)(27.0)} \leq 1.0$$

$$0.41 + 0.12 + 0.27 = 0.80 < 1.0 \qquad \textbf{O.K.}$$

9.    Checking ASDS Equation (H1-2) yields

$$\frac{f_a}{0.6F_y} + \frac{f_{bx}}{F_{bx}} + \frac{f_{by}}{F_{by}} \leq 1.0$$

$$\frac{5.63}{21.6} + \frac{3.85}{21.6} + \frac{7.92}{27.0} \leq 1.0$$

$$0.26 + 0.18 + 0.29 = 0.73 < 1.0 \qquad \textbf{O.K.}$$

The W8 × 35 is satisfactory.

---

## 6-4

## EFFECTIVE LENGTH FACTOR *K*

In the design of steel structures one is often concerned with continuous frames of various types consisting of beams and columns that are rigidly connected. When this occurs, the columns are subjected to the combined action of compression and bending and may be categorized as beam-columns.

As discussed previously, the terms $F_a$ and $F_e'$ in the ASDS interaction formula are functions of slenderness ratio. This in turn is a function of the effective length factor $K$, which when multiplied by an actual length of a member will result in an effective length $(K\ell)_x$ or $(K\ell)_y$. The effective length of the member depends on the restraints against relative rotation and lateral movement (sidesway) imposed at the ends of the member.

Consider a simple portal frame, as shown in Figure 6-8. The beam is rigidly connected to the supporting columns. When sidesway (lateral movement) of the frame is effectively prevented by some means, its deformed shape under vertical load may be observed. The effective length factor $K$ for the columns can have values that range from 0.5 for ends fixed against rotation to 1.0 for pinned ends. When the frame depends on its own stiffness for resistance to sidesway, as shown in Figure 6-9, $K$ will have a value larger than 1.0. Figure 6.9a shows the deformed

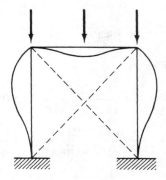

**FIGURE 6-8**  Loaded braced frame.

shape due to vertical load. The frame will deflect to the side (sidesway) so as to equalize the moments at the tops of the columns. Sidesway may also be caused by a laterally applied force, as shown in Figure 6-9b. As a rule, columns free to translate in a sidesway mode are appreciably weaker than columns of equal length braced against sidesway. Also of importance is that the magnitude of the sidesway of a column is directly affected by the stiffness of the other members in the frame; or, the magnitude of joint rotation is directly affected by the stiffness of the members framing into the joint. The problem can easily become very complex.

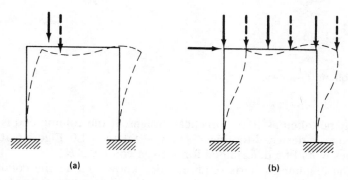

**FIGURE 6-9**   Loaded unbraced frame.

To simplify the determination of the effective length factor $K$, alignment charts are furnished in the ASDM, Part 3 (Figure 1), for the two cases of *sidesway prevented* and *sidesway not prevented*. These charts afford a means of obtaining more precise values for $K$ than those offered by Table C-C2.1 in the ASDS Commentary and discussed in Chapter 3. The use of the charts requires an evaluation of the relative stiffness of the members of the frame at each end of the column. The stiffness ratio or *relative stiffness* of the members rigidly connected at each joint may be expressed as

$$G_A = \frac{\Sigma(I_c/\ell_c)_A}{\Sigma(I_g/\ell_g)_A}$$

and

$$G_B = \frac{\Sigma(I_c/\ell_c)_B}{\Sigma(I_g/\ell_g)_B}$$

where $A$ and $B$ subscripts refer to the joint at which the relative stiffness is being determined; the $c$ and $g$ subscripts refer to *column* and *girder* or beam, respectively; $I$ is the moment of inertia; and $\ell$ is the unsupported length of member. The $I/\ell$ terms are taken with respect to an axis normal to the plane of buckling under consideration. Having determined $G_A$ and $G_B$, the appropriate chart may be used to determine $K$. The points representing the values of $G_A$ and $G_B$ are connected with a straight line, and the value of $K$ is read at the intersection of this line with the central vertical $K$ reference line.

For column ends supported by, but not rigidly connected to, a footing or foundation, $G_B$ is theoretically infinity, but unless the joint is designed as a true friction-

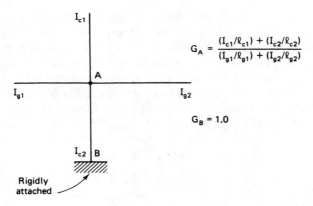

$$G_A = \frac{(I_{c1}/\ell_{c1}) + (I_{c2}/\ell_{c2})}{(I_{g1}/\ell_{g1}) + (I_{g2}/\ell_{g2})}$$

$$G_B = 1.0$$

**FIGURE 6-10**   Column $G$ values.

free pin, $G_B$ may be taken as 10 for practical designs. If the column end is rigidly attached to a properly designed footing, $G_B$ may be taken as 1.0. Figure 6-10 shows how the $G$ values would be determined for a given column $AB$.

The use of the alignment charts requires prior knowledge of the column and beam sizes. In other words, before the charts can be used, a trial design has to be made of each of the members. To start the design, it is therefore necessary to assume a reasonable value of $K$, choose a column section to support the axial load and moments, and then determine the actual value of $K$. In addition, trial beam sizes must be reasonably estimated.

**Example 6-5** _____

Compute the effective length factor $K$ for each of the columns in the frame shown in Figure 6-11 using ASDM alignment charts. Preliminary sizes of each

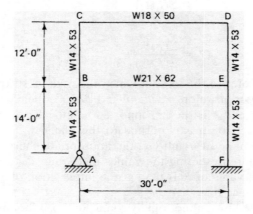

**FIGURE 6-11**   Two-story rigid frame.

member are furnished. Sidesway is *not* prevented. Webs of the wide-flange shapes are in the plane of the frame.

**Solution:**

| Member | Shape | $I$ (in.⁴) | $\ell$ (in.) | $I/\ell$ |
|--------|-------|-----------|-------------|----------|
| AB | W14 × 53 | 541 | 168 | 3.22 |
| BC | W14 × 53 | 541 | 144 | 3.76 |
| CD | W18 × 50 | 800 | 360 | 2.22 |
| BE | W21 × 62 | 1330 | 360 | 3.69 |
| DE | W14 × 53 | 541 | 144 | 3.76 |
| EF | W14 × 53 | 541 | 168 | 3.22 |

$G$ factors for each joint are determined as follows:

| Joint | $\dfrac{\Sigma(I_c/\ell_c)}{\Sigma(I_g/\ell_g)}$ | | $G$ |
|-------|------------------|---|-----|
| A | Pinned end | = | 10.0 |
| B | $\dfrac{3.76 + 3.22}{3.69}$ | = | 1.89 |
| C | $\dfrac{3.76}{2.22}$ | = | 1.69 |
| D | $\dfrac{3.76}{2.22}$ | = | 1.69 |
| E | $\dfrac{3.76 + 3.22}{3.69}$ | = | 1.89 |
| F | Fixed end | = | 1.0 |

Column $K$ factors from the chart (sidesway uninhibited) are

| Column | $G$ values at column ends | | $K$ |
|--------|---------|------|------|
| AB | 10.0 | 1.89 | 2.08 |
| BC | 1.89 | 1.69 | 1.54 |
| DE | 1.69 | 1.89 | 1.54 |
| EF | 1.89 | 1.0 | 1.44 |

Because of the smaller effective length factors used for frames where sidesway is prevented, it is advisable to provide lateral support wherever possible. This may be accomplished with diagonal bracing, shear walls, or attachment to an adjacent structure having adequate lateral stability or by floor slabs or roof decks secured horizontally by walls or bracing systems.

As discussed previously, the determination of the $K$ factors utilizing the alignment charts is based on several assumptions. Two of the principal assumptions are that all columns in a story buckle simultaneously and that all column behavior is purely elastic. Either or both of these conditions may not exist in an actual structure, and as a result, the use of the alignment charts will produce overly conservative designs.

The ASDM, Part 3, contains a design procedure to reduce the $K$ factor value by multiplying the elastic $G$ value by a stiffness reduction factor. Then, using the alignment charts as discussed previously, an *inelastic K* factor is obtained. The stiffness reduction factor is obtained from Table A in Part 3 of the ASDM.

**Example 6-6**

Compute the *inelastic K* factors for columns $BC$ and $DC$ of Example 6-5, as shown in Figure 6-12. Consider behavior in the plane of the frame only. Use Figure 1 and Table A of the ASDM, Part 3, for the inelastic $K$ factor procedure. Preliminary sizes are shown. Sidesway in the plane of the frame is not prevented. Webs of the W-shape members are in the plane of the frame. Use A36 steel. Assume that the columns support loads of 245 kips, as shown in Figure 6-12.

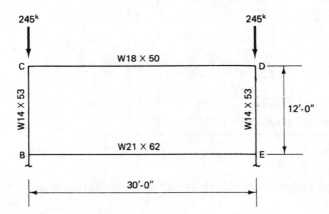

**FIGURE 6-12** Inelastic $K$ factor determination.

**Solution:**

1.   The trial column size is W14 × 53. $A = 15.6$ in.²

2.  Compute $f_a$:

$$f_a = \frac{245}{15.6} = 15.7 \text{ ksi}$$

3.  From Table A, the stiffness reduction factor $f_a/F'_e$ is 0.621.
4.  From Example 6-5, the elastic stiffness ratios are 1.89 at joints $B$ and $E$ (bottom) and 1.69 at joints $C$ and $D$ (top).
5.  Calculate $G_{\text{inelastic}}$:

$$G_{\text{inelastic (top)}} = 0.621(1.69) = 1.05$$

$$G_{\text{inelastic (bottom)}} = 0.621(1.89) = 1.17$$

6.  Determine $K$ from Figure 1 of the ASDM, Part 3:

$$K = 1.35$$

This compares with a $K$ of 1.54 as determined in Example 6-5, indicating a greater column capacity if inelastic behavior is considered.

## PROBLEMS

*Note: The ASDS applies in all the following problems.*

**6-1.**   A W12 × 53 column of A36 steel supports a vertical load of 40 kips at an eccentricity of 12 in. with respect to the strong axis. The length is 12 ft, and $K = 1.0$ with respect to both axes. Determine whether the member is adequate.

**6-2.**   In Problem 6-1, determine the maximum *moment* (bending about the strong axis) that the beam-column can safely support. The vertical *load* remains 40 kips.

**6-3.**   In Problem 6-1, determine whether the beam-column is adequate if the 40-kip load acts at an eccentricity of 12 in. with respect to *each* axis.

**6-4.**   A single-story W14 × 109 of A36 steel is to support a load of 280 kips at an eccentricity of 10 in. Bending is to be about the strong axis. Sidesway is prevented. The length is 24 ft, and $K = 1.0$. Determine whether the member is adequate.

**6-5.**   A single-story W10 × 45 of A36 steel supports a beam reaction of 70 kips. Bending is about the *weak* axis and eccentricity is 3 in. Sidesway is prevented. The length is 16 ft, and $K = 1.0$. Determine whether the member is adequate.

**6-6.** An A36 W10 × 39 is subjected to the loads shown. Sidesway is prevented by bracing. The member length is 24 ft, and $K = 1.0$. Determine whether the member is adequate.

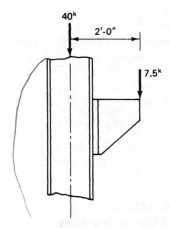

**PROBLEM 6-6**

**6-7.** A W8 × 40 column has a length of 16 ft. The bottom is fixed and the top is pinned. Sidesway is prevented by bracing. A36 steel is used. Determine the maximum load that may be safely supported by this column, assuming a strong-axis eccentricity of 6 in.

**6-8.** The built-up column shown supports a vertical load $P$ of 100 kips at an eccentricity of 8 in. The length is 16 ft. Sidesway is prevented. $K = 1.0$. Determine whether the column is adequate.

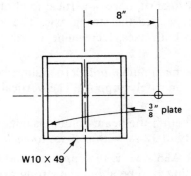

**PROBLEM 6-8**

**6-9.** Select the lightest W shape of A588 steel to support an axial load of 200 kips and a moment of 50 ft-kips with respect to the strong axis. The column length is 18 ft, and $K = 1.0$ in both directions. Sidesway is prevented.

**6-10.** Select the lightest W shape of A36 steel to support an axial load of 280 kips and a weak-axis moment of 60 ft-kips. The member is part of a frame in which sidesway is not prevented. The length is 14 ft, and $K$ has been estimated to be 1.3.

**6-11.** Rework Problem 6-10 using A572 steel ($F_y = 50$ ksi).

**6-12.** Select the lightest W shape of A36 steel to support an axial load of 600 kips and a lateral uniformly distributed load of 0.25 kips/ft. The lateral load causes bending with respect to the weak axis of the column. The member is part of a frame in which sidesway is *not* prevented. The length of the column is 20 ft, and $K$ has been estimated to be 1.20.

**6-13.** Rework Problem 6-12, assuming the uniformly distributed load to cause bending with respect to the strong axis of the column.

**6-14.** Rework Problem 6-12 using A572 steel ($F_y = 50$ ksi).

**6-15.** Select the lightest W12 for the beam-column shown. Bending is about the strong axis. Sidesway is not prevented, and $K = 1.18$. Use A36 steel. $L = 14$ ft.

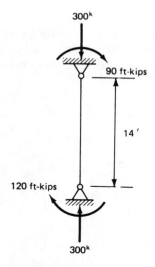

**PROBLEM 6-15**

**6-16.** Select the lightest W shape of A572 steel ($F_y = 50$ ksi) to support an axial load of 500 kips and a lateral uniformly distributed load of 0.30 kip/ft with respect to the strong axis. The member length is 25 ft, and $K = 1.0$. Sidesway is prevented.

**6-17.** In Problem 6-16, select the lightest W shape if the lateral load occurs with respect to the weak axis.

**6-18.** The corner column shown is part of a frame that is braced against sidesway and has $K = 1.0$. The length of the column is 20 ft. The column supports girts at its midheight. The girts, which support exterior wall panels, induce a maximum horizontal load of 2 kips on the column in each direction. Select the lightest W12 shape using A36 steel.

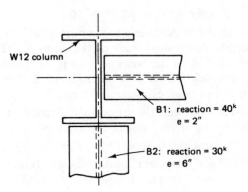

**PROBLEM 6-18**

**6-19.** Determine the effective length factor $K$ for each of the two columns in the rigid frame shown. The column bases are to be considered pinned. The webs of the members are in the plane of the frame. Sidesway is not prevented.

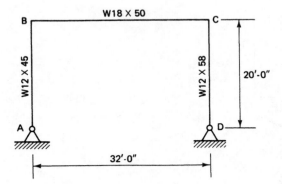

**PROBLEM 6-19**

**6-20.** Determine the effective length factor $K$ for each column in the frame shown. Sidesway is not prevented. The webs of the members are in the plane of the frame.

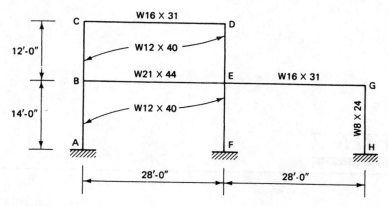

**PROBLEM 6-20**

# CHAPTER 7

# Bolted Connections

## 7-1

### INTRODUCTION

The preceding chapters have covered the three fundamental structural members—bending members, tension members, and compression members—of which every structure must in part be composed, no matter how simple or how complex. A structure may be considered to be an assemblage of these various members that must be fastened together to make the finished product.

Irrespective of how scientifically or efficiently the basic structural members may have been designed, if the necessary connections are inadequate, the result could be catastrophic collapse. The importance of economical and structurally adequate connections cannot be overemphasized.

Connection behavior is so complex that numerous simplifying assumptions must be made so that connection design is brought to a practical level. It is generally agreed among designers that the design of the basic structural members is simple compared with the design of the connections between those members.

The most common types of structural steel connections currently being used are bolted connections and welded connections. For many years rivets were the predominant type of fasteners in structures. Because of their low strength, high installation costs, and other disadvantages, however, they have been superseded and may be considered obsolete. Despite this, the 9th edition of the ASDM continues to include rivet data.

Several types of bolts can be used for connecting structural steel members. The two types generally used in structural applications are unfinished bolts and high-strength bolts. Proprietary bolts incorporating ribbed shanks, end splines, and slotted ends are also available, but in reality these are only modifications of the high-strength bolt. The use of these bolts is allowed subject to the approval of the responsible engineer. Unfinished bolts are also known as *machine, common,* or *ordinary bolts.* They are designated in the ASDM as *ASTM A307 bolts,* conforming to the requirements of ASTM A307, *Specifications for Carbon Steel Bolts and Studs, 60,000 psi Tensile Strength.* Permissible loads on these bolts are significantly less than those permitted on high-strength bolts. Their application should be limited to secondary members not subjected to vibrations or dynamic and fatigue loading.

## 7-2

### TYPES OF BOLTED CONNECTIONS

Connections serve primarily to transmit load from or to intersecting members; hence, the design of connections must be based on structural principles. This involves creating a detail that is both structurally adequate and economical as well as practical.

The simplest form of bolted connection is the ordinary *lap connection* shown in Figure 7-1a. Some connections in structures are of this general type, but it is not a commonly used detail due to the tendency of the connected members to deform.

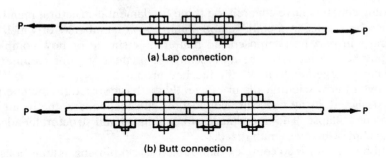

(a) Lap connection

(b) Butt connection

**FIGURE 7-1**   Types of connections.

A more common type of connection, the *butt connection,* is shown in Figure 7-1b. It is a type that may be used for tension member splices, in effect replacing the member at the point where it is cut. Other commonly used bolted connections are shown in Figure 7-2.

An understanding of the behavioral aspect of connections is important since the connections establish the support conditions of the connected members. The design of the members, which always precedes the design of the connections, must necessarily be based on assumed support conditions. There is a vast number of connection types, but a series of relatively standard connections has been developed over the years and categorized in a behavioral sense. These connections are primarily beam-to-column and beam-to-beam building connections.

There are three AISC basic types of construction and associated design assumptions:

Type 1, commonly designated as *rigid frame* (continuous frame), assumes that beam-to-column connections have sufficient rigidity to hold, virtually unchanged, the original angles between intersecting members.

Type 2, commonly designed as *simple framing* (unrestrained, pin-connected), assumes that insofar as gravity loading is concerned, the ends of the beams and girders are connected for shear only and are free to rotate under gravity load.

Type 3, commonly designated as *semirigid framing* (partially restrained), assumes that the connections of beams and girders possess a dependable and known moment capacity intermediate in degree between the rigidity of Type 1 and the flexibility of Type 2.

Therefore, in the design of a steel frame building, the type of construction must be established prior to the design of any of the structural members. After the structural members (beams and columns) are designed, the connections must be

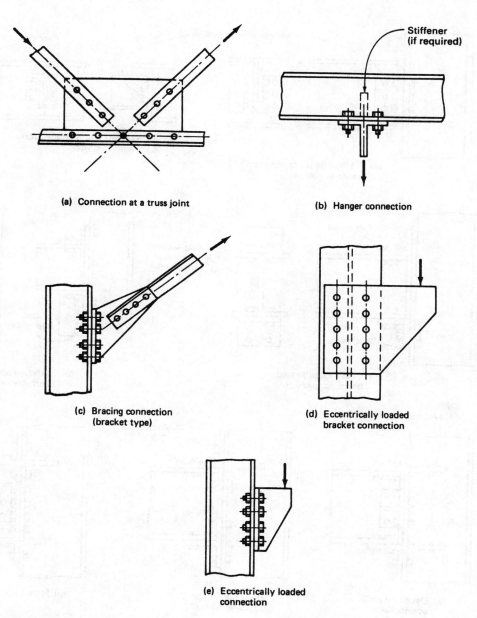

(a)  Connection at a truss joint

(b)  Hanger connection

(c)  Bracing connection
(bracket type)

(d)  Eccentrically loaded
bracket connection

(e)  Eccentrically loaded
connection

**FIGURE 7-2**   Common types of bolted connections.

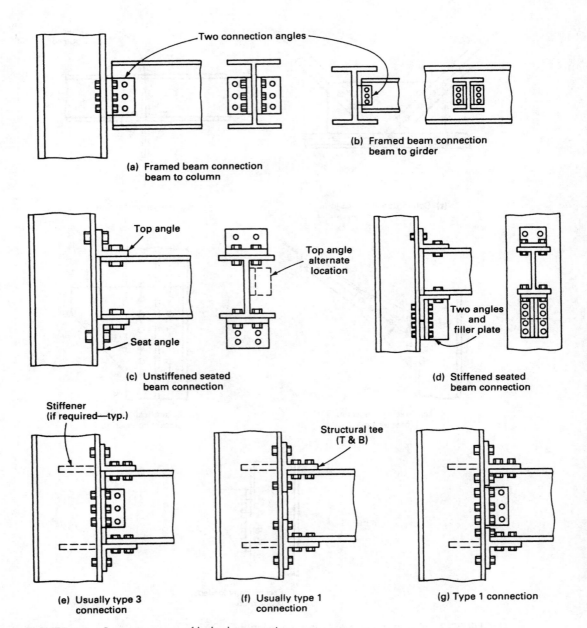

**FIGURE 7-3**    Common types of bolted connections.

designed consistent with the type of construction. Some of the common types of bolted building connections are shown in Figure 7-3.

## 7-3

### HIGH-STRENGTH BOLTS

The high-strength bolt is undoubtedly the most commonly used mechanical fastener for structural steel. Bolting with high-strength bolts has become the primary means of connecting steel members in the field as well as in the shop.

The two primary types are the A325 high-strength bolt and the A490 high-strength bolt. Specifications covering the chemical and mechanical requirements of the bolt *materials* are the latest ASTM A325 and ASTM A490 standards [1, 2]. The A490 bolt has the higher material strength. Both fasteners may be described as heavy hex structural bolts and are used with heavy hex nuts. They have shorter thread lengths than comparable bolts used for other applications.

High-strength bolts are available in several *types*. The type specified depends on the condition of use or performance desired, such as use at elevated temperatures, enhanced corrosion resistance, or certain weathering characteristics. In general, the A325 and A490 bolts are available in diameters ranging from $\frac{1}{2}$ to $1\frac{1}{2}$ in. inclusive, with $\frac{3}{4}$-in., $\frac{7}{8}$-in., and 1-in. diameters the most commonly used sizes. For details on types and sizes, see References 1 and 2.

With the profusion of different bolts available, it became necessary to establish a means of identification. It is required that the top of the bolt head be marked (either A325 or A490) along with a symbol identifying the manufacturer. Additional markings may be required to identify the type of bolt. The *Specification for Structural Joints Using ASTM A325 or A490 Bolts,* ASDM, Part 5, contains further details on markings and many other aspects of high-strength bolting. For simplicity, this specification will be referred to as the *SSJ.* The SSJ, together with its accompanying *commentary,* is relatively short and should be read in its entirety. Additionally, for a concise compilation of authoritative data on the science of high-strength structural bolting, the reader is referred to Reference 3. Part 5 of the ASDM will assist the reader in the proper identification and designation of the bolts, nuts, and washers.

## 7-4

### INSTALLATION OF HIGH-STRENGTH BOLTS

When the first A325 bolting procedure was approved for structural use, bolts were merely substituted for like numbers of hot-driven ASTM A141 steel rivets of the same diameter. Bolts were always torqued to high initial tension to ensure adequate clamping of the connected part. It was also recognized that the clamping force prevented the connected parts from movement relative to each other and allowed much of or all the load to be transmitted by friction between the parts. Further, properly torqued bolts did not loosen under dynamic or cyclic loading conditions. Until recently, the installation of all high-strength bolts required that the bolts be tightened in such a manner that the tension induced into the bolt be equal to or greater than 70% of the specified minimum tensile strength for that steel. This specified minimum *fastener tension* is found in Table 4 of the SSJ (or ASDS, Table J3.7).

The current specification recognizes that in many cases, some small movement or slip between the connected parts may occur. As discussed in Section C8 of the SSJ commentary, this slip would be extremely small and not detrimental to the performance of the connection. Therefore, depending on the category of the connection, a high-strength bolt may be installed by tightening either to the full pretensioning load or to the *snug tight* condition. "Snug tight" is defined as the tightness that exists when all plies in a joint are in firm contact. This tightness may be attained by a few impacts of an impact wrench or by the full effort of a person using an ordinary spud wrench. (A spud wrench is an open-end wrench, about 15 to 18 in. long, that has the handle end formed into a tapered pin. The pin is used to align the holes of members being connected.) Categories of connections and their specific applications are discussed in Section 7-5.

When tightening to the full pretensioning load is required, the SSJ requires that the bolts be installed and properly tightened by one of the following four methods:

1.  Turn-of-nut tightening
2.  Calibrated wrench tightening
3.  Alternate design bolt installation
4.  Direct tension indicator tightening

The objective of the tightening method is the same in all cases, namely, to induce the minimum required tension into the bolts. Tightening may be accomplished by turning either the bolt head or the nut while preventing the other element from rotating. The actual tightening is almost always accomplished by using a pneumatic or electric *impact wrench.*

In the *turn-of-nut tightening method,* enough bolts are brought to a snug-tight condition to ensure that the parts of the joint are brought into good contact with each other. Bolts must then be placed in any remaining holes in the connection and also brought to snug tightness. From this snug point all bolts must be tightened additionally by some amount of turning element rotation, varying anywhere from one-third to one full turn, depending on the length and diameter of the bolts. During this operation no rotation of the part *not* turned by the wrench is permissible. The precise amount of turning element rotation, as well as the need for washers when using this tightening method, is indicated in the SSJ.

In the *calibrated wrench tightening method,* calibrated wrenches that must be set to provide a tension at least 5% in excess of the prescribed minimum bolt tension are used. A hardened washer must be used under the element turned in tightening. Wrenches must be calibrated at least once each working day for each bolt diameter, length, and used grade, using fastener assemblies that are being installed in the work. Calibration must be accomplished by tightening, in a device capable of indicating actual bolt tension, three typical bolts of each diameter from the bolts being installed. Even though the wrench is adjusted to stall at a designated bolt tension, another check should be made by verifying during the actual installation that the turned element rotation from snug position is not in excess of that prescribed for the turn-of-nut method. The tightening procedure for this method is identical to the turn-of-nut tightening procedure. In addition, it is necessary that the wrench be returned to touch up bolts previously tightened, which may have been loosened by the tightening of subsequent bolts, until all are tightened to the prescribed amount.

*Alternate design bolt installation* involves fasteners incorporating a design feature that will directly or indirectly indicate the bolt tension. The manufacturer's installation procedure must be followed carefully to achieve proper installation. Calibration tests are required to ensure that each bolt develops a tension not less than 5% greater than the tension required in the SSJ, Table 4. A bolt that falls into this category is termed a *Load Indicator Bolt* by Bethlehem Steel Corporation. It is a high-strength bolt with a splined portion added to the threaded end. In combination with the use of a special wrench, the installation process has been simplified and reduced to a one-person operation. The principle is quite simple (see Figure 7-4). The wrench, by virtue of its construction, grasps both the nut and the spline, applying a clockwise turning force to the nut and a counterclockwise turning force to the spline. When the fastener assembly reaches a predetermined torque, the splined end will be twisted off.

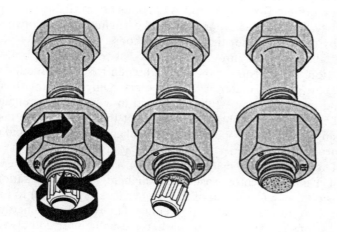

**FIGURE 7-4** Load indicator bolt. (Copyright by Bethlehem Steel Corporation. Reproduced by permission.)

**PHOTO 7-1**   Framed beam connection that is shop-bolted to the beam web using load-indicator bolts. The bolts have not yet been tightened. Note also the cope in the beam, which is made to allow clearance for the flange of the supporting member.

Although this would seem to indicate that the bolt has been brought to the specified minimum tension, there is no guarantee that it has. In fact, this type of fastener has been found to produce widely varying installed tensions depending on the conditions during installation. Because of this, inspection procedures for these bolts should be specified when they are used in connections where the specified minimum tension is required. Both A325 and A490 bolts are available with splined ends. These bolts may be reused (if approved by the engineer) as a conventional high-strength bolt by using the other tightening methods.

The required installation procedure as per the SSJ is to install the bolts in all holes of the connection. All bolts must then be tightened to an intermediate level of tension adequate to pull all materials into contact. Only after this has been accomplished should the fasteners be fully tensioned in a systematic manner and the splined ends sheared off. Tests have shown that if the fasteners are installed and tensioned in a single continuous operation, they will give a misleading indication that the bolts are properly tightened.

One device that falls under the category of *direct tension indicator tightening* involves the use of a load indicator washer (Bethlehem Steel Corporation). Its installation depends on the deformation of the washer (see Figure 7-5) to indicate the induced bolt tension during and after tightening. The round hardened washer has a series of protrusions on one face. The washer is usually inserted between the bolt head and the gripped plates with the protrusions bearing against the underside

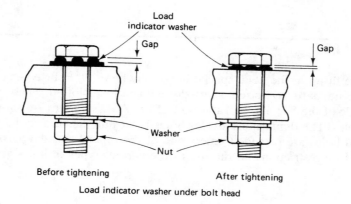

**FIGURE 7-5**   Direct tension indicator.

of the bolt head, leaving a gap. Upon tightening, the protrusions are partially flattened and the gap is reduced. Bolt tension is evaluated by measurement of the gap closure. When the gap is reduced to a prescribed dimension, the bolt has been properly tightened. Load indicator washers are produced for both A325 and A490 bolts and are identified appropriately to avoid any errors. Installation by this method still requires the use of wrenches. The calibration of the wrench, or establishing the nut rotation from a snug-tight condition, however, is not necessary when using the load indicator washer. In addition, the use of the load indicator washer furnishes simplified and economical inspection that may be performed by one person using a simple metal feeler gage.

   With respect to the reuse of all high-strength bolts, it is stipulated in the SSJ that A490 bolts and galvanized A325 bolts shall not be reused. Other A325 bolts may be reused once or twice if approved by the engineer responsible. A properly installed high-strength bolt will have experienced some deformation, however. Proper reinstallation of the bolt will be difficult if not impossible. According to the SSJ, retightening previously tightened bolts that may have been loosened by the tightening of adjacent bolts should not be considered as a reuse.

7-5

## HOLE TYPES

To expedite fit-up in the field, special holes may be used. Four types of holes are recognized: standard, oversized, short-slotted, and long-slotted. These are defined in Section 3c of the SSJ and in Table J3.1 of the ASDS. Figure 7-6 shows the four types of holes. The dimensions are for holes for a $\frac{3}{4}$-in.-diameter bolt.

Standard holes are provided unless one of the other types is approved by the designer. More restrictions on the use of the various types of holes are discussed shortly.

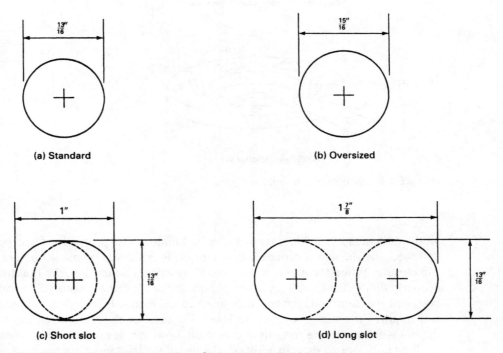

FIGURE 7-6    Four hole types (for a $\frac{3}{4}$-in.-diameter bolt).

7-6

## STRENGTH AND BEHAVIOR OF HIGH-STRENGTH BOLTED CONNECTIONS

In Chapter 2 we briefly discussed the strength of bolted connections and how the details of connections affect the design and analysis of tension members. We now investigate the rationale on which the strength of high-strength bolted connections is based. In determining the strength of high-strength bolted connections, one must

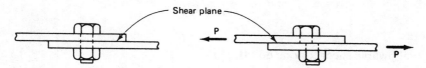

**FIGURE 7-7**   Bolt in single shear.

consider the aspects of *shear, bearing,* and *tension* with regard to both the fasteners and the connected materials.

In most structural connections the bolt is required to prevent the movement of connected material in a direction perpendicular to the length of the bolt, as in Figure 7-7. In such cases the bolt is said to be loaded in *shear.* In the lap connection shown, the bolt has a tendency to shear off along the single contact plane of the two plates. Since the bolt is resisting the tendency of the plates to slide past one another along the contact surface and is being sheared on a single plane, the bolt is said to be in *single shear.*

In a butt connection such as that shown in Figure 7-8, there are two contact planes; therefore, the bolt is offering resistance along two planes and is said to be in *double shear.*

It is easy to visualize the plates slipping in the direction of the applied force until they bear against the bolt. This is called a *bearing-type* connection. It is sometimes referred to as a *shear bearing* connection. For this type of connection to exist, a small amount of movement (slip) must take place to bring the bolts into bearing. In many applications, a small amount of slip in a connection is not detrimental and is sometimes even desirable. Bolts used in a bearing connection need only be tightened to the snug-tight condition. In other connections even the smallest amount of slip is undesirable. These connections are called *slip-critical* connections, and the high-strength bolts used in them must be tightened to the full pretensioning load. The ASDS, Section J1.12, is specific as to where the latter type of connection must be used. The list, in part, includes (a) connections of all beams and girders to columns and of any other beams and girders on which the bracing of columns is dependent, in structures over 125 ft in height; and (b) connections for supports of running machinery or of other live loads that produce impact or reversal of stress. The SSJ also adds, among others, connections subject to fatigue loadings.

The load-carrying capacity, or working strength, of a bolt in single shear is equal to the product of the cross-sectional area of its shank and an allowable shear stress:

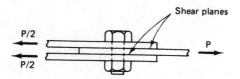

**FIGURE 7-8**   Bolt in double shear.

$$r_v = A_b F_v$$

where

   $r_v$ = allowable shear for one bolt (kips)

$A_b$ = cross-sectional area of one bolt (in.$^2$)

 $F_v$ = allowable shear stress (ksi)

   The allowable shear stress depends on the type of high-strength bolt, whether the connection is bearing-type or slip-critical, and the type of hole. Values for allowable shear stress $F_v$ can be found in Table J3.2 of the ASDS.

   When a bolt is subjected to more than one plane of shear, such as in double shear (Figure 7-8), the allowable shear for the one bolt will be $r_v$ multiplied by the number of shear planes.

   Although the bolts in a connection may be adequate to transmit the applied load in shear, the connection will fail unless the *material joined* is capable of transmitting the load *into* the bolts. This capacity is a function of the bearing (or crushing) strength of the connected material, as shown in Figure 7-9. The bearing failure mode is shown in Figure 2-15. The true distribution of the bearing pressure on the material around the perimeter of the hole is unknown; therefore, the resisting contact area has been taken as the nominal diameter of the bolt multiplied by the thickness of the connected material. This assumes a uniform pressure acting over a rectangular area.

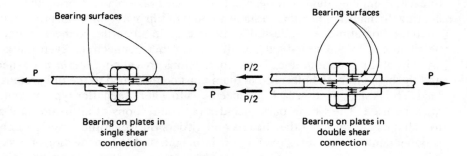

**FIGURE 7-9**   Bearing pressures.

   The strength of one bolt in bearing may be expressed as

$$r_v = dt F_p$$

where

   $r_v$ = allowable bearing for one bolt (kips)

   $d$ = nominal bolt diameter (in.)

    $t$ = thickness of plate or connected part (in.)

$F_p$ = allowable bearing stress (ksi)

The allowable bearing stress at the bolt hole depends on the type of hole in which the bolts are placed, the spacing between the centers of the bolts, and the distance from the center of a bolt to the nearest edge (called *edge distance;* defined in Section 2-2 of this text). The allowable bearing stress $F_p$ (ksi) on the projected area of bolts in shear connections with the end distance in the line of force not less than $1.5d$ and the center-to-center distance of bolts not less than $3d$ is to be taken as follows:

1.  In standard or short-slotted holes with two or more bolts in the line of force,

$$F_p = 1.2F_u \qquad \text{ASDS Eqn. (J3-1)}$$

2.  In long-slotted holes with the axis of the slot perpendicular to the direction of the load and with two or more bolts in the line of force,

$$F_p = 1.0F_u \qquad \text{ASDS Eqn. (J3-2)}$$

On the projected area of the bolt *closest to the edge* in standard or short-slotted holes with the edge distance less than $1.5d$ and in all connections with a single bolt in the line of force,

$$F_p = \frac{L_e F_u}{2d} \le 1.2F_u \qquad \text{ASDS Eqn. (J3-3)}$$

where

$L_e$ = distance from the free edge to the center of the bolt (in.), as shown in Figure 7-10

$F_u$ = the specified minimum tensile strength of the connected part (ksi)

Therefore, for a group of bolts in a line (parallel to the line of force), if the edge distance in the line of force is less than $1.5d$, $F_p$ need be reduced for only the bolt closest to the edge.

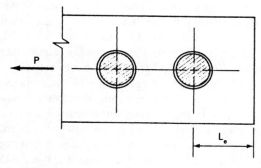

**FIGURE 7-10**   Explanation of $L_e$.

In the slip-critical connection, it is assumed that the load is transmitted from one connected part to another entirely by the friction that results from the high tension in the bolt. The theory behind the slip-critical connection is that *no slippage* occurs between the connected parts and that the bolts are not actually loaded in shear or bearing. *For design purposes,* however, it is assumed that the bolts *are* in shear and bearing and, therefore, allowable stress values are furnished. Even though no slippage is expected to occur, the bearing consideration will result in an acceptable design should the remotely possible slippage take place. In addition to the strength calculations based on allowable shear and bearing stresses, the SSJ requires that slip-critical connections be checked to ensure that slip will not occur at working loads. The force on the connection must not exceed the allowable resistance $P_s$ calculated from

$$P_s = F_s A_b N_b N_s$$

where

$F_s$ = allowable slip load per unit area of bolt from the SSJ, Table 3

$A_b$ = area corresponding to the nominal body area of the bolt

$N_b$ = number of bolts in the joint

$N_s$ = number of slip planes

Table 3 of the SSJ provides $F_s$ as a function of hole type, direction of load application, and the condition of the contact surfaces of the connected parts. With regard to the latter, the contact surface can be categorized as class A, B, or C. Each class has an associated minimum *slip coefficient*. The slip coefficient $k_s$ is analogous to (but not quite the same as) a coefficient of friction. When determining $P_s$, if coatings are used on the contact surfaces, it will be necessary to ensure that the minimum slip coefficients of Table 3 are provided. Appendix A of the SSJ contains a standardized test method that can be used by any certified testing agency to evaluate the slip coefficient for a particular coating.

For the purposes of this text, we assume that all slip-critical joints have class A contact surfaces. This implies a minimum slip coefficient of 0.33 (clean mill scale or blast-cleaned surfaces with class A coatings). Therefore, the allowable shear stresses of ASDS, Table J3.2, apply directly, and $P_s$ will not be calculated.

In the bearing-type connection, it is accepted that the bolt is actually in shear and that the load is transmitted by the shearing resistance of the bolt as well as bearing of the connected parts on the bolt. The frictional resistance between the connected parts is of no concern, nor is the surface condition of the connected parts. Bearing-type connections may be designed with the bolt threads in, or out of, the shear plane. The allowable shearing stresses reflects the selected condition. Depending on the type of connection and location of the threads, the ASDM uses the following bolt designations:

| Bolt designation | Type of application |
|---|---|
| A325SC, A490SC | Slip-critical connection |
| A325N, A490N | Bearing-type connection, threads in the shear plane |
| A325X, A490X | Bearing-type connection, threads excluded from the shear plane |

A high-strength bolted connection, where the bolts are subjected to pure tensile loads, is shown in Figure 7-2b. This is a hanger-type connection. The permissible tension in a bolt may be taken as the product of its nominal cross-sectional area and its allowable unit tensile stress. This may be expressed as

$$r_t = A_b F_t$$

where

$r_t$ = allowable tension for one bolt (kips)

$A_b$ = nominal cross-sectional area of one bolt (in.$^2$)

$F_t$ = allowable unit tensile stress (ksi)

The allowable tensile stresses are furnished in Table J3.2 of the ASDS and Table 1-A in Part 4 of the ASDM. Although tensile stresses exist in high-strength bolts before any external tensile load is applied, it has been found that the external tensile load does not substantially affect the stresses in the bolt until the externally applied load exceeds the tension initially induced into the bolt by one of the tightening methods discussed previously. Therefore, the bolt is permitted to develop its full allowable tensile strength to resist the load.

In the design of hanger-type connections, prying action must be considered. In Figure 7-11 this is a tendency of the flanges of the structural tee to act as cantilever beams under the action of the downward load $P$ and the upward force of the restraining bolt. Actually, prying forces are present to some extent in nearly all connections employing bolts loaded in tension. The effect of the prying action is to increase the tension in the bolts. The magnitude of this increased tensile load is a function of the supporting member, bolt type and size, and the connection geometry. A design example considering prying forces is presented later in this chapter. Reference 4 contains some background.

Connections subjected simultaneously to shear and tension loads require special analysis to assure conformance to specification provisions for values of $F_v$ and $F_t$. Connections for this type of combined loading occur frequently at the end of diagonal bracing members, as shown in Figure 7-2c. Interaction formulas permit the computation of a new allowable tensile stress $F_t$ for bearing-type connections that must exceed the actual tensile stress induced by the applied loads. The new $F_t$ is based on $f_v$, which is the actual bolt shear stress induced by the applied loads.

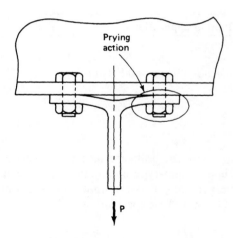

**FIGURE 7-11**    Prying action.

According to the ASDS, Table J3.3, for bearing-type connections, the allowable tensile stress may be determined as follows:

| Bolt | Expression for $F_t$ (ksi) |
|------|----------------------------|
| A325N | $\sqrt{(44)^2 - 4.39 f_v^2}$ |
| A325X | $\sqrt{(44)^2 - 2.15 f_v^2}$ |
| A490N | $\sqrt{(54)^2 - 3.75 f_v^2}$ |
| A490X | $\sqrt{(54)^2 - 1.82 f_v^2}$ |

In all cases the shear stress $f_v$ cannot exceed the allowable shear stresses as furnished in Table J3.2 of the ASDS. In slip-critical connections (A325SC and A490SC), the maximum shear stress allowed by Table J3.2 shall be multiplied by the reduction factor

$$1 - \frac{f_t A_b}{T_b}$$

where

$f_t$ = the average tensile stress due to a direct load applied to all the bolts in a connection

$T_b$ = the pretension load of the bolt specified in Table J3.7

The design and detailing of bolted connections requires proper edge distance (as previously discussed) as well as proper *bolt spacing*. The ASDS, Section J3.8, requires that the distance between centers of standard, oversized, or slotted fastener holes (bolt spacing) must not be less than $2.67d$, where $d$ is the nominal diameter of the bolt. A distance of $3d$ is preferred. For standard holes, however, bolt spacing $s$ along a line of transmitted force shall be such that

$$s \geq \frac{2P}{F_u t} + \frac{d}{2} \qquad \text{ASDS Eqn. (J3-5)}$$

where

$P$ = force transmitted by one bolt to the critical connected part (kips)

$F_u$ = lowest specified minimum tensile strength of the critical connected part (ksi)

$t$ = thickness of the critical connected part (in.)

For *oversized and slotted holes,* this distance is increased by an increment $C_1$ given in Table J3.4 of the ASDS. This provides the same clear distance between holes as for standard holes.

The edge distance must not be less than that shown in Table J3.5 of the ASDS. The edge distance $L_e$ (see Figure 7-10) along a line of transmitted force, in the direction of the force, cannot be less than $1.5d$ when $F_p$ is determined by Equation (J3-1) or (J3-2), however. Otherwise, the edge distance shall not be less than

$$L_e \geq \frac{2P}{F_u t} \qquad \text{ASDS Eqn. (J3-6)}$$

where $P$, $F_u$, and $t$ are as defined previously. For oversized and slotted holes, this distance is increased by an increment $C_2$ given in Table J3.6 of the ASDS. This provides the same clear distance from the edge of the hole as for a standard hole. The maximum distance from the center of a bolt hole to the nearest edge of any member shall be 12 times the thickness of the connected part under consideration, but shall not exceed 6 in.

Another mode of failure to be considered is block shear, also referred to as *web tear-out.* This type of failure may be critical at a beam connection if the end of the beam is *coped.* It may also be critical in tension member connections discussed in Chapter 2 of this text. ASDS, Section J4, discusses this type of failure. As described in Section 2-2, block shear is a combination of shear failure along a plane through a line of bolts (or a weld) and simultaneous tension failure along a perpendicular plane. For a coped beam, this failure mode is shown in Figure 7-12. (Also refer to Figure 4-2.)

Recall that block shear strength is calculated from the summation of net shear area $A_v$ times the allowable shear stress $F_v$ and net tension area $A_t$ times the allowable tensile stress $F_t$, where $F_v = 0.30F_u$ and $F_t = 0.50F_u$. Mathematically, this is stated as

$$P = A_v F_v + A_t F_t$$

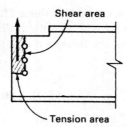

**FIGURE 7-12**   Block shear for a coped beam.

Also recall that, for purposes of block shear strength calculations, hole diameters for the net area determination are taken as the *fastener* diameter plus $\frac{1}{8}$ in. Table I-G in Part 4 of the ASDM may be used for the block shear strength determination if conditions match those for which the table is set up.

In the design or analysis of a connection, it may be necessary to check the tensile capacity of the connected member itself since the original design of the tension member was based on an assumed connection. The tensile capacity of the connected member will be denoted $P_t$, following the convention of Chapter 2 of this text. Naturally, the final tensile capacity (or allowable load) of the member may be controlled either by the member itself or by the connection.

**Example 7-1** _____

Compute the tensile capacity, $P_t$, for the single-shear lap connection shown in Figure 7-13. The plates are A36 steel ($F_u = 58$ ksi), and the high-strength

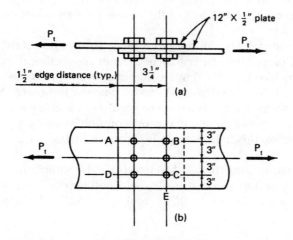

**FIGURE 7-13**   Single-shear lap connection.

bolts are $\frac{7}{8}$-in.-diameter A325SC in standard holes. Assume a class A contact surface and a slip-critical connection.

**Solution:**

(*Note:* All table references are to the ASDM, Part 4.)

*Bolt shear:*   The allowable shear stress (Table I-D) is

$$F_v = 17.0 \text{ ksi}$$

The allowable shear load per bolt ($A_b$ from Table I-D) is

$$r_v = A_b F_v$$
$$= 0.6013 \, (17.0) = 10.2 \text{ kips}$$

Note that the allowable shear per bolt could also be obtained directly without computations from Table I-D.
The allowable shear load per connection (six bolts) is

$$P_t = 10.2(6) = 61.2 \text{ kips}$$

*Bearing* (on $\frac{1}{2}$-in.-thick plate):   The allowable bearing stress $F_p$ depends on the edge distance and the center-to-center spacing of the bolts. The edge distance is 1.5 in.

$$\frac{1.5}{0.875} = 1.71$$

Therefore, the edge distance is $1.71d$, which exceeds $1.5d$. Distance center-to-center of bolts is 3.25 in. This is $3.71d$, which exceeds $3d$. Therefore, $F_p$ is calculated from Equation (J3-1):

$$F_p = 1.2F_u = 1.2(58) = 69.6 \text{ ksi}$$

The allowable bearing load per bolt is

$$r_v = dtF_p = 0.875(0.5)(69.6) = 30.5 \text{ kips}$$

Note that the allowable bearing load per bolt could also be obtained directly without computations from Table I-E, which is based on $F_p = 1.2 \, F_u$.
The allowable bearing load per connection is

$$P_t = 30.5(6) = 183.0 \text{ kips}$$

*Tension:*   This is a check of the tensile capacity of the plates (bolts are *not* in tension) and is based on the principles of Sections 2-2 and 2-3 of this text.

$$A_n = A_g - A_{\text{holes}} = \left(12 \times \frac{1}{2}\right) - 3\left(\frac{7}{8} + \frac{1}{8}\right)\frac{1}{2} = 4.50 \text{ in.}^2$$

Based on $A_g$ ($F_t = 0.60 \ F_y = 22$ ksi),

$$P_t = A_g F_t = \left(12 \times \frac{1}{2}\right)(22) = 132 \text{ kips}$$

Based on $A_n$ [$F_t = 0.50 F_u = 0.50(58) = 29.0$ ksi; shear lag consideration is not applicable],

$$P_t = A_n F_t = 4.50(29.0) = 130.5 \text{ kips}$$

Next, check the block shear strength (also termed *shear rupture* or *web tearout*) in accordance with the ASDS, Section J4. Two possible cases are investigated. The block shear strength is written

$$P_t = A_v F_v + A_t F_t = A_v(0.30 \ F_u) + A_t(0.50 F_u)$$

where $A_v$ and $A_t$ are the *net* shear area and *net* tension area, respectively. Note that the hole diameter is taken as

$$\frac{7}{8} + \frac{1}{8} = 1.00 \text{ in.}$$

*Case I—failure line ABCD* (see Figure 7-13b):

$$A_v = 2(0.50)(4.75 - 1.5(1.00)) = 3.25 \text{ in.}^2$$
$$A_t = 0.50(6.00 - 2(1.00)) = 2.00 \text{ in.}^2$$
$$P_t = 3.25(0.30)(58) + 2.00(0.50)(58) = 114.6 \text{ kips}$$

*Case II—failure line ABCE* (see Figure 7-13b):

$$A_v = 0.50(4.75 - 1.5(1.00)) = 1.625 \text{ in.}^2$$
$$A_t = 0.50(9.00 - 2.5(1.00)) = 3.25 \text{ in.}^2$$
$$P_t = 1.625(0.30)(58) + 3.25(0.50)(58) = 122.5 \text{ kips}$$

Note that Case I is the more critical of the two block shear cases. Therefore, bolt shear governs the strength of the lap connection. The connection has a tensile capacity of 61.2 kips (272 kN).

---

**Example 7-2** _____

Rework Example 7-1 assuming that the bolts are $\frac{7}{8}$-in.-diameter A325X (bearing-type connection with threads excluded from the shear plane) in standard holes.

**Solution:**

(*Note:* All table references are to the ASDM, Part 4.)

*Bolt shear:*   The allowable shear stress $F_v$ from (Table I-D) is 30 ksi. Therefore, the allowable shear load per bolt is

$$r_v = A_b F_v$$

$$= 0.6013(30) = 18 \text{ kips}$$

Note that the allowable shear load per bolt could also be obtained directly, without computations from Table I-D.

The allowable shear load per connection (six bolts) is

$$P_t = 18(6) = 108 \text{ kips}$$

Bearing, tension, and block shear strengths are the same as for Example 7-1: 183.0 kips, 130.5 kips, and 114.6 kips, respectively. Therefore, the tensile capacity = 108 kips (480 kN).

---

## Example 7-3

A single-angle tension member in a roof truss is attached to a $\frac{3}{8}$-in.-thick gusset plate with A325 $\frac{3}{4}$-in.-diameter high-strength bolts (A325N) in standard holes, as shown in Figure 7-14. The gusset plate and the angle are A36 steel ($F_u = 58$ ksi). The tension member is to support a 42-kip load. Determine if the member and the connection are adequate.

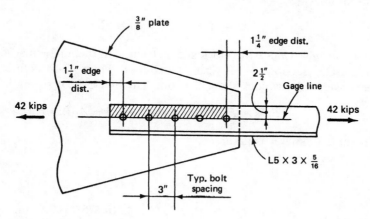

**FIGURE 7-14**  Single-angle tension member.

### Solution:

(*Note:* All table references are to the ASDM, Part 4.)

*Bolt shear* (S-single shear):   See Table I-D.

$$P_t = 9.3(5) = 46.5 \text{ kips}$$

*Bearing* (on the thinnest material, which is the $\frac{5}{16}$-in.-thick angle):   Edge distance in the line of force is 1.25 in., which is $1.67d$ ($> 1.5d$). Center-to-center spacing of bolts is 3 in., which is $4.0d$ ($> 3.0d$). Both of these are O.K.; therefore, Table I-E is applicable. The capacity of one bolt is 16.3 kips. Therefore,

$$P_t = 16.3(5) = 81.5 \text{ kips}$$

*Tension:*   In the design of this truss, the angle tension member was designed based on an assumed end connection of one gage line and one hole per cross section. The actual connection agreed with the assumption. Under these conditions, it would not be necessary to check the angle in tension. To be complete, however, we will determine the tensile capacity of the single angle. Calculate the net area ($A_n$):

$$A_n = A_g - A_{\text{holes}}$$
$$= 2.40 - 0.875(0.3125) = 2.13 \text{ in.}^2$$

For calculation of effective net area $A_e$, $U = 0.85$ (see Table 2-1 of this text). Based on $A_e$ [$F_t = 0.50F_u = 0.50(58.0) = 29.0$ ksi],

$$P_t = A_e F_t = U A_n F_t = 0.85(2.13)(29.0) = 52.5 \text{ kips}$$

Based on $A_g$ ($F_t = 0.60F_y = 22$ ksi),

$$P_t = A_g F_t = 2.40(22) = 52.8 \text{ kips}$$

*Block shear:*   Check the block shear strength in the angle in accordance with the ASDS, Section J4. The cross-hatching in Figure 7-14 defines the area to be considered for block shear tear-out. The block shear strength is written as

$$P_t = A_v F_v + A_t F_t = A_v(0.30F_u) + A_t(0.50F_u)$$

The hole diameter is taken as

$$\frac{3}{4} + \frac{1}{8} = 0.875 \text{ in.}$$

$$A_v = 0.3125[4(3.0) + 1.25 - 4.5(0.875)] = 2.91 \text{ in.}^2$$
$$A_t = 0.3125[2.50 - 0.5(0.875)] = 0.645 \text{ in.}^2$$
$$P_t = 2.91(0.30)(58) + 0.645(0.50)(58) = 69.3 \text{ kips}$$

The tensile capacity of the angle is taken as the lowest of the four capacities calculated. Bolt shear, $P_t = 46.5$ kips, controls. This is greater than the actual load of 42 kips. Therefore, the connection and the member are O.K.

**Example 7-4** _____

Compute the tensile capacity $P_t$ for the double-shear butt connection shown in Figure 7-15. The plates are A36 steel ($F_u = 58$ ksi). The high-strength bolts are $\frac{7}{8}$-in.-diameter A325 in standard holes. Assume that the connection is

(a)  Slip-critical (A325SC, class A).
(b)  Bearing-type, with threads excluded from the shear plane (A325X).
(c)  Bearing-type, with threads in the shear plane (A325N).

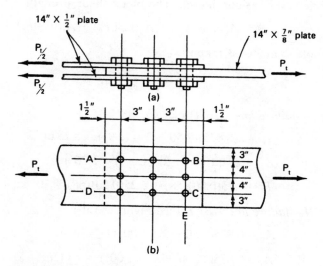

**FIGURE 7-15**   Double-shear butt connection.

**Solution:**

(*Note:* All table references are to the ASDM, Part 4).

(a)  *A325SC connection:* Bolt shear (D; double shear):

$$P_t = 20.4(9) = 183.6 \text{ kips (Table I-D)}$$

*Bearing* (on $\frac{7}{8}$-in. plate): Edge distance = $1.71d$; bolt spacing (in line of force) = $3.43d$. Therefore, Table I-E applies. No value for allowable bearing per bolt is tabulated, however, since this value exceeds the double-shear capacity of an A490X bolt and bearing would not control. Alternatively, we could multiply the allowable bearing per bolt value for a 1-in. thickness of material by the thickness of *this* plate ($\frac{7}{8}$ in.) to obtain

$$\frac{7}{8}(60.9) = 53.3 \text{ kips/bolt}$$

This exceeds the allowable shear per bolt of 20.4 kips; thus we see that bearing does not control.

The tensile capacity based on tensile fracture of the net area is

$$P_t = A_n F_t = (14 - 3)(0.875)(29) = 279 \text{ kips}$$

The tensile capacity based on yielding of the gross area is

$$P_t = A_g F_t = 14(0.875)(22) = 270 \text{ kips}$$

The block shear strength is based on failure lines in the $\frac{7}{8}$-in. plate. Two possible cases are considered. The block shear strength is written as

$$P_t = A_v F_v + A_t F_t = A_v(0.30F_u) + A_t(0.50F_u)$$

The hole diameter is taken as

$$\frac{7}{8} + \frac{1}{8} = 1.00 \text{ in.}$$

*Case I—failure line ABCD* (see Figure 7-15b):

$$A_v = 2(0.875)(7.50 - 2.5(1.00)) = 8.75 \text{ in.}^2$$

$$A_t = 0.875(8.00 - 2.0(1.00)) = 5.25 \text{ in.}^2$$

$$P_t = 8.75(0.30)(58) + 5.25(0.50)(58) = 305 \text{ kips}$$

*Case II—failure line ABCE* (see Figure 7-15b):

$$A_v = 0.875(7.50 - 2.5(1.00)) = 4.38 \text{ in.}^2$$

$$A_t = 0.875(11.00 - 2.50(1.00)) = 7.44 \text{ in.}^2$$

$$P_t = 4.38(0.30)(58) + 7.44(0.50)(58) = 292 \text{ kips}$$

Note that Case II is the more critical of the two block shear cases. Comparing all the preceding values, it is seen that bolt shear controls and the tensile capacity $P_t$ is 183.6 kips (817 kN).

(b)  *A325X connection:* Bolt shear (D):

$$P_t = 36.1(9) = 324.9 \text{ kips (Table I-D)}$$

Bearing, tension, and block shear strengths are as in part (a): bearing is not critical, tension strength is 270 kips, and block shear strength is 292 kips. Tension on the gross area controls and the tensile capacity $P_t$ is 270 kips (1201 kN).

(c)  *A325N connection:* Bolt shear (D):

$$P_t = 25.3(9) = 228 \text{ kips (Table I-D)}$$

Bearing, tension, and block shear are as in part (a). Bolt shear controls and the tensile capacity $P_t$ is 228 kips (1014 kN).

**Example 7-5**

A tension member made up of a pair of angles is connected to a column with eight $\frac{3}{4}$-in.-diameter A325 high-strength bolts in standard holes as shown in Figure 7-16. Note that the 70.7-kip forces are components of the 100-kip force. All structural steel is A36. Determine if the connection to the column is satisfactory. (Assume that the connection between the angles and the structural tee is satisfactory.) Consider

(a)  A bearing-type connection with threads excluded from the shear plane (A325X)

(b)  A slip-critical connection (A325SC)

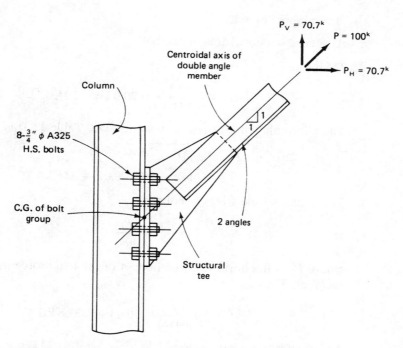

**FIGURE 7-16**  Bracing connection.

**Solution:**

In this connection, the bolts are subjected to combined shear and tension. Note that the connection is detailed so that the centroidal axis of the tension member intersects the center of the bolt group. This will eliminate eccentricity, and each bolt is considered equally loaded. The problem of combined stresses and the associated allowable stresses is discussed in the ASDS, Sections J3.5

and J3.6. In this solution, actual stresses will be compared with allowable stresses.

(a)   *A325X connection:* The allowable bolt shear stress is

$$F_v = 30 \text{ ksi (Table I-D, ASDM, Part 4)}$$

The actual bolt shear stress is

$$f_v = \frac{P_V}{nA_b}$$

where

$P_V$ = vertical component of the axial force in the tension member

$n$ = number of bolts

$A_b$ = cross-sectional area of one bolt

Therefore,

$$f_v = \frac{70.7}{8(0.4418)} = 20.0 \text{ ksi} < 30 \text{ ksi} \qquad \textbf{O.K.}$$

The allowable tensile stress in bolts (ASDS, Table J3.3) is

$$F_t = \sqrt{(44)^2 - 2.15f_v^2}$$
$$= \sqrt{(44)^2 - 2.15(20.0)^2} = 32.8 \text{ ksi}$$

The actual tensile stress in bolts is

$$f_t = \frac{P_H}{nA_b}$$

where $P_H$ is the horizontal component of the axial force in the tension member. Thus

$$f_t = \frac{70.7}{8(0.4418)} = 20.0 \text{ ksi} < 32.8 \text{ ksi} \qquad \textbf{O.K.}$$

The connection is satisfactory.

(b)   *A325SC connection:* The average tensile stress in bolts is

$$f_t = \frac{70.7}{8(0.4418)} = 20.0 \text{ ksi}$$

The allowable bolt shear stress, where bolts are subjected to both shear and tension in a slip-critical connection (ASDS, Section J3.6), is

$$F_v = 17.0 \left( 1 - \frac{f_t A_b}{T_b} \right)$$

$$= 17.0 \left[ 1 - \frac{20(0.4418)}{28} \right]$$

$$= 11.63 \text{ ksi}$$

$T_b$, the minimum bolt tension applied in the tightening process, may be found in the ASDS, Table J3.7, or in the SSJ, Table 4.

The actual bolt shear stress is

$$f_v = \frac{70.7}{8(0.4418)} = 20.0 \text{ ksi}$$

Since $f_v > F_v$, the connection is *not satisfactory* as a slip-critical type.

---

### Example 7-6

A W12 × 30 is used as a tension member in a truss. The flanges of the member are connected to $\frac{3}{8}$-in. gusset plates as shown in Figure 7-17. All structural steel is A36 ($F_u = 58$ ksi). A connection is to be designed to develop the full tensile capacity of the member. Determine the required number of $\frac{3}{4}$-in.-diameter high-strength bolts. Assume standard holes and a slip-critical condition. Establish the pitch and edge distance.

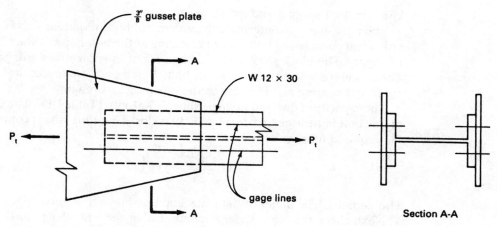

**FIGURE 7-17**  Truss member connection.

### Solution:

(*Note:* Table references are to the ASDM, Part 4, unless noted otherwise.)

The tensile capacity of the wide-flange section will be determined first. Assume two gage lines per flange with 3 or more bolts per line. Thus

$$A_n = A_g - A_h$$

$$= 8.79 - 4(0.875)(0.440) = 7.25 \text{ in.}^2$$

Based on $A_e$, using $U = 0.85$ (since $b_f < \frac{2}{3}d$) (see Table 2-1 in this text) and using $F_t = 0.50F_u = 29$ ksi,

$$P_t = A_e F_t = UA_n F_t = 0.85(7.25)(29) = 178.7 \text{ kips}$$

Based on $A_g$ ($F_t = 0.60F_y = 22$ ksi),

$$P_t = A_g F_t = 8.79(22) = 193.4 \text{ kips}$$

Therefore, the tensile capacity for which the connection will be designed is 178.7 kips as controlled by the effective net area consideration.

Next we establish bolt pitch and edge distances so that $F_p = 1.2F_u$ (Equation J3-1). Table I-E will be applicable. The minimum edge distance (in the line of force) is

$$1.5d = 1.5(0.75) = 1.125 \text{ in.}$$

Use $1\frac{1}{4}$ in. for all edge distances (alternatively, refer to Table J3.5). The minimum bolt spacing in the line of force is

$$3d = 3(0.75) = 2.25 \text{ in.}$$

Use 3-in. bolt spacing and a 4-in. gage.

The next step is to determine the number of bolts required. This is an easy task when considering bolt shear or bearing at the bolt holes. Tables I-D and I-E apply. The block shear consideration is not as straightforward, however. Therefore, we select the number of bolts based on the more critical of bolt shear and bearing and then check the block shear strength.

The capacity of one bolt in single shear is 7.51 kips (Table I-D). The capacity of one bolt in bearing on $\frac{3}{8}$-in. plate is 19.6 kips. Therefore, shear controls and the required number of bolts $n$ is

$$n = \frac{178.7}{7.51} = 23.8 \text{ bolts}$$

Use 24 bolts. The detail of the connection is shown in Figure 7-18.

Next, check the block shear strength. Either the wide-flange or the gusset plates may control. The perimeters to be considered are shown cross-hatched in Figure 7-18. The hole diameter is taken as $\frac{3}{4} + \frac{1}{8} = 0.875$ in. For the wide flange:

$$A_v = 4(0.44)[5(3) + 1.25 - 5.5(0.875)] = 20.1 \text{ in.}^2$$

$$A_t = 4(0.44)[1.25 - 0.5(0.875)] = 1.430 \text{ in.}^2$$

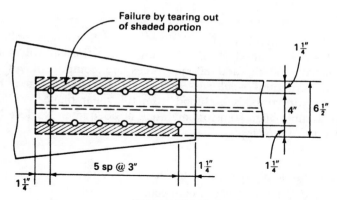

W 12 × 30 block shear

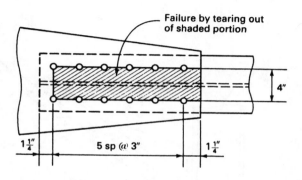

Gusset plate block shear

**FIGURE 7-18**    Connection detail.

$$P_t = A_v(0.3F_u) + A_t(0.5F_u)$$

$$= 20.1(0.3)(58) + 1.430(0.5)(58) = 391 \text{ kips}$$

For the gusset plates (considering two plates),

$$A_v = 4(0.375)[5(3) + 1.25 - 5.5(0.875)] = 17.16 \text{ in.}^2$$

$$A_t = 2(0.375)(4.00 - 0.875) = 2.34 \text{ in.}^2$$

$$P_t = 17.16(0.30)(58) + 2.34(0.50)(58) = 366 \text{ kips}$$

Both conditions of block shear strength are in excess of the design load (178.7 kips). Therefore, the connection detail, as shown in Figure 7-18, is satisfactory.

**Example 7-7**

Determine the number of $\frac{3}{4}$-in.-diameter high-strength bolts (A325) in standard holes required for the bracing connection shown in Figure 7-19. Note that the 60-kip and 80-kip forces are components of the 100-kip force. All structural steel is A36 ($F_u = 58$ ksi). Design a bearing-type connection with threads excluded from the shear plane (A325X). Assume that the double-angle tension member is adequate. Neglect block shear. Consider the following:

(a) Tension member-to-structural tee connection
(b) Structural tee-to-column connection

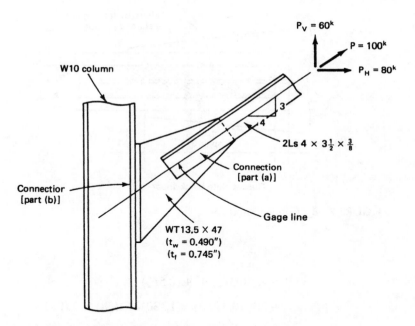

**FIGURE 7-19** Bracing connection design.

**Solution:**

(*Note:* All table references are to the ASDM, Part 4, unless noted otherwise.)

Establish the bolt pitch and edge distance. From the ASDS Table J3.5, the minimum edge distance is $1\frac{1}{4}$ in. Gage $g$ for a 4-in. angle leg is $2\frac{1}{2}$ in. (from the ASDM table Usual Gages for Angles). Use of this gage would provide an edge distance of $1\frac{1}{2}$ in. Use a $1\frac{1}{2}$-in. edge distance in the line of force, also.

(a)  *Connection of the angles to the structural tee* (bolts are in double shear; the bearing is on the web of the structural tee): The capacity of one bolt in double shear (Table I-D) is 26.5 kips. From Table I-E, the capacity of one bolt in bearing on the web of the structural tee ($t_w = 0.490$ in.) is

$$0.490(52.2) = 25.6 \text{ kips}$$

Bearing controls. The required number of bolts is

$$n = \frac{100}{25.6} = 3.91 \text{ bolts}$$

Use 4 bolts with a 3-in. pitch.

(b)  *Connection of the structural tee to the flange of the column:* Bolts are in combined shear and tension (A325X). The design of this connection is a trial-and-error procedure. Assume eight bolts, with four in each of two vertical rows, and assume that the tensile load passes through the center of gravity of the bolt group.

The allowable bolt shear stress is

$$F_v = 30 \text{ ksi (Table I-D)}$$

The average bolt shear stress (single shear) is

$$f_v = \frac{P_V}{nA_b} = \frac{60}{8(0.4418)} = 17.0 \text{ ksi} < 30 \text{ ksi} \qquad \textbf{O.K.}$$

The allowable tensile stress in bolts (ASDS, Table J3.3) is

$$F_t = \sqrt{(44)^2 - 2.15f_v^2}$$
$$= \sqrt{(44)^2 - 2.15(17.0)^2} = 36.3 \text{ ksi}$$

The actual tensile stress in bolts is

$$f_t = \frac{P_H}{nA_b} = \frac{80}{8(0.4418)} = 22.6 \text{ ksi} < 36.3 \text{ ksi} \qquad \textbf{O.K.}$$

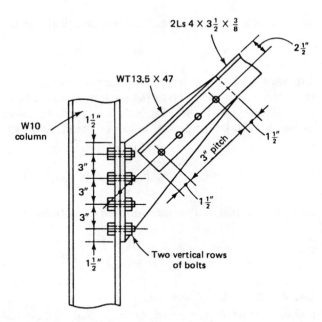

**FIGURE 7-20**   Design sketch.

**Use eight bolts,** as shown in Figure 7-20. The prying force on these bolts should also be considered (reference is made to the ASDM, Part 4, for example design calculations).

## 7-7

## FRAMED BEAM CONNECTIONS

Framed beam connections are probably the most commonly used type of beam-to-column and beam-to-girder connections. These connections are portrayed in Figure 7-3. This type of connection is used in Type 2 construction, where lateral loads (such as wind) are neglected in the design or where other systems in the structure will resist these loads. The framed connection is categorized as a simple beam connection. It is unrestrained and flexible, and it permits freedom of rotation at the supports. In reality a certain amount of capacity to resist moment is developed by these connections, but it is disregarded and the connections are usually designed to resist shear only.

The analysis of a framed beam connection using angles on the beam web follows. In addition to the possible failure modes previously discussed (bolt shear, bearing on the connected material, and block shear), a check will also be made on the *net shear area* of the connecting angles. In this situation an allowable shear stress of $0.3F_u$ is used, and the hole diameter will conservatively be taken as the fastener diameter plus $\frac{1}{8}$ in.

## Example 7-8

Compute the load-carrying capacity of the framed connection shown in Figure 7-21. The structural steel is A36 ($F_u = 58$ ksi). The bolts are $\frac{3}{4}$-in.-diameter A325 high-strength bolts in standard holes. Assume a slip-critical (class A) connection.

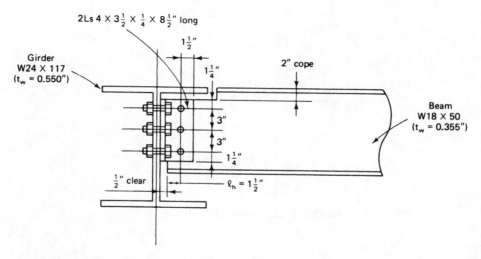

**FIGURE 7-21**   Framed beam connection analysis.

## Solution:

(*Note:* All table references are to the ASDM, Part 4.)

A.  Consider bolts through the web of the W18 × 50. These bolts are in double shear.

1.  The capacity of the connection based on *bolt shear* may be calculated to be 45.1 kips or it may be read directly from Table II-A.

2.  Consider the capacity of the connection based on bearing on the web of the beam. The bolt spacing is 3 in. = 4d (> 3d), and the edge distance parallel to the line of force is $1\frac{1}{4}$ in. = 1.67d (> 1.5d); therefore, Table I-E applies. Bearing may be critical on either (a) the beam web (3 bolts, $t_w = 0.355$ in.) or (b) the angles (three bolts, $t = \frac{1}{4} + \frac{1}{4} = \frac{1}{2}$ in.). The beam web is the more critical of the two:

$$0.355(52.2)(3) = 55.6 \text{ kips}$$

3.   The capacity of the connection based on *shear on the net area* of the connecting angles is calculated (using $F_v = 0.30\,F_u$) with reference to Figure 7-22:

$$\text{hole diameter} = \frac{3}{4} + \frac{1}{8} = 0.875 \text{ in.}$$

$$A_v = 2(0.25)(8.5 - 3(0.875)) = 2.94 \text{ in.}^2$$

$$\text{capacity} = A_v F_v = 2.94(0.30)(58) = 51.2 \text{ kips}$$

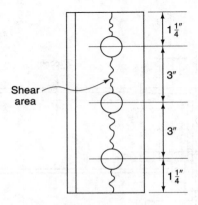

**FIGURE 7-22**   Shear on the net area.

Alternatively, Table II-C can be used. It is based on the hole diameter being taken as the fastener diameter plus $\frac{1}{16}$ in., however.

4.   The capacity of the connection based on *web tear-out* (block shear) is calculated as follows (refer to Figure 7-23):

$$\text{hole diameter} = \frac{3}{4} + \frac{1}{8} = 0.875 \text{ in.}$$

$$P_t = A_v(0.30)F_u + A_t(0.50)F_u$$

$$A_v = 0.355(7.25 - 2.5(0.875)) = 1.797 \text{ in.}^2$$

$$A_t = 0.355(1.5 - 0.5(0.875)) = 0.377 \text{ in.}^2$$

$$P_t = 1.797(0.30)(58) + 0.377(0.50)(58) = 42.2 \text{ kips}$$

The controlling value for this part of the connection is 42.2 kips.

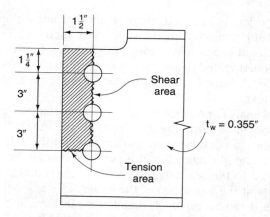

**FIGURE 7-23**   Web tear-out (block shear).

B.  Consider the bolts through the supporting member (W24 × 117). These bolts are in single shear. The following items should be checked:
1.  Capacity in single shear from Table II-A: 45.1 kips.
2.  Capacity in bearing on the 0.550-in.-thick web of the supporting member, based on 3-in. bolt spacing, from Table I-E:

$$0.550(52.2)(6) = 172.3 \text{ kips}$$

3.  Capacity in bearing on the $\frac{1}{4}$-in.-thick angles:

$$6(13.1) = 78.6 \text{ kips}$$

4.  Capacity based on shear on the net area of the connection angles is the same as for the other part of the connection: 52.7 kips.
5.  Web tear-out is not applicable for this part of the connection.

Hence the controlling value for the total connection is 42.2 kips (187.7 kN).

The design of this type of connection is usually made by the structural detailer as opposed to the structural designer. Hence the design of a typical connection is furnished in Chapter 11.

## 7-8

## UNSTIFFENED SEATED BEAM CONNECTIONS

An alternative type of "simple beam" connection that theoretically will behave as does the framed beam connection with respect to end rotation is the unstiffened seated beam connection. The seated connection has an angle under the beam that is usually shop-connected to the supporting member. The supporting member may be a column or a girder. In addition, there is another angle (generally on top of the beam) that is field-connected to the supporting member. The top angle may be placed at an optional location as shown in Figure 7-3 if space is restricted. The design of an unstiffened seated connection is generally accomplished by the use of tables in the ASDM. A sample design problem using these tables is furnished in Chapter 11 of this text.

Example 7-9 is an analysis problem of a given unstiffened seated beam connection that will demonstrate how the allowable loads of the ASDM tables have been determined. Reference 5 contains some background on the calculation method.

In the design of a seated connection, it is assumed that all the end reaction from the beam is delivered to the seat angle. The top angle is added to provide lateral support and stability at the top flange of the beam. This angle is assumed to carry no load. Therefore, it can be relatively small and can have as few as two bolts in each of its legs. The common size for a top angle would be L$3\frac{1}{2}$ × $3\frac{1}{2}$ × $\frac{1}{4}$ or L4 × 3 × $\frac{1}{4}$. The size is generally based on judgment and practical considerations.

There are two main considerations in the analysis or design of the seat angle. First, the seat, or the outstanding leg (OSL) must be sufficiently long to satisfy the web yielding and web crippling restrictions for the beam. The allowable load tables (Tables V-A and V-B) incorporate the web yielding consideration but leave the web crippling to be checked separately. Second, the seat angle must be thick enough so that it can support the beam reaction without exceeding the allowable bending stress. The OSL is analyzed and designed as a cantilever beam having a critical section for bending moment at the toe of the fillet of the seat angle. The location of the toe of the fillet is taken as $t_a + \frac{3}{8}$ in. from the back of the vertical leg of the angle, where $t_a$ is the thickness of the angle.

With reference to Figure 7-24c, the distance from the end of the beam to the face of the support, called the *setback*, is nominally $\frac{1}{2}$ in. To allow for beams that are cut short (but still within tolerance), a setback of $\frac{3}{4}$ in. is assumed for analysis and design purposes. The location of the beam reaction $R$, in this case, is assumed to be at the center of some bearing length $b$. Note that $b$ is not the actual bearing length of the beam on the seat angle but is the length of the bearing when the web is stressed to its maximum allowable stress value of $0.66F_y$ over a distance $b + 2.5k$ in the plane of the toe of the fillet of the beam.

The allowable loads of Tables V-A and V-B are based on fractional (rather than decimal) beam web thicknesses $t_w$ and approximate values for the distance $k$ from the outside of the flange to toe of the fillet in the beam. This means that the tabulated allowable loads are approximations for most beams, since the beams do not quite fit the table values of $t_w$ and $k$. Note also that the allowable loads are based on the seat angle having an outstanding leg of 4 in. Therefore, the actual bearing length $N$ is $3\frac{1}{4}$ in.

**Example 7-9** _____

Calculate the allowable load for the seat angle/beam combination shown in Figure 7-24. All steel is A36, and the bolts are $\frac{3}{4}$-in.-diameter A325N in standard holes.

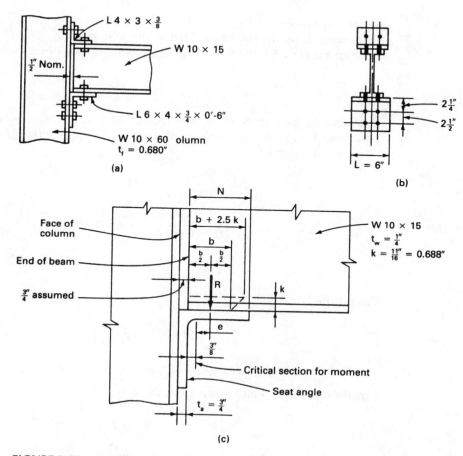

**FIGURE 7-24**  Unstiffened seated beam connection.

## Solution:

(*Note:* All units are kips and inches.)

Based on web yielding and substituting beam bearing length $b$ for $N$, Equation (K1-3) may be written as

$$\frac{R}{t_w(b + 2.5k)} = 0.66F_y \qquad \text{Eqn. 1}$$

Based on bending at the assumed critical section in the angle, where the maximum bending stress is the allowable bending stress $(0.75F_y)$ and $S$ is the section modulus (rectangular section) of the OSL,

$$\frac{M}{S} = \frac{Re}{\left(\dfrac{Lt_a^2}{6}\right)} = 0.75F_y \qquad \text{Eqn. 2}$$

where the moment $M$ is the product of the beam reaction and the eccentricity $e$ measured from the toe of the fillet in the seat angle.

From Figure 7-24, the following relationship can be observed:

$$\frac{b}{2} + \frac{3}{4} = t_a + \frac{3}{8} + e \qquad \text{Eqn. 3}$$

Note that we have three unknowns ($R$, $b$, and $e$) and three equations. Solve Equations 1 and 3 for $b$, equate, and write a new expression for $e$.

From Equation 1,

$$b = \frac{R}{t_w(0.66F_y)} - 2.5k$$

$$= \frac{R}{0.25(0.66)(36)} - 2.5(0.688)$$

$$= \frac{R}{5.94} - 1.72 \qquad \text{Eqn. 4}$$

From Equation 3,

$$b = 2\left(t_a + \frac{3}{8} + e - \frac{3}{4}\right)$$

$$b = 0.75 + 2e \qquad \text{Eqn. 5}$$

Equate Equations 4 and 5 and solve for $e$:

$$0.75 + 2e = \frac{R}{5.94} - 1.72$$

$$e = \frac{R}{11.88} - 1.235 \qquad \text{Eqn. 6}$$

Solve Equation 2 for $e$:

$$e = \frac{0.75F_y}{R}\left(\frac{Lt_a^2}{6}\right)$$

$$= \frac{0.75(36)}{R}\left(\frac{6(0.75)^2}{6}\right)$$

$$= \frac{15.19}{R} \qquad \text{Eqn. 7}$$

Equate Equations 6 and 7:

$$\frac{R}{11.88} - 1.235 = \frac{15.19}{R}$$

Multiplying by $11.88R$ results in a quadratic equation in $R$:

$$R^2 - 14.67R - 180.5 = 0$$

which can be solved for the positive root of $R$ to yield $R = 22.6$ (kips). Note that this differs slightly from the tabular value of 21.9 kips since the table is based on approximate values of $k$.

Next, check to see if $b$ falls within the acceptable range for our assumptions to be valid. Note that the bearing length $b$ may be computed using Equation 4:

$$b = \frac{R}{5.94} - 1.72 = \frac{22.6}{5.94} - 1.72 = 2.08 \text{ in.}$$

$$2.5k = 2.5(0.688) = 1.72 \text{ in.}$$

Therefore,

$$2.5k < b < N = 3.25 \qquad\qquad \textbf{O.K.}$$

If the resulting value of $b$ were less than $2.5k$, the assumption made for the location of the reaction would not be valid and modifications would have to be made to the previous equations. If the value of $b$ were greater than $N$ (where $N = 3.25$ in. for an OSL of 4 in.), web yielding would control, the allowable load could not exceed

$$R = 0.66F_y t_w(3.25 + 2.5k)$$

and the position of the resultant for the analysis would be assumed at $N/2$ or 1.625 in. from the end of the beam.

Since only web yielding has been considered (Equation 1), web crippling should be checked to establish the maximum allowable end reaction $R$. Using the constants from the ASDM Allowable Uniform Load tables,

$$R = R_3 + NR_4$$

$$= 11.7 + 3.25(2.76) = 20.7 \text{ kips}$$

Therefore, the allowable load based on web crippling is 20.7 kips.

Last, check the bolted connection of the vertical leg of the seat angle to the column flange.

*Shear:*   Bolts are in single shear with a capacity of 9.3 kips/bolt (Table I-D).

*Bearing:*   The column flange thickness is 0.680 in., which is less than the seat angle thickness of 0.75 in. Therefore, check bearing on the column flange. There are four bolts in two rows with two bolts per row in the line of force:

$$\text{edge distance} = 1\tfrac{1}{4} \text{ in.} = 1.67d > 1.5d$$

$$\text{bolt spacing} = 2\tfrac{1}{2} \text{ in.} = 3.3d > 3.0d$$

Therefore, use Table I-E. The allowable load in bearing is

$$\frac{0.680}{1.0}(52.2) = 35.5 \text{ kips/bolt}$$

Therefore, shear controls, and the capacity of the connection based on the bolts is

$$4(9.3) = 37.2 \text{ kips}$$

Therefore, web crippling is the controlling factor and the allowable load for the unstiffened seated beam-to-column connection is 20.7 kips (92.1 kN).

---

As previously mentioned (and as the reader can appreciate), analysis or design of this type of connection should be accomplished using the available tables. A complete design of an unstiffened seated connection is presented in Chapter 11.

When end reactions are relatively large, the required thickness of the seat angle becomes excessive and it becomes necessary to use a *stiffened* seated beam connection as shown in the diagrams accompanying Table VII of the ASDM, Part 4. This type of connection is similar to the unstiffened type, but in this type, stiffener angles are placed under the outstanding leg of the seat angle and fitted to bear on the underside, thereby relieving the bending in the horizontal leg. The design is usually accomplished through the use of the standard tables in the ASDM, Part 4, Table VII.

**7-9**

## END-PLATE SHEAR CONNECTIONS

A relatively recent type of flexible connection that has evolved is the end-plate shear connection. This type is generally economical, and it has been used successfully where beam end reactions are relatively light. It consists of a rectangular plate welded to the end of the *beam web* and generally bolted to the supporting member. Fabrication of this type of connection requires close control in cutting the beam to length as well in squaring the beam ends. For adequate end rotation capacity, the end-plate thickness should range from $\frac{1}{4}$ to $\frac{3}{8}$ in. inclusive, the plate depth is limited to the beam $T$ dimension, and only two vertical lines of bolts are permitted. The gage, $g$, between lines of bolts should be between $3\frac{1}{2}$ and $5\frac{1}{2}$ in., and the edge distance should be $1\frac{1}{4}$ in.

The design of this type of connection is usually accomplished through the use of the standard tables in the ASDM, Part 4, Table IX. At this point, we assume that the end plate is adequately welded to the end of the beam. The welded connection is considered in Chapter 8. The remaining problem is to design the bolted connection to the supporting member. The design assumes no eccentricity and considers only shear and bearing.

## Example 7-10

Design the end-plate shear connection as shown in Table IX, Part 4 of the ASDM, for a W18 × 60 beam framing to the flange of a W8 × 31 column. All structural steel is A36 ($F_u = 58$ ksi). The end reaction is 40 kips. Use $\frac{3}{4}$-in.-diameter A325N bolts (threads in the shear plane) in standard holes. Neglect welding of the plate to the end of beam. Use $\frac{5}{16}$-in.-thick plate and assume that the welding is adequate.

### Solution:

(*Note:* All table references are to the ASDM, Part 4.)

From Table I-D, the capacity of one bolt in single shear is 9.3 kips. For bearing capacity, assume an edge distance of $1\frac{1}{4}$ in. and a 3-in. bolt spacing. The $\frac{5}{16}$-in. plate will control since it is thinner than the 0.435-in. flange thickness of the column. Check applicability of Table I-E. The edge distance is

$$\frac{1.25}{0.75} = 1.67d > 1.5d \qquad\qquad \textbf{O.K.}$$

The bolt spacing is

$$\frac{3.0}{0.75} = 4.0d > 3.0d \qquad\qquad \textbf{O.K.}$$

Therefore, using Table I-E, the allowable bearing load is

$$P = 16.3 \text{ kips/bolt}$$

Therefore, shear controls, and the required number of bolts is

$$n = \frac{40.0}{9.3} = 4.30 \text{ bolts} \qquad (\text{use six bolts})$$

The *required* plate length (vertically) is

$$2(3) + 2(1.25) = 8.5 \text{ in.}$$

The *maximum allowable* plate length is the beam $T$ dimension, which is $15\frac{1}{2}$ in. from ASDM, Part 1. Therefore, for this end-plate connection, use a plate

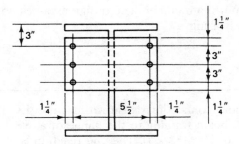

**FIGURE 7-25**   End-plate connection.

$8 \times 8\frac{1}{2} \times \frac{5}{16}$ with six $\frac{3}{4}$-in.-diameter A325N high-strength bolts. The connection is shown in Figure 7-25.

## 7-10

## SEMIRIGID CONNECTIONS

The framed beam, seated beam, and end-plate shear connections discussed previously are designed primarily to transfer shear only. These connections are used in Type 2 construction. In essence, these are the connections that permit end rotation, thereby making the bending members behave as simply supported members. When members are to be continuous or when resistance to wind or other lateral forces is to be provided by the joints, connections that offer predictable moment resistance must be used.

Connections for Type 1 and Type 3 construction are categorized as rigid and semirigid connections, respectively. Various types are shown in Figure 7-3. One of the simplest semirigid connection types is shown in Figure 7-3e. The web angles are designed to resist the shear as in the framed beam connection. The top and bottom angles (sometimes called *clip angles*) are designed to resist the moment. A modification of this connection is the use of a pair of structural tees instead of the top and bottom angles. The tendency of the end of the beam to rotate is resisted by a couple produced by a horizontal tensile force on the top flange and a horizontal compressive force on the bottom flange. The magnitude of these horizontal forces may be determined by dividing the end moment by the nominal depth of the beam

$$T = C = \frac{M}{d}$$

Assuming the use of angles on top and bottom of the beam, the horizontal force can be developed by transferring the force in the beam flange to the angles by single shear in the connecting bolts. This force is then transmitted by bending in the angles to the column flange by tension in the top connecting bolts and bearing between the column flange and the bottom clip angle. The design for the end moment involves determining how many bolts are required through top and bottom angles as well as the size and thickness of the angles themselves. In the classical approach to this design problem, the angle thickness may be determined by assuming some bending behavior of the angle and applying the beam flexure formula. The following example demonstrates this approach. Example 7-12 demonstrates the current AISC approach.

**Example 7-11** _____

Design a connection of the type shown in Figure 7-26 to resist a moment of 50 ft-kips and a shear (reaction) of 35 kips. Use $\frac{3}{4}$-in.-diameter A325N high-strength bolts in standard holes. All structural steel is A36 ($F_u = 58$ ksi).

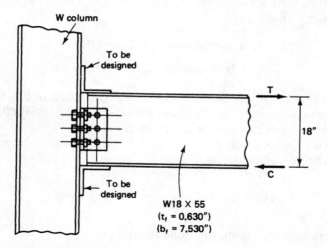

**FIGURE 7-26**  Semirigid connection.

**Solution:**

(*Note:* All table references are to the ASDM, Part 4.)

1. The connection on the beam web to resist the shear has been designed using the ASDM and consists of 2L5 × $3\frac{1}{2}$ × $\frac{5}{16}$ × $8\frac{1}{2}$ in. with three rows of bolts.

2. Consider the bolts through the *beam* flanges and the top and bottom angles. The capacity of one bolt in single shear is 9.3 kips, from Table I-D. The capacity of one bolt in bearing is based on bolt spacing and edge distance. Assume the angle-to-flange connection as shown in Figure

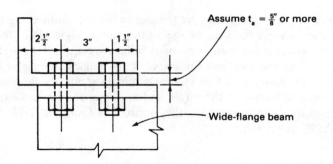

**FIGURE 7-27**    Angle-to-beam flange connection.

7-27. Assume a horizontal leg of 7 in. with two bolts per row in the line of force. Thus

$$\text{edge distance} = 1\tfrac{1}{2} \text{ in.} = 2.0d > 1.5d$$

$$\text{bolt spacing} = 3 \text{ in.} = 4.0d > 3.0d$$

Therefore, $F_p = 1.2F_u$. Assuming an angle thickness of $\tfrac{5}{8}$ in. or more, the minimum bearing capacity (from Table I-E) is 32.6 kips/bolt. Therefore, shear controls this part of the connection.

The force in each flange (tension in the top and compression in the bottom) is calculated as

$$T = C = \frac{50(12)}{18} = 33.3 \text{ kips}$$

Therefore, the number of bolts required is

$$n = \frac{33.3}{9.3} = 3.6 \text{ bolts} \qquad \text{(use four bolts)}$$

3.  Consider the bolts through the *column* flanges and the top and bottom angles. The capacity of one bolt in tension is 19.4 kips, from Table I-A. Conservatively use the applied force $T = C = 33.3$ kips. The force is actually less than that because the distance between the bolts is in excess of 18 in. The number of bolts required is

$$n = \frac{33.3}{19.4} = 1.72 \text{ bolts} \qquad \text{(use two bolts)}$$

4.  Using an L7 × 4 to accommodate the required bolts and assuming a $\tfrac{3}{4}$-in. angle thickness, the design will be accomplished based on an assumed

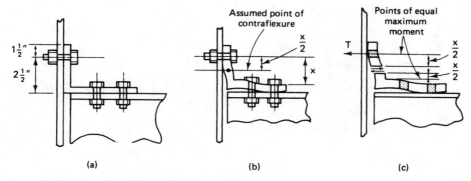

**FIGURE 7-28**    Top-angle analysis.

point of contraflexure located as shown in Figure 7-28b. This point is a point of zero moment, a point at which only shear exists. Therefore, the calculation of moment at the assumed points of equal maximum moment, shown in Figure 7-28c, becomes possible. Assume that the angle is 8 in. long (to match approximately the width of the beam flange), and compute the thickness required based on an allowable bending stress $F_b$ of $0.75F_y$.

The distance from the bolts in tension to the top of the horizontal leg of the angle is (see Figure 7-28b)

$$x = 2.5 - 0.75 = 1.75 \text{ in.}$$

Determine the tensile force using the actual vertical distance between forces $T$ and $C$ of 23 in.:

$$T = C = \frac{50(12)}{23} = 26.1 \text{ kips}$$

Therefore, the moment at the assumed critical section is

$$M = 26.1\left(\frac{1.75}{2}\right) = 22.8 \text{ in.-kips}$$

Solving for the required thickness, recall that for a rectangular shape,

$$S = \frac{bt^2}{6}$$

and that

$$\text{required } S = \frac{M}{F_b}$$

Therefore,

$$\text{required } t = \sqrt{\frac{6(M)}{F_b(b)}}$$

$$= \sqrt{\frac{6(22.8)}{0.75(36)(8)}} = 0.80 \text{ in.} > 0.75 \text{ in.} \qquad \textbf{N.G.}$$

Recalculate, assuming an L8 × 4 × 1 × 8 in.:

$$x = 2.5 - 1 = 1.5 \text{ in.}$$

$$T = C = 26.1 \text{ kips (unchanged)}$$

$$M = 26.1\left(\frac{1.50}{2}\right) = 19.6 \text{ in.-kips}$$

$$\text{required } t = \sqrt{\frac{6(19.6)}{0.75(36)(8)}} = 0.74 \text{ in.}$$

Therefore, use top and bottom angles L8 × 4 × 1 × 8 in.

---

The preceding example does not take into account *prying action*. The actual distribution of stress in the clip angle is very complex. The stiffness of the clip angle, rather than its bending strength, is usually more critical since large deformation would be detrimental. The ASDM provides an empirical design method in Part 4, entitled Hanger Type Connections. Reference 6 provides some background. The following example demonstrates the method. Reference should be made to the ASDM.

## Example 7-12

Rework Example 7-11 using the ASDM empirical design method.

### Solution:

(*Note:* The steps of the ASDM solution procedures, Method 1 (Design), are followed.)

Given from Example 7-11:

$$B = \text{allowable tension per bolt} = 19.4 \text{ kips}$$

$$F_y = 36 \text{ ksi}$$

$$T = \text{applied tension (total)} = 26.1 \text{ kips}$$

$$p = \text{length of angle tributary to each bolt}$$

$$p = \frac{8}{2} = 4 \text{ in.}$$

1. The number of bolts required is

$$n = \frac{26.1}{19.4} = 1.35 \qquad \text{(use 2 bolts)}$$

Therefore,

$$T = \frac{26.1}{2} = 13.1 \text{ kips/bolt} < 19.4 \text{ kips/bolt}$$

2. From the preliminary selection table for hanger type connections in the ASDM, Part 4, based on a load per inch of

$$\frac{26.1}{8} = 3.26 \text{ kips/inch}$$

and an estimated required $b$ of $1\frac{3}{4}$ in. (see ASDM, Part 4, Assembling Clearances for Threaded Fasteners), try a preliminary angle thickness of $\frac{9}{16}$ in., tentatively select an L8 × 4 × $\frac{9}{16}$, use a $2\frac{1}{2}$-in. gage on the vertical leg. With reference to Figure 7-29 and the expressions in the ASDM procedure, the following constants are applicable:

$$t = 0.563 \text{ in.}$$

$$b = 2.5 - 0.563 = 1.937 \text{ in.}$$

$$a = 4.0 - 2.5 = 1.5 \text{ in.}$$

$$b' = b - \frac{d}{2} = 1.937 - 0.375 = 1.562 \text{ in.}$$

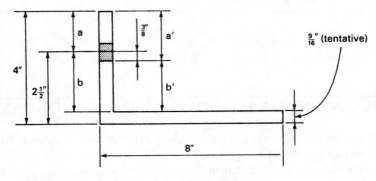

**FIGURE 7-29**   Clip angle design.

$$a' = a + \frac{d}{2} = 1.5 + 0.375 = 1.875 \text{ in.}$$

$$d' = \frac{13}{16} = 0.813 \text{ in.}$$

$$\rho = \frac{b'}{a'} = \frac{1.562}{1.875} = 0.833$$

$$\delta = 1 - \frac{d'}{p} = 1 - \frac{0.813}{4} = 0.797$$

3.
$$\beta = \frac{1}{\rho}\left(\frac{B}{T} - 1\right) = \frac{1}{0.833}\left(\frac{19.4}{13.1} - 1\right) = 0.577 < 1.0$$

Therefore, $\alpha'$ is taken as the lesser of

$$\frac{1}{\delta}\left(\frac{\beta}{1-\beta}\right) = \frac{1}{0.797}\left(\frac{0.577}{1-0.577}\right) = 1.712$$

or 1.0.

Therefore, $\alpha' = 1.0$.

4. The required thickness is then calculated from

$$\text{required } t = \sqrt{\frac{8Tb'}{pF_y(1 + \delta\alpha')}}$$

$$= \sqrt{\frac{8(13.1)(1.562)}{4(36)[1 + 0.797(1.0)]}} = 0.795 \text{ in.}$$

Since 0.795 in. $> \frac{9}{16}$ in., a thicker angle must be chosen and steps 2, 3, and 4 repeated. This iterative process will show that an angle thickness of $\frac{3}{4}$ in. will be satisfactory.

**7-11**

## ECCENTRICALLY LOADED BOLTED CONNECTIONS

When bolt groups are loaded by some external load that does not act through the center of gravity of the group, the load is said to be *eccentric*. It will tend to cause a relative rotation and translation of the connected parts, and the individual bolts will have unequal loads induced in them.

If the eccentrically applied load lies in the plane of the connection as shown in Figure 7-30 and elastic behavior is assumed, the bolt group may be analyzed by resolving the eccentric load $P$ into a concentric load acting through the centroid of the bolt group and a torsional moment $M$ (where $M = Pe$). The moment acts with respect to the centroid of the bolt group as a center of rotation. Hence, the forces acting on the bolts will be made up of two components: $Q_v$ due to the axial effect of the eccentric load and $Q_m$ due to the torsional moment effect, as shown in Figure 7-31.

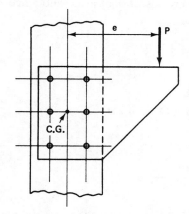

**FIGURE 7-30**   Eccentrically loaded connection.

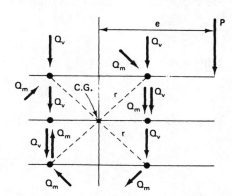

**FIGURE 7-31**   Bolt group forces.

$Q_v$ will be the same for all the bolts and may be taken as the load $P$ divided by the number of bolts ($n$):

$$Q_v = \frac{P}{n}$$

$Q_m$ will vary with the distance $r$ from the center of gravity of the bolt group to the bolt and will act in a direction normal to a line from the bolt to the center of gravity. Therefore, the connection must be designed so that the resultant of these two components acting on any bolt does not exceed the maximum permissible bolt capacity as determined by shear or bearing.

The torsional load $Q_m$ may be determined by applying the classic torsional stress formula for circular members to the bolt group:

$$f_v = \frac{Mr}{J}$$

where

$f_v$ = shear stress in any bolt

$M$ = torsional moment $(Pe)$

$r$ = radial distance from center of gravity of bolt group to any bolt

$J$ = polar moment of inertia = $\Sigma Ar^2$

In the expression for polar moment of inertia (taken from any engineering mechanics text), $A$ represents the cross-sectional area of one bolt. Since all bolts in any given connection will have the same cross-sectional area, and since $r^2$ may be expressed as $x$ and $y$ coordinates,

$$J = \Sigma A(x^2 + y^2)$$

This can be written as

$$J = A\Sigma(x^2 + y^2) = A(\Sigma x^2 + \Sigma y^2)$$

The torsional stress formula then becomes

$$f_v = \frac{Mr}{A(\Sigma x^2 + \Sigma y^2)}$$

If we multiply both sides of the equation by $A$, we obtain the torsional load $Q_m$ on any bolt:

$$f_v A = Q_m = \frac{Mr}{\Sigma x^2 + \Sigma y^2}$$

If we assume $r$ to be a *unit distance* from the center of gravity, the expression becomes

$$Q_{mu} = \frac{M}{\Sigma x^2 + \Sigma y^2}$$

This represents a force or load acting at a unit distance from the center of gravity. By knowing this force we can compute any force acting on any bolt normal to a radial line from the center of gravity to the bolt by multiplying $Q_{mu}$ by the radial distance $r$ to that bolt.

It is generally more convenient to resolve the $Q_m$ force into vertical and horizontal components and then vectorially add the $Q_v$ force to obtain the resultant force on the bolt. Usually, the bolt most remote from the group center of gravity and on the load side will be subjected to the critical (greatest) load and will determine the adequacy of the connection.

**Example 7-13** _____

For the eccentrically loaded bolted connection shown in Figure 7-32, determine the force acting on the most critical bolt. Assume $\frac{3}{4}$-in.-diameter A325N bolts in standard holes. The applied load is 10,000 lb.

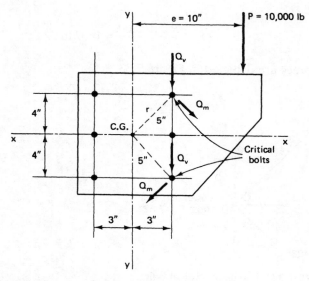

**FIGURE 7-32**    Eccentrically loaded bracket.

**Solution:**

1.    The torsional moment is

$$M = Pe = 10,000(10) = 100,000 \text{ in.-lb}$$

2.    The polar moment of inertia (divided by $A$) is

$$\Sigma x^2 = 6(3)^2 = 54 \text{ in.}^2$$

$$\Sigma y^2 = 4(4)^2 = 64 \text{ in.}^2$$

$$\frac{J}{A} = \Sigma x^2 + \Sigma y^2 = 118 \text{ in.}^2$$

3.    The torsional load ($Q_{mu}$) acting at a unit distance from the center of gravity of the bolt group is

$$Q_{mu} = \frac{M}{\Sigma x^2 + \Sigma y^2} = \frac{100,000}{118} = 847 \text{ lb/in.}$$

The torsional load on the critical bolt is

$$Q_m = 847(5) = 4235 \text{ lb}$$

4. The horizontal component of $Q_m = 4235(4/5) = 3388$ lb; the vertical component of $Q_m = 4235(3/5) = 2541$ lb.

5. The force on the bolt due to the 10,000-lb load applied at the center of gravity is

$$Q_v = \frac{P}{n} = \frac{10,000}{6} = 1667 \text{ lb}$$

The forces are depicted in Figure 7-33.

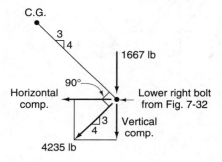

**FIGURE 7-33**  Forces on critical bolt.

6. The resultant force $R$ on the critical bolt is determined with reference to Figure 7-34:

$$R^2 = (1667 + 2541)^2 + 3388^2$$

$$R = 5402 \text{ lb } (24.0 \text{ kN})$$

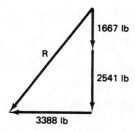

**FIGURE 7-34**  Vector diagram.

**Example 7-14**

Find the maximum load $P$ that can be supported by the bracket shown in Figure 7-35. The column and bracket are A36 steel. Use $\frac{7}{8}$-in.-diameter A325SC (class A) high-strength bolts in standard holes. Assume that the column flange and bracket are thick enough that single shear in the bolts will control.

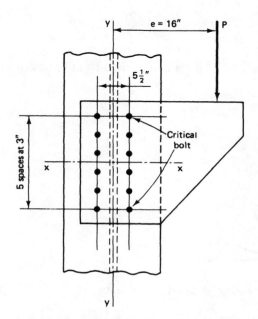

**FIGURE 7-35**    Bracket analysis.

**Solution:**

1.    The torsional moment is

$$M = Pe = 16P \text{ in.-lb} \quad \text{(when } P \text{ is in pounds)}$$

2.    The polar moment of inertia (divided by $A$) is

$$\Sigma x^2 = 12(2.75)^2 = 90.75 \text{ in.}^2$$
$$\Sigma y^2 = 4(1.5)^2 + 4(4.5)^2 + 4(7.5)^2 = 315 \text{ in.}^2$$
$$\frac{J}{A} = \Sigma x^2 + \Sigma y^2 = 405.8 \text{ in.}^2$$

3.   The torsional load $(Q_{mu})$ at unit distance from the center of gravity of the bolt group is

$$Q_{mu} = \frac{M}{\Sigma x^2 + \Sigma y^2} = \frac{16P}{405.8}$$

The torsional load on the critical bolt is

$$Q_m = \frac{16P(7.99)}{405.8} = 0.315P$$

where the radius of 7.99 in. is determined from Figure 7-36.

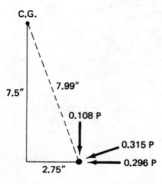

**FIGURE 7-36**   Torsional loads on critical bolts.

4.   The horizontal component of $Q_m$ is

$$\frac{7.5}{7.99}(0.315P) = 0.296P$$

and the vertical component of $Q_m$ is

$$\frac{2.75}{7.99}(0.315P) = 0.108P$$

5.   The force on the bolt due to $P$ load applied at the center of gravity is

$$Q_v = \frac{P}{n} = \frac{P}{12} = 0.083P$$

6.   The resultant force $R$ on the critical bolt cannot exceed the capacity of one bolt in single shear: 10.2 kips (ASDM, Part 4, Table I-D). The vector diagram for the determination of $R$ is shown in Figure 7-37. Thus

$$R^2 = (0.191P)^2 + (0.296P)^2$$

$$R = 0.352P$$

from which

$$0.352P = 10,200 \text{ lb}$$

$$P = 28,980 \text{ lb } (128.9 \text{ kN})$$

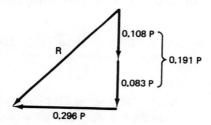

**FIGURE 7-37**   Vector diagram.

The two examples for eccentrically loaded bolt groups are based on a so-called elastic method that is currently designated *Alternate Method 1—Elastic*. This method, although providing a simplified and conservative solution, does not result in a design having a consistent factor of safety.

As a result, the ultimate-strength method was developed and is the current method used in the ASDM (9th edition) tables of Eccentric Loads on Fastener Groups. Discussion of this method is provided in Part 4 of the ASDM. Since these tables are intended for eccentric loads that are vertical, they are not applicable directly for the case where eccentric loads are inclined at some angle from the vertical. For this condition, *Alternate Method 1—Elastic* could be used. In addition, a new method designated *Alternate Method 2* has been developed to overcome the tabular limitations. Discussion of the new method is also provided in Part 4 of the ASDM. As a means of comparison, Example 7-14 will be recalculated using the ultimate-strength method.

**Example 7-15** _____

Rework Example 7-14 using an ultimate-strength method.

**Solution:**

Refer to the discussion of the method in the ASDM, Part 4, and use Tables XI–XVIII. When entering Table XIII, use the following values:

$$n = 6 \text{ (number of bolts in one vertical row)}$$

$$r_v = 10.2 \text{ kips (permissible load on one bolt in single shear)}$$

$$l = e = 16 \text{ in. (actual eccentricity)}$$

$$D = 5\tfrac{1}{2} \text{ in. (distance between gage lines)}$$

$$b = 3 \text{ in. (bolt pitch)}$$

From Table XIII, the coefficient $C$ is 3.55. Therefore, the allowable load is calculated as

$$P = Cr_v = 3.55(10{,}200) = 36{,}210 \text{ lb (161.1 kN)}$$

---

These methods are discussed further in the ASDM on the pages preceding the tables. The use of the tables expedites and simplifies design and analysis of eccentrically loaded bolted connections. In fact, with the use of tables, the process of design is probably best accomplished by assuming a number and arrangement of bolts and then checking the group capacity. Revising and rechecking may also be quickly accomplished.

When an eccentrically applied load lies outside the plane of the connection, as shown in Figure 7-38, the eccentric load tends to separate the bracket from the column flange at the top and press the bracket against the flange at the bottom.

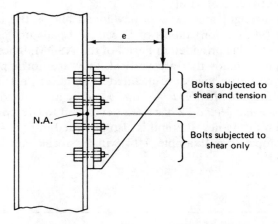

**FIGURE 7-38** Eccentrically loaded connection.

Therefore, the bolts are subjected to a varying and decreasing tensile force from the top down to the neutral axis. The bolt is also placed in shear by the vertical effect of the eccentric load.

Simplifying assumptions have been made with respect to analyses of this type of connection and are based on an assumption that the tension induced in any bolt as a result of the applied eccentric load will not be in excess of the initial tension induced by tightening.

An assumption is made that the neutral axis lies at the middepth or center of gravity of the bolt group. Further, it is assumed that the tensile stress distribution above the neutral axis varies linearly from zero at the neutral axis to a maximum at the bolt farthest from the neutral axis and may be obtained from the flexure formula:

$$f_t = \frac{Mc}{I} = \frac{Pec}{I}$$

where

$P$ = applied eccentric load

$e$ = eccentricity of load from face of column

$c$ = distance from the neutral axis to the center of the most distant bolt

$I$ = moment of inertia of the bolt areas

In addition, the shear stress is

$$f_v = \frac{P}{nA_b}$$

where $n$ is the total number of bolts, and $A_b$ is the cross-sectional area of one bolt.

As discussed previously, the allowable tensile stress must be modified by an interaction formula that is based on an actual bolt shear stress $f_v$, which in turn cannot be in excess of an allowable shear stress (see the ASDS, Table J3.3, and the SSJ, Table 2). In slip-critical connections the allowable shear stress must be modified by an expression based on an actual tensile stress, which in turn cannot exceed the allowable tensile stress as shown in the ASDS, Table J3.2, or the ASDM, Part 4, Table I-A. The modification factor is from the ASDS, Section J3.6.

**Example 7-16** _____

For the connection shown in Figure 7-38, the bracket is a structural tee, $P$ = 25 kips, and $e$ = 12 in. Determine if the connection is satisfactory assuming $\frac{3}{4}$-in.-diameter A325N high-strength bolts in two vertical rows. The bolts are in standard holes spaced 3 in. vertically. Assume that the column and bracket are adequate.

**Solution:**

1.   The moment of inertia of the bolts about the neutral axis is

$$I = \Sigma(A_b d^2)$$
$$I = 0.4418[4(1.5)^2 + 4(4.5)^2] = 39.8 \text{ in.}^4$$

2.   The actual tensile stress in the top bolt is

$$f_t = \frac{Pec}{I} = \frac{25[12(4.5)]}{39.8} = 33.9 \text{ ksi}$$

3.   The allowable shear stress $F_v$ is 21 ksi (from the ASDS, Table J3.2). The actual shear stress in the bolts is

$$f_v = \frac{25}{8(0.4418)} = 7.07 \text{ ksi} < 21 \text{ ksi} \qquad \textbf{O.K.}$$

4.   The allowable tensile stress (from the ASDS, Table J3.3) is

$$F_t = \sqrt{(44)^2 - 4.39 f_v^2}$$
$$= \sqrt{(44)^2 - 4.39(7.07)^2} = 41.4 \text{ ksi}$$

Since the actual tensile stress is less than the allowable tensile stress, the connection is satisfactory.

## REFERENCES

[1]   *Standard Specification for High-Strength Bolts for Structural Steel Joints,* ASTM A325, The American Society for Testing and Materials, 1916 Race Street, Philadelphia, PA 19103.

[2]   *Standard Specification for Heat-Treated Steel Structural Bolts, 150 ksi Min. Tensile Strength,* ASTM A490, The American Society for Testing and Materials, 1916 Race Street, Philadelphia, PA 19103.

[3]   *High Strength Bolting for Structural Joints,* Bethlehem Steel Corp., Bethlehem, PA 18016.

[4]   W. A. Thornton, "Prying Action—A General Treatment," *AISC Engineering Journal,* Vol. 22, No. 2, 1985, pp. 67–75.

[5]   J. H. Garrett, Jr., and R. L. Brockenbrough, "Design Loads For Seated-Beam Connections in LRFD," *AISC Engineering Journal,* Vol. 23, No. 2, 1986, pp. 84–88.

[6]   A. Astaneh, "Procedure For Design and Analysis of Hanger-Type Connections," *AISC Engineering Journal,* Vol. 22, No. 2, 1985, pp. 63–66.

## PROBLEMS

*Note: In the following problems, use $F_u = 58$ ksi for A36 steel and assume a class A contact surface for all slip-critical connections.*

**7-1.**   Compute the tensile capacity $P_t$ for the single-shear lap connection shown. The plates are A36 steel, the high-strength bolts are $\frac{3}{4}$-in.-diameter A325 in standard holes. Assume that the connection is

**(a)**   A325SC.

**(b)**   A325X.

**(c)**   A325N.

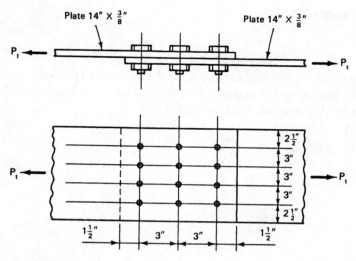

**PROBLEM 7-1**

**7-2.** Compute the tensile capacity $P_t$ for the connection shown. The gusset plate and the angle are A36 steel. The high-strength bolts are $\frac{3}{4}$-in.-diameter A325 in standard holes. Assume that the connection is

   **(a)** A325X.

   **(b)** A325N.

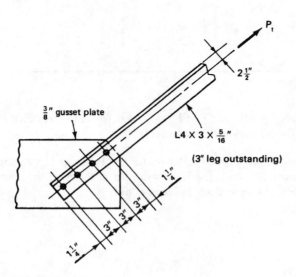

**PROBLEM 7-2**

**7-3.** Find the tensile capacity $P_t$ for the double-shear butt connection shown. Bolts are $\frac{3}{4}$-in.-diameter high-strength bolts in standard holes. All structural steel is A36. Assume that the connection is

   **(a)** A325SC.

   **(b)** A325N.

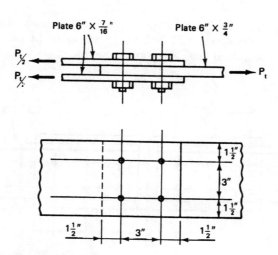

**PROBLEM 7-3**

**7-4.** Rework Problem 7-3 using $\frac{7}{8}$-in.-diameter bolts and changing the $\frac{7}{16}$-in.-thick plates to $\frac{5}{16}$-in.-thick plates.

**7-5.** Compute the tensile capacity $P_t$ for the double-shear butt connection shown. The plates are A36 steel. The $\frac{3}{4}$-in.-diameter high-strength bolts are in standard holes. Assume that the connection is

**(a)** A325SC.

**(b)** A325X.

**(c)** A325N.

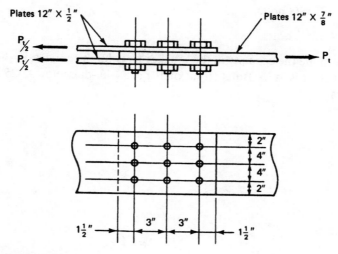

**PROBLEM 7-5**

**7-6.** Compute the tensile capacity $P_t$ for the double-angle tension member shown. Fasteners are $\frac{7}{8}$-in.-diameter A325X bolts in standard holes. All structural steel is A36. Use standard gage lines in the angles.

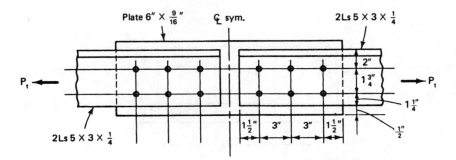

**PROBLEM 7-6**

**7-7.** A tension member made up of a pair of angles is connected to a column with eight $\frac{7}{8}$-in.-diameter A325 high-strength bolts in standard holes similar to that shown in Figure 7-14. All structural steel is A36. The angles are oriented 30° with the horizontal and are subjected to an axial load of 120 kips. Assume that the connection between the angles and the tee is satisfactory and that the angles are satisfactory for the applied load. Determine whether the connection to the column is satisfactory. Consider the following high-strength bolts:

**(a)** A325SC bolts.

**(b)** A325X bolts.

**7-8.** Determine the capacity of the bracket-to-column connection shown. All structural steel is A36. Bolts (total of 10) are $\frac{7}{8}$-in.-diameter A325SC in standard holes.

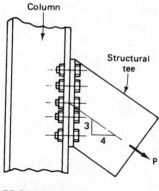

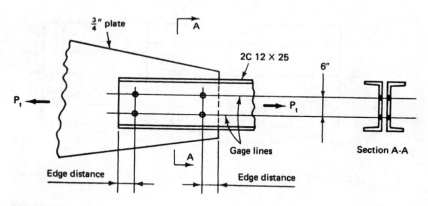

**PROBLEM 7-8**

**7-9.** A truss member made of two C12 $\times$ 25 channels is connected to a $\frac{3}{4}$-in.-thick gusset plate, as shown. Structural steel is A36. Determine the number of $\frac{7}{8}$-in.-diameter A490SC high-strength bolts in standard holes required to develop the full tensile capacity of the channels. Assume that the edge distance is $1\frac{1}{2}$ in., the pitch is 3 in., and the gage is 6 in.

**PROBLEM 7-9**

**7-10.** A single-shear lap connection is made of two $14 \times \frac{3}{4}$ in. A36 plates. The plates are to be connected with $\frac{3}{4}$-in.-diameter A490 high-strength bolts in standard holes. Determine the number of bolts required, the pitch, and the edge distance to transfer a tensile load of 150 kips (assume three gage lines) if the connection is designed as

**(a)** A490SC.

**(b)** A490X.

**(c)** A490N.

**7-11.** Determine the number of $\frac{3}{4}$-in.-diameter A325SC high-strength bolts in standard holes required to develop the full tensile capacity of the member that contains the butt connection shown. Determine the edge distance to be used. All structural steel is A36.

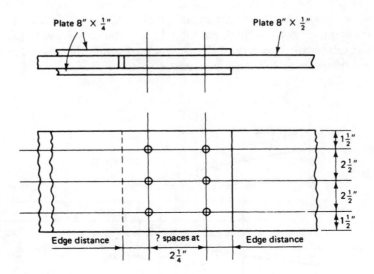

**PROBLEM 7-11**

**7-12.** A roof truss member is made of 2L6 $\times$ 4 $\times$ $\frac{1}{2}$ (long legs back to back) of A36 steel and is subjected to an axial load of 180 kips. If the angles frame into a $\frac{3}{8}$-in.-thick gusset plate, determine the number of $\frac{7}{8}$-in.-diameter A325 high-strength bolts in standard holes required for its end connection using a slip-critical connection (A325SC). Assuming two gage lines, establish the pitch and the edge distance.

**7-13.** A truss tension member is made up of two C10 $\times$ 30 channels laced together and connected to two $\frac{3}{8}$-in.-thick gusset plates, as shown. The channels and gusset plates are A36 steel. How many $\frac{7}{8}$-in.-diameter A325X high-strength bolts in standard holes are required to develop the full tensile capacity of the channels? Establish the pitch and the edge distance.

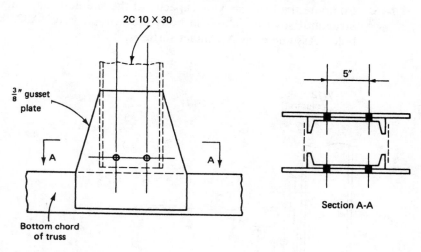

**PROBLEM 7-13**

**7-14.** The single-angle tension member shown, part of a bracing system, is to support a design tensile load of 36 kips. The $\frac{3}{8}$-in.-thick gusset plate will be shop-welded to the column and field-bolted to the angle. Structural steel is A36.

**(a)** Using $\frac{3}{4}$-in.-diameter A325N bolts in standard holes, design the connection. Include the number of bolts, gage line location, edge distances, and bolt spacing.

**(b)** Design the gusset plate. (Neglect weld design.)

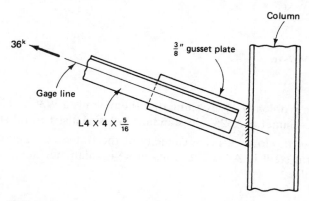

**PROBLEM 7-14**

**7-15.** In Problem 7-8, assume that the applied tensile load is 80 kips. Determine the number of $\frac{3}{4}$-in.-diameter A325SC bolts required for the bracket-to-column connection.

**7-16.** Compute the load-carrying capacity of the simple-beam connection shown. Structural steel is A36. The bolts are $\frac{7}{8}$-in.-diameter A325SC in standard holes. Assume class A contact surfaces.

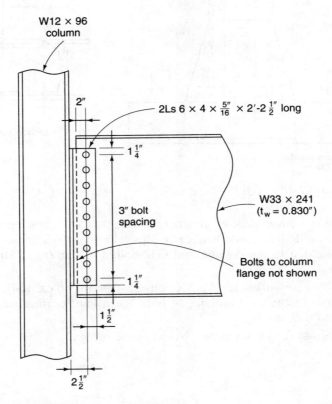

**PROBLEM 7-16**

**7-17.** Rework Problem 7-16. Assume that the beam is a W36 × 170 and the bolts are $\frac{7}{8}$-in.-diameter A325X in standard holes. All structural steel is A36.

**7-18.** Compute the load-carrying capacity of the framed connection shown. The structural steel is A36. The bolts are $\frac{3}{4}$-in.-diameter A325N bolts in standard holes.

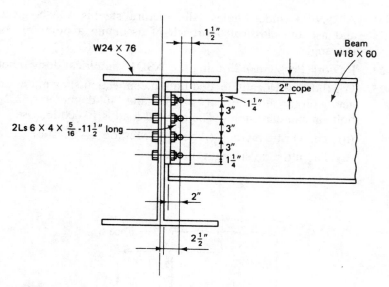

**PROBLEM 7-18**

**7-19.** Refer to Figure 7-24 and assume the following: the beam is a W21 × 57, the column is a W12 × 40, and the seat angle is an L7 × 4 × $\frac{3}{4}$ × 8 in. (4-in. leg is horizontal) connected to the column web with four $\frac{7}{8}$-in.-diameter A490X bolts. Assume an actual setback of $\frac{3}{4}$ in. Only one beam frames to the column at this point. All structural steel is A36. Determine the maximum beam reaction that can be transmitted by this connection.

**7-20.** An unstiffened seated connection is to be used for the connection of a W12 × 40 beam to the flange of a W12 × 65 column. The seat angle is an L7 × 4 × $\frac{3}{4}$ × 0′–8. The bolt pattern is Type B, and the bolts are $\frac{7}{8}$-in.-diameter A325N in standard holes. All steel is A36. Sketch the connection and determine the allowable load. (*Hint:* Make use of any appropriate tables.)

**7-21.** Rework Problem 7-20 with a W16 × 89 beam and $\frac{3}{4}$-in.-diameter A325SC bolts.

**7-22.** Design an end-plate shear connection for a W14 × 38 beam framing to the flange of a W8 × 35 column. All structural steel is A36. The end reaction is 38 kips. Use $\frac{3}{4}$-in.-diameter A325N high-strength bolts in standard holes. Neglect the welding of the plate to the end of the beam. Use $\frac{5}{16}$-in.-thick plate and assume that the welding is adequate.

**7-23.** Design the top and bottom clip angles and bolts for a semirigid connection similar to that shown in Figure 7-26. Assume that the beam is a W21 × 62, the reaction is 60 kips, and the end moment to be transmitted is 75 ft-kips. Neglect the connection on the beam web. Bolts are $\frac{7}{8}$-in.-diameter

A325N in standard holes. All structural steel is A36. Neglect prying action and use the classical approach of assuming a point of contraflexure in the angle.

**7-24.** Rework Problem 7-23 using the ASDM empirical design method.

**7-25.** For the eccentrically loaded bolted connection shown, determine the force acting on the most critical bolt. Assume $\frac{3}{4}$-in.-diameter A325N high-strength bolts in standard holes. The applied load is 10,000 lb. Use

    **(a)** The elastic method (Alternate Method 1—Elastic).

    **(b)** The ultimate strength method.

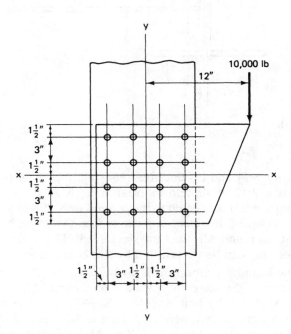

**PROBLEM 7-25**

**7-26.** For the eccentrically loaded bolted connection shown, determine the maximum allowable load that can be supported by the $\frac{3}{8}$-in. plate connected to the W8 × 24 column. The plate and column are A36 steel. Use $\frac{3}{4}$-in.-diameter A325SC high-strength bolts in standard holes. Assume that the plate and column are adequate. Use

    **(a)** The elastic method (Alternate Method 1—Elastic).

    **(b)** The ultimate strength method.

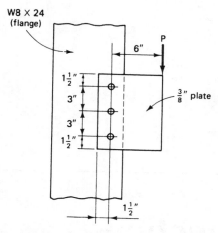

**PROBLEM 7-26**

**7-27.** Determine if the connection shown is adequate. The angle thickness is $\frac{5}{16}$ in., and both angle and column are A36 steel. Bolts are $\frac{7}{8}$-in.-diameter A490SC high-strength bolts in standard holes. Assume that the angle and column are adequate. Use the elastic method (Alternate Method 1—Elastic).

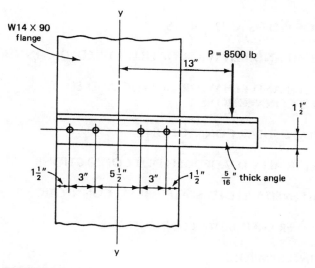

**PROBLEM 7-27**

**7-28.** Refer to Figure 7-38. Determine if the connection is satisfactory if $P = 20$ kips, $e = 15$ in., and the bolts are $\frac{3}{4}$-in.-diameter A325SC in two rows at 3-in. pitch.

# CHAPTER 8

# Welded Connections

## 8-1

### INTRODUCTION

Welding is a process in which two pieces of metal are fused together by heat to form a joint. In structural welding this process is usually accompanied by the addition of filler metal from an electrode. Structural welds are usually made either by the *manual shielded metal-arc process* or by the *submerged arc process.*

The manual shielded metal-arc welding process, commonly called stick welding, is designed primarily for manual application and is used both in the shop and in the field. An electric arc is formed between the end of a coated metal electrode and the steel components to be welded. This arc generates an approximate temperature of 6500°F, which melts a small area of the base metal. The tip of the electrode also melts, and this metal is forcibly propelled across the arc. The small pool of molten metal that is formed is called a *crater.* As the electrode is moved along the joint, the crater follows it, solidifying rapidly as the temperature of the pool behind it drops below the melting point. Figure 8-1 illustrates this process. During the welding process, as the electrode coating decomposes, it forms a gas shield to prevent absorption of impurities from the atmosphere. In addition, the coating contains a material (commonly called *flux*) that will prevent or dissolve oxides and other undesirable substances in the molten metal, or which will facilitate removal of these substances from the molten metal.

The submerged arc welding process is primarily a shop-welding process performed by either an automatic or a semiautomatic method. The principle is similar to manual shielded metal-arc welding, but a bare metal electrode is used instead

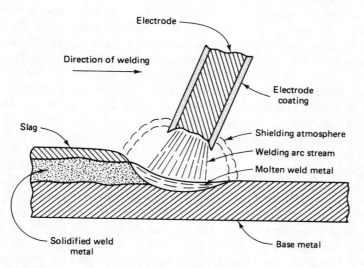

**FIGURE 8-1**   Manual shielded metal-arc welding process.

of a coated electrode. Loose flux is supplied separately in granular form and is placed over the joint to be welded. The electrode is pushed through the flux, and as the arc is formed, part of the flux melts to form a shield that coats the molten metal. This welding process is faster and results in deeper weld penetration. In the automatic process an electrically controlled machine supplies the flux and metal electrode through separate nozzles as it moves along a track. Figure 8-2 illustrates this process.

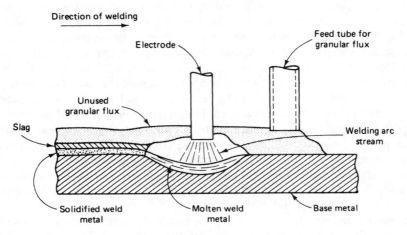

**FIGURE 8-2**   Submerged arc welding process.

The chemical and mechanical properties of the deposited weld metal (electrodes) should be as similar as possible to those of the base metal. Hence a variety of electrodes are needed to satisfy the requirements of the various steels. As a result, the American Welding Society (AWS), in cooperation with the ASTM, has established an electrode numbering system that classifies welding electrodes. The system utilizes a prefix letter, E, to indicate an electrode, followed by four or five digits. In the manual shielded metal-arc welding process, the first two (or three) digits designate the minimum tensile strength (in ksi) of the deposited metal. The third (or fourth) digit indicates the welding position in which the electrode is capable of making sound welds. These positions are illustrated in Figure 8-9. A 1 in the electrode numbering system indicates all positions: flat, horizontal, vertical, and overhead; a 2 indicates flat and horizontal; and a 3 indicates the flat position only. The fourth (or fifth) digit refers to the current supply and the type of coating on the electrode. For example, E7014 indicates an electrode with a minimum tensile strength of 70 ksi, usable in all positions with either ac or dc current, and with an iron powder added to the electrode covering so that the arc can easily be maintained. Initially, for structural design, the item of significance is the minimum tensile strength of the electrode material, since the designer is concerned that the weld metal be of adequate strength with respect to the base metal.

The American Welding Society *Structural Welding Code—Steel* (AWS D1.1) (see Reference 1) specifies the electrode classes and welding processes that can be used to achieve matching weld metal for a proper electrode-base metal combination. The code stipulates that both E60XX and E70XX electrodes may be used with A36 steel. The authors recommend the use of the E70XX. The minimum yield strength of the E70 electrode is 60 ksi, which is approximately 65% higher than the minimum yield strength of A36 steel. Thus weld metal is invariably stronger than the metals it connects.

In the submerged arc process, the electrode numbering system is somewhat different, as it includes a combination of flux and electrode designations, such as F7X-E7XX. The first portion of the designation is pertinent to the flux and the second portion to the electrode. F represents flux, and the first digit after F represents the tensile strength requirement of the resulting weld. The second digit indicates the impact strength requirement. The second portion, E7XX, indicates an E for electrode, with the first digit after the E indicating the minimum tensile strength of the weld metal. The last two digits classify the electrode. For this process the American Welding Society *Structural Welding Code—Steel* (AWS D1.1) stipulates that both F6X and F7X-EXXX flux-electrode combinations may be used with A36 steel. The most widely used electrode for structural work at present is the E70, since it is compatible with all grades of steel with a yield stress $F_y$ up to 60 ksi.

## 8-2

## TYPES OF WELDS AND JOINTS

The two types of welds that predominate in structural applications are the fillet weld and the groove weld. Other structural welds are the plug weld and slot weld, which generally are used only under circumstances in which fillet welds lack adequate load-carrying capacity. These four weld types are illustrated in Figure 8-3. The most commonly used weld for structural connections is the fillet weld. Groove welds may require extensive edge preparation as well as precise fabrication and, as a result, are more costly.

In any given welded structure the adjoining members may be situated with respect to each other in several ways. These joints may be categorized as butt, tee, corner, lap, and edge. They are illustrated in Figure 8-4.

Fillet welds are welds of theoretically triangular cross section joining two surfaces approximately at right angles to each other in lap, tee, and corner joints. The cross section of a typical fillet weld is a right triangle with equal legs. Figure 8-5 illustrates a typical fillet weld together with its pertinent nomenclature. The leg size designates the size of the weld. The *root* is the vertex of the triangle, or the point at which the legs intersect.

The *face of weld* is a *theoretical* plane, since weld faces will be either convex or concave, as shown in Figure 8-6. The convex fillet weld is the more desirable of

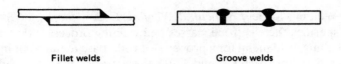

Fillet welds                              Groove welds

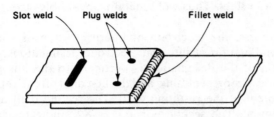

**FIGURE 8-3**  Weld types.

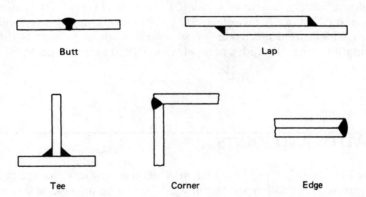

Butt                                          Lap

Tee                    Corner                    Edge

**FIGURE 8-4**  Joint types.

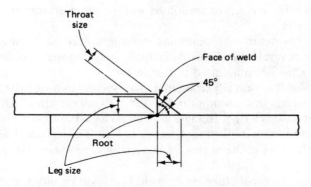

**FIGURE 8-5**  Typical fillet weld.

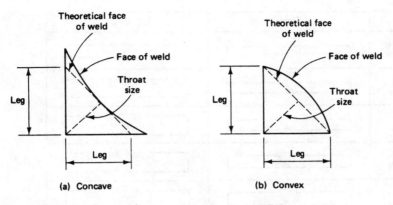

FIGURE 8-6   Fillet welds.

the two, since it has less tendency to crack as a result of shrinking while cooling. The distance from the theoretical face of weld to the root is called the *throat size*. Variations of this fillet weld are permitted and may be necessary. Leg sizes may be unequal. If the pieces to be joined do not intersect at right angles, the welds are considered to be *skewed fillets,* as shown in Figure 8-7. If the intersection is not within the angular limits shown in Figure 8-7, the welds are considered to be *groove welds.*

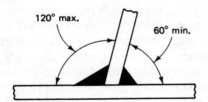

FIGURE 8-7   Fillet weld limitations.

Groove welds are welds made in a groove between adjacent ends, edges, or surfaces of two parts to be joined in a butt, tee, or corner joint. The weld configuration in these joints can be made in various ways. A welded butt joint can be made square, double-square, single-bevel, double-bevel, single V, double V, single J, double J, single U, or double U, as illustrated in Figure 8-8. With the exception of the square groove weld, some edge preparation is required for either one or both of the members to be connected.

Groove welds are further classified as either complete penetration or partial penetration welds. A *complete penetration weld* is one that achieves fusion of weld and base metal throughout the depth of the joint. It is made by welding from both sides of the joint, or from one side to a backing bar. The throat dimension of a full

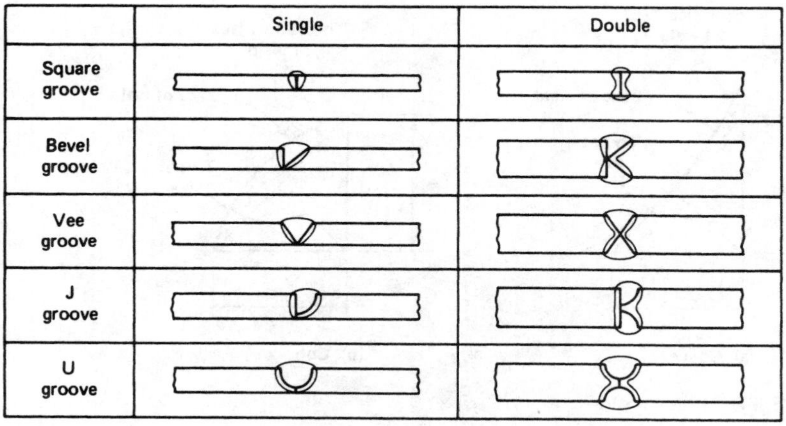

| | Single | Double |
|---|---|---|
| Square groove | | |
| Bevel groove | | |
| Vee groove | | |
| J groove | | |
| U groove | | |

**FIGURE 8-8**   Groove welds.

penetration groove weld is considered to be the full thickness of the thinner part joined, exclusive of weld *reinforcement.* Reinforcement is added weld metal over and above the thickness of the welded material.

*Partial penetration groove welds* are used when load requirements do not require full penetration or when welding must be done from one side of a joint only without the use of a backing bar. For more detailed information regarding groove penetration, welding processes, and positions, the reader is referred to the ASDM, Part 4, Welded Joints, as well as the *Structural Welding Code* of the AWS. Where possible, groove welds should be avoided because of their high cost compared with fillet welds. Where groove welds are necessary, the type required—a complete joint penetration weld or a partial joint penetration weld—should be simply stated. This permits the fabricator to use the most economical groove weld for the particular situation and equipment. With respect to groove weld design, if the proper electrode is used with the parent metals, allowable stresses in the weld will be the same as those for the parent material.

Plug and slot welds are used in lap joints (see Figure 8-3). Round holes or slotted holes are punched or otherwise formed through one of the members to be joined (before assembly). Weld metal is deposited in the openings. The openings may be partially or completely filled, depending on the thickness of the punched material. A variation of the slot weld is the use of a fillet weld in the slotted hole.

The AWS has established certain joints used in structural welding as *prequalified.* These may be found in Part 4 of the ASDM. These joints may be made by manual shielded metal-arc or submerged arc welding and used without the need for performing welding procedure qualification tests, provided that the electrodes and flux conform to AWS specifications and the fabricator meets certain workmanship standards. Prequalified joints are possible because such joints and the procedures for making them have a long history of satisfactory performance. Joints that are not prequalified under AWS codes and specifications may be used in structural welding, but they must first pass a procedure qualification test as prescribed by the AWS. This is usually costly. Therefore, whenever possible, only prequalified joints should be used.

Welds are also classified as flat, horizontal, vertical, and overhead. These four positions are illustrated in Figure 8-9. The position of the joint when welding is performed is of economic significance. The flat weld is the most economical, and the overhead weld the most expensive. Therefore, the flat position is preferred in all types of welding.

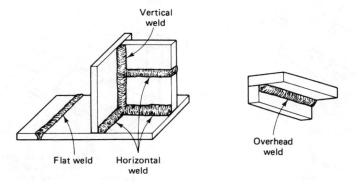

**FIGURE 8-9**   Weld types.

## 8-3

## STRENGTH AND BEHAVIOR
## OF FILLET WELDED CONNECTIONS

A fillet weld is a surface weld, and its shape or size is not restricted by the shape and size of a groove. Hence it is necessary to establish the size and length of the fillets to avoid overwelding or underwelding. Since tests have shown that fillet welds are stronger in tension and compression than they are in shear, the controlling fillet weld stresses are considered to be shear stresses on an effective (theoretical) throat area. This throat area establishes the strength of a fillet weld and is defined as the shortest distance from the root of the joint to the theoretical face of weld (as shown in Figure 8-5). In a fillet with equal leg sizes, where the cross-sectional shape of the weld is theoretically a 45° right triangle, the effective throat distance is

$$\sin 45° \times \text{leg size} = 0.707 \times \text{leg size}$$

If weld metal exists outside the theoretical right triangle, this additional weld metal is considered to be *reinforcement* and is assumed to add no strength.

The strength of a fillet depends on the direction of the applied load, which may be parallel or perpendicular to the axis of the weld. In parallel loading the applied load is transferred parallel to the weld from one leg face to the other, as shown in Figure 8-10a. The minimum resisting area in the fillet occurs at the throat and is equal to 0.707 times the leg size. (This assumes equal leg sizes for the weld, which is generally the case.) The strength of the weld is computed by multiplying the allowable shear stress of the weld by the throat area.

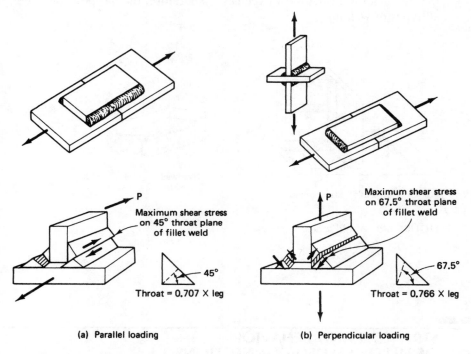

(a) Parallel loading                                    (b) Perpendicular loading

**FIGURE 8-10**   Fillet weld loading.

Tests indicate that a fillet weld loaded perpendicular to the fillet (transverse loading as shown in Figure 8-10b) is approximately one-third stronger than when loaded in a parallel direction. The ASDS, however, does not permit this to be considered when designing welds. The strengths of all fillets are based on the values calculated for loads applied in a parallel direction. The fillet weld loaded in a perpendicular direction has greater strength because the failure plane develops at an angle other than 45°; hence the resisting area of the fillet is greater than the throat area, which is perpendicular to the theoretical face of the weld. In addition, transverse fillet welds are more uniformly stressed than parallel loaded fillet welds.

The allowable shear stress for the weld metal is (ASDS, Table J2.5)

$$F_v = 0.3F_u$$

where $F_u$ is the specified minimum tensile strength of the electrode. Therefore, the strength of a fillet weld *per linear inch of weld* is

$$P = F_v(0.707)(\text{leg size})$$

or

$$P = 0.3F_u(0.707)(\text{leg size}) = 0.212F_u(\text{leg size})$$

We now calculate strength per inch for a fillet weld having a leg size of $\frac{1}{16}$ in. (which is a hypothetical value, since a minimum weld size according to the ASDS is $\frac{1}{8}$ in.). The strength or load-carrying capacity of other leg-size fillet welds can then be obtained by multiplying by the number of sixteenths in the leg size. For an E70XX electrode ($F_u = 70$ ksi),

$$P = 0.212(70)\left(\frac{1}{16}\right) = 0.928 \text{ kip/in.}$$

This value is generally rounded to 0.925 kip/in. Using 0.925 kip/in. as a basic value, the strength of other sizes of fillet welds may be computed and tabulated. For example, the strength of a $\frac{3}{16}$-in. fillet weld would be

$$0.925(3) = 2.78 \text{ kips/in.}$$

A similar approach could be used for E60XX electrodes, where $F_u = 60$ ksi.

When the submerged arc process is used, greater heat input produces a deeper weld penetration, and as a result, the ASDS stipulates that the effective throat distance for welds larger than $\frac{3}{8}$ in. may be taken equal to the theoretical throat plus 0.11 in. In addition, for welds of $\frac{3}{8}$ in. or less, the strength of the weld is based on leg size rather than throat distance. These values are shown in Table 8-1.

**TABLE 8-1    Strength of Welds (kips per linear inch)**

| Weld size (in.) | E70XX SMAW[a] | E60XX SMAW | E70XX SAW[b] | E60XX SAW |
|---|---|---|---|---|
| $\frac{1}{16}$ | 0.925 | 0.795 | 1.31 | 1.13 |
| $\frac{1}{8}$ | 1.85 | 1.59 | 2.63 | 2.25 |
| $\frac{3}{16}$ | 2.78 | 2.39 | 3.94 | 3.38 |
| $\frac{1}{4}$ | 3.70 | 3.18 | 5.25 | 4.50 |
| $\frac{5}{16}$ | 4.63 | 3.98 | 6.56 | 5.63 |
| $\frac{3}{8}$ | 5.55 | 4.77 | 7.88 | 6.75 |
| $\frac{7}{16}$ | 6.48 | 5.57 | 8.81 | 7.55 |
| $\frac{1}{2}$ | 7.40 | 6.36 | 9.73 | 8.34 |
| $\frac{9}{16}$ | 8.33 | 7.16 | 10.66 | 9.14 |
| $\frac{5}{8}$ | 9.25 | 7.95 | 11.59 | 9.93 |
| $\frac{11}{16}$ | 10.18 | 8.75 | 12.52 | 10.73 |
| $\frac{3}{4}$ | 11.10 | 9.54 | 13.45 | 11.52 |
| $\frac{13}{16}$ | 12.03 | 10.34 | 14.37 | 12.32 |
| $\frac{7}{8}$ | 12.95 | 11.13 | 15.30 | 13.12 |

[a]Shielded metal-arc welding.

[b]Submerged arc welding.

As discussed in Chapters 2 and 7 of this text, another factor that must be considered with respect to the strength of a connection is block shear (or shear rupture). As with bolted connections, block shear must be checked for some types of fillet welded connections. The ASDS, Section J4, states that the minimum net failure path on the periphery of welded connections shall be checked. In essence, this is a shear and tension tearout of either the supporting or supported member.

The equation for block shear strength remains the same as for bolted connections:

$$P_t = A_v F_v + A_t F_t$$
$$= A_v(0.30F_u) + A_t(0.50F_u)$$

In addition to the strength criteria, the ASDS furnishes design requirements with respect to minimum and maximum sizes and lengths of fillet welds. *Minimum* leg sizes for various thickness of members being joined are shown in the ASDS, Table J2.4. Note that the minimum size of a fillet weld allowed in structural work is $\frac{1}{8}$ in. Also, the minimum size is based on the *thicker* of the two parts being joined, except that the weld size need not exceed the thickness of the thinner part. The minimum size limitation is based on the fact that the heat generated in depositing a small weld is not enough to heat a much thicker member beyond the immediate vicinity of the weld. As a result, the weld cools rapidly, with subsequent cracks developing.

The *maximum size* of fillet welds against the edges of connected parts of a joint is limited so that the weld does not overstress the parts it connects. This means that the fillet weld capacity cannot exceed the capacity of the connected material either in tension or shear. The maximum permissible leg size is $\frac{1}{16}$ in. less than the thickness of the material for material thickness of $\frac{1}{4}$ in. or over. Along edges of material less than $\frac{1}{4}$ in. thick, the weld leg size may equal the thickness of the material (see Figure 8-11).

Within the constraints of the minimum and maximum criteria for fillet welds, economy can best be achieved by using welds that require a minimum amount of metal and can be deposited in the least amount of time. As shown previously, the strength of a fillet weld is directly proportional to its size; the volume of deposited metal, and hence the cost of the weld, however, increases as the square of the weld

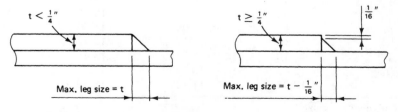

**FIGURE 8-11** Maximum leg sizes for fillet welds.

size. Hence it is generally preferable to use a long small-leg-size weld rather than a short large-leg-size weld. In addition, a weld deposited in a single pass is cheaper than welds made in multiple passes. The largest weld that can be made by a welder in a single pass is a $\frac{5}{16}$-in. fillet weld. Multiple passes require appreciably more time and weld metal, and as a result are more expensive.

In addition, the ASDS, Section J2.2, imposes limitations on the lengths of fillet welds. The minimum effective length of a fillet weld must not be less than four times the nominal size, or else the size of the weld must be considered not to exceed one-fourth its effective length. This also applies to intermittent fillet welds (see Figure 8-3), with the added requirement that each weld length not be less than $1\frac{1}{2}$ in. If longitudinal fillet welds are used alone (without transverse welds) in end connections of flat bar tension members, the length of each fillet weld cannot be less than the perpendicular distance between them. Nor can the transverse spacing of longitudinal fillet welds used in end connections exceed 8 in. unless specific design provisions are utilized.

Side or end fillet welds terminating at ends or sides, respectively, of parts or members should, whenever practicable, be returned continuously around the corners for a distance not less than two times the nominal size of the weld. This weld detail is called an *end return* (see Figure 8-14). The ASDS, Section J2.2a, states that the effective length of fillet welds includes the length of the end returns used.

Where lap joints are used, the minimum amount of lap should be five times the thickness of the thinner part joined but not less than 1 in. Lap joints joining plates or bars subjected to an axial load should be fillet welded along the end of both lapped parts (see the ASDS, Section J2.2).

## Example 8-1

Determine the allowable tensile load that may be applied to the connection shown in Figure 8-12. The steel is A36, and the electrode used was E70 (manual shielded metal-arc welding). The weld is a $\frac{7}{16}$-in. fillet weld.

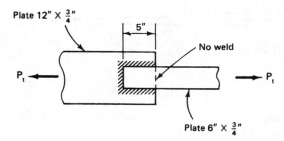

**FIGURE 8-12**  Welded lap joint.

**Solution:**

The total length of $\frac{7}{16}$-in. weld is 16 in. From Table 8-1, the capacity of a $\frac{7}{16}$-in. weld is 6.48 kips/in. Thus

$$\text{weld capacity} = 6.48(16) = 103.7 \text{ kips}$$

The tensile capacity of the plate (using $F_t = 22$ ksi) is

$$P_t = 6(0.75)(22) = 99 \text{ kips}$$

Therefore, the allowable tensile load is 99 kips (440 kN).
    Check block shear:

$$A_v = 2(5)\left(\frac{3}{4}\right) = 7.5 \text{ in.}^2$$

$$A_t = 6\left(\frac{3}{4}\right) = 4.5 \text{ in.}^2$$

$$P_t = A_v(0.30F_u) + A_t(0.5F_u)$$

$$= 7.5(0.30)(58) + 4.5(0.50)(58) = 261 \text{ kips}$$

$$261 \text{ kips} > 99 \text{ kips} \qquad\qquad \textbf{O.K.}$$

**Example 8-2**

Determine the allowable tensile load that may be applied to the connection shown in Figure 8-13. The steel is A36, and the electrode used was E70. The fillet weld is $\frac{5}{16}$ in., and the shielded metal-arc welding (SMAW) process was used.

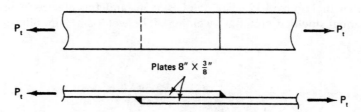

**Plates 8″ × $\frac{3}{8}$″**

**FIGURE 8-13**   Welded lap joint.

**Solution:**

The length of $\frac{5}{16}$-in. weld is 16 in. The capacity of a $\frac{5}{16}$-in. weld per linear inch, from Table 8-1, is 4.63 kips/in. Thus

$$\text{weld capacity} = 4.63(16) = 74.1 \text{ kips}$$

The tensile capacity of the plate (using $F_t = 22$ ksi) is

$$P_t = 8(0.375)(22) = 66 \text{ kips}$$

Therefore, the allowable tensile load is 66 kips (294 kN). Block shear is not applicable.

## Example 8-3

Design longitudinal fillet welds to develop the tensile capacity of the plate shown in Figure 8-14. The steel is A36, and the electrode is E70 (SMAW).

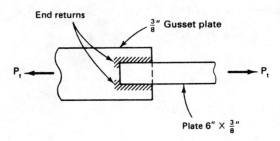

**FIGURE 8-14**   Parallel loaded welded lap joint.

**Solution:**

Calculate the tensile capacity of the plate. Based on gross area,

$$P_t = 6(0.375)(0.60)(36) = 48.6 \text{ kips}$$

Based on effective net area (see ASDS Section B3),

$$\text{assume } U = 0.75 \qquad (1.5\,w > \ell > w)$$

$$P_t = F_t A_e = F_t U A_g = 0.50(58)(0.75)(6)(0.375) = 48.9 \text{ kips}$$

Use $P_t = 48.6$ kips.

$$\text{Maximum weld size} = \frac{3}{8} - \frac{1}{16} = \frac{5}{16} \text{ in.}$$

(see the ASDS, Section J2.2). From Table 8-1, the capacity of a $\frac{5}{16}$-in. weld per linear inch is 4.63 kips/in. The length of the weld required is

$$\frac{48.6}{4.63} = 10.5 \text{ in.}$$

Use end returns with a minimum length of 2 × (leg size):

$$2\left(\frac{5}{16}\right) = \frac{5}{8} \text{ in.} \qquad \text{(use 1 in.)}$$

The length of the longitudinal welds required is

$$10.5 - 1 - 1 = 8.5 \text{ in.}$$

$$\frac{8.5}{2} = 4.25 \text{ in. each side of plate}$$

The ASDS, however, stipulates that the minimum length of longitudinal fillet welds must not be less than the perpendicular distance between them. Therefore, use a minimum length of 6 in. on each side of the plate.

Check the assumed value of $U = 0.75$:

$$w = 6 \text{ in.} \qquad 1.5 \, w = 9 \text{ in.}$$

Therefore,

$$1.5 \, w > \ell \geq 6 \qquad\qquad \textbf{O.K.}$$

Check block shear in the gusset plate:

$$A_v = 2(6)\left(\frac{3}{8}\right) = 4.5 \text{ in.}^2$$

$$A_t = 6\left(\frac{3}{8}\right) = 2.25 \text{ in.}^2$$

$$P_t = A_v(0.30F_u) + A_t(0.5F_u)$$

$$= 4.5(0.30)(58) + 2.25(0.50)(58) = 143.6 \text{ kips}$$

$$143.6 \text{ kips} > 49.5 \text{ kips} \qquad\qquad \textbf{O.K.}$$

---

The welds connecting the plates in Example 8-3 could also have included an end transverse weld together with the longitudinal welds. With this type of end connection, the criterion for the minimum length of longitudinal fillet weld used in Example 8-3 does not apply (see the ASDS, Section J2.2). In the special case where single or double angles subject to static tensile loads are welded to plates, as in a welded truss, the ASDS permits their connections to be designed using procedures similar to those of the previous examples.

**Example 8-4** _____

Design an end connection using longitudinal welds and an end transverse weld to develop the full tensile capacity of the angle, shown in Figure 8-15.

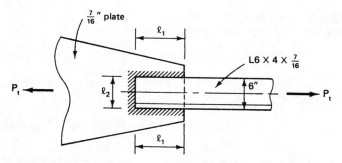

**FIGURE 8-15**   Welded connection for angle.

Use A36 steel and E70 electrodes (SMAW). The angle is an L6 × 4 × $\frac{7}{16}$ with the long leg connected to the plate.

**Solution:**

Find the tensile capacity of the angle. Based on gross area $A_g$,

$$P_t = F_t A_g$$
$$= 0.60 F_y A_g$$
$$= 0.60(36)(4.18)$$
$$= 90.3 \text{ kips}$$

Based on effective net area $A_e$,

$$P_t = F_t A_e$$
$$= 0.50 F_u A_e$$

where $A_e = U A_g$ and $U = 0.85$ from Table 2-1, Case II, of this text. Therefore,

$$P_t = 0.50(58)(0.85)(4.18) = 103.0 \text{ kips}$$

The tensile capacity $P_t$ of 90.3 kips controls.

$$\text{Maximum weld size} = \frac{7}{16} - \frac{1}{16} = \frac{3}{8} \text{ in.}$$

From Table 8-1, the capacity of a $\frac{3}{8}$-in. weld per linear inch is 5.55 kips/in. The length of the weld required is

$$\frac{90.3}{5.55} = 16.3 \text{ in.}$$

End weld $\ell_2$ has length of 6 in. Therefore, the side welds must furnish

$$16.3 - 6 = 10.3 \text{ in.} \qquad \text{(say 11 in.)}$$

Use a length $\ell_1$ of $5\frac{1}{2}$ in. for each side of the angle.
Check block shear in the gusset plate:

$$A_v = 2(5.5)(7/16) = 4.81 \text{ in.}^2$$

$$A_t = 6(7/16) = 2.63 \text{ in.}^2$$

$$P_t = A_v(0.30F_u) + A_t(0.50F_u)$$

$$= 4.81(0.30)(58) + 2.63(0.50)(58) = 160.0 \text{ kips}$$

$$160.0 \text{ kips} > 90.3 \text{ kips} \qquad \textbf{O.K.}$$

---

Where the angle tension member is subject to repeated variation of stress, such as stress reversal that may occur with moving loads, the placement of the welds must conform to the distribution of the angle area, with the centroid of the resisting welds collinear with the centroidal axis of the angle. In effect, the welding pattern will not be symmetrical if the member is not symmetrical.

**Example 8-5** _____

Design an end connection using longitudinal welds and an end transverse weld to develop the full tensile capacity of the angle shown in Figure 8-16. The member is subjected to repeated stress variations. Use A36 steel and an E70 electrode (SMAW). The angle is an L6 × 4 × $\frac{1}{2}$ with the long leg connected to a gusset plate.

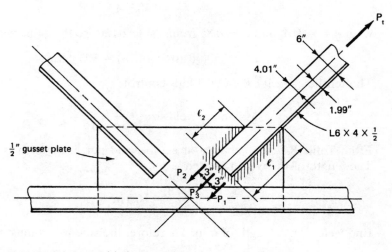

**FIGURE 8-16**   Welded truss joint.

**Solution:**

Find the tensile capacity of the angle. Based on gross area $A_g$,

$$P_t = F_t A_g$$
$$= 0.60 F_y A_g$$
$$= 0.60(36)(4.75)$$
$$= 102.6 \text{ kips}$$

Based on effective net area $A_e$,

$$P_t = F_t A_e$$
$$= 0.50 F_u A_e$$

where $A_e = U A_g$ and $U = 0.85$ from Table 2-1, Case II, of this text. Therefore,

$$P_t = 0.50(58)(0.85)(4.75) = 117.1 \text{ kips}$$

The tensile capacity $P_t$ of 102.6 kips controls.

$$\text{Maximum weld size} = \frac{1}{2} - \frac{1}{16} = \frac{7}{16} \text{ in.}$$

The capacity of a $\frac{7}{16}$-in. weld per linear inch is 6.48 kips (Table 8-1). Different-sized welds could be used along the back and toe of the angle; for both economic and practical reasons, however, the weld will be kept the same size. In addition, a single-pass weld ($\frac{5}{16}$ in. maximum) could be used, but it would require a greater length of weld. Therefore, a $\frac{7}{16}$-in. fillet weld will be used for this connection.

Since the force in the angle is assumed to act along its centroidal axis, the centroid of the resisting welds must be collinear to eliminate any eccentricity.

The resistance of the transverse end weld is

$$P_3 = 6(6.48) = 38.9 \text{ kips}$$

$P_3$ acts along the centerline of the attached leg.

Taking moments about line $\ell_1$, resisting force $P_2$ may be determined:

$$102.6(1.99) = P_2(6) + 38.9(3)$$
$$P_2 = 14.6 \text{ kips}$$

$P_1$ may be determined by a summation of forces parallel to the length of the angle:

$$P_1 = 102.6 - 38.9 - 14.6$$
$$= 49.1 \text{ kips}$$

The length of longitudinal weld required is based on the weld capacity per linear inch, 6.48 kips. Thus

$$\ell_2 = \frac{14.6}{6.48} = 2.3 \text{ in.} \qquad \text{(use 3 in.)}$$

$$\ell_1 = \frac{49.1}{6.48} = 7.6 \text{ in.} \qquad \text{(use 8 in.)}$$

Check block shear in the gusset plate:

$$A_v = (8 + 3)(1/2) = 5.5 \text{ in.}^2$$

$$A_t = 6(1/2) = 3.0 \text{ in.}^2$$

$$P_t = A_v(0.30F_u) + A_t(0.50F_u)$$

$$= 5.5(0.30)(58) + 3.0(0.50)(58) = 182.7 \text{ kips}$$

$$182.7 \text{ kips} > 102.6 \text{ kips} \qquad\qquad\qquad \textbf{O.K.}$$

### 8-4

## STRENGTH AND BEHAVIOR OF PLUG AND SLOT WELDED CONNECTIONS

Where inadequate length of fillet welds exists to resist the applied load in a given connection, additional strength may be furnished through the use of plug or slot welds. The ASDS states that plug or slot welds may be used to transmit shear in a lap joint or to prevent buckling of lapped parts and to join component parts of built-up members.

The terms *plug weld* and *slot weld* are used with reference to circular holes, or slotted holes with circular ends, that are filled with weld metal completely or to such depth as prescribed by the ASDS. The effective area for such welds is assumed to be the nominal area of the hole or slot in the plane of the contact surfaces of the elements being joined.

Plug and slot welds are not usually used for strength. These welds, especially plug welds, often serve a useful purpose in stitching together elements of a member. Joints or connections made with these welds have exhibited poor fatigue resistance, however.

Fillet welds in holes or slots also may be used to transmit shear in lap joints; they are not to be considered plug or slot welds, however. Since large plug and slot welds may exhibit excessive shrinkage, it is usually more desirable to use fillet welds in large holes or slots. Also, special care and special procedures are necessary for sound plug or slot welds, whereas the making of a fillet weld in a hole or slot is a normal procedure.

The ASDS requirements for plug and slot welds may be summarized as follows:

1. The width of slot or diameter of hole cannot be less than material thickness plus $\frac{5}{16}$ in. (rounded to the next greater odd $\frac{1}{16}$ in.) and cannot exceed 2.25 times the thickness of the weld. Additionally, for a plug weld, the diameter may not exceed the minimum diameter plus $\frac{1}{8}$ in.

2. For material up to $\frac{5}{8}$ in. thick, the weld thickness must equal the material thickness.

3. For material greater than $\frac{5}{8}$ in. thick, the weld thickness may not be less than $\frac{1}{2}$ times the material thickness or less than $\frac{5}{8}$ in.

4. The maximum length of slot is 10 times the weld thickness.

5. Minimum center-to-center spacing of plug welds is four times the hole diameter.

6. Minimum spacing of lines of slot welds transverse to their length is four times the width of the slot.

7. Minimum center-to-center spacing in a longitudinal direction on any line is two times the length of the slot.

8. The ends of the slot must be semicircular or be rounded to a radius not less than the thickness of the material (except for ends that extend to the edge of the material).

**Example 8-6** _____

Design the connection of a C10 × 30 to a $\frac{3}{8}$-in. gusset plate, as shown in Figure 8-17, to develop the full tensile capacity of the channel. Welding is not permitted on the back of the channel. All steel is A36. Use E70 electrode (SMAW). The maximum length of lap on the gusset plate is 10 in. (space limitations). Neglect block shear.

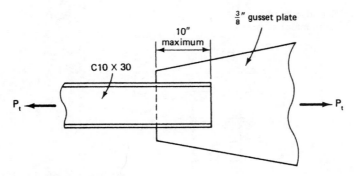

**FIGURE 8-17**   Channel connection.

**Solution:**

1. Find the tensile capacity of the C10 × 30. Based on gross area $A_g$,

$$P_t = F_t A_g$$
$$= 0.60 F_y A_g$$
$$= 0.60(36)(8.22)$$
$$= 177.6 \text{ kips}$$

Based on effective net area $A_e$,

$$P_t = F_t A_e$$
$$= 0.50 F_u A_e$$

where $A_e = U A_g$ and $U = 0.85$ from Table 2-1, Case II, of this text. Therefore,

$$P_t = 0.50(58)(0.85)(8.22) = 202.6 \text{ kips}$$

The tensile capacity $P_t$ of 177.6 kips controls.

2. The web thickness of the channel is 0.673 in.; therefore, a $\frac{9}{16}$-in. fillet weld could be used. For economic reasons, however, a $\frac{5}{16}$-in. fillet weld will be tried. This is the maximum size for a single-pass weld. The capacity of a $\frac{5}{16}$-in. fillet weld per linear inch is 4.63 kips, from Table 8-1.

3. The required length of weld is

$$\frac{177.6}{4.63} = 38.4 \text{ in.}$$

4. The length available for welding is

$$10(2) + 10 = 30 \text{ in.}$$

There is insufficient length available for a $\frac{5}{16}$-in. fillet weld. Although the size of the fillet weld could simply be increased to provide the required strength with the available length, we will design a slot weld for illustrative purposes. Figure 8-18 shows the layout of the slot weld.

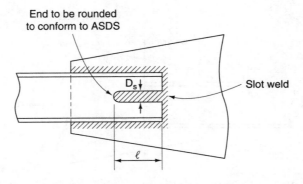

**FIGURE 8-18**   Slot weld notation.

5.  For the slot weld, use a weld thickness equal to the thickness of the channel web.

6.  The minimum width of slot is

$$\text{minimum } D_s = 0.673 + \frac{5}{16} = 0.986 \text{ in.}$$

The maximum width of slot is

$$\text{maximum } D_s = 2.25(0.673) = 1.51 \text{ in.}$$

Use a slot width $D_s = 1\frac{1}{16}$ in.

7.  The total length of the longitudinal and transverse end fillet weld is reduced by the width of the slot. The length of the $\frac{5}{16}$-in. fillet weld is

$$\ell = 30 - 1\frac{1}{16} = 28\frac{15}{16} \text{ in.}$$

$$= 28.94 \text{ in.}$$

8.  The capacity of the $\frac{5}{16}$-in. fillet weld is

$$P = 28.94(4.63) = 134 \text{ kips}$$

9.  The load to be resisted by the slot weld is

$$177.6 - 134 = 43.6 \text{ kips}$$

10. The required length of the slot weld $\ell$ may be determined from the relationships

$$P = AF_v = D_s\ell F_v$$

where

$P$ = load to be resisted by the slot weld

$A$ = area of the slot in the plane of the contact surfaces

$F_v$ = allowable shear stress in the weld (see Section 8-3 of this text or the ASDS, Table J2.5); $F_v = 0.30F_u$

$D_s$ and $\ell$ are as shown in Figure 8-18($A = D_s\ell$). Substituting yields

$$43.6 = \left(1\frac{1}{16}\right)(\ell)(0.3)(70)$$

from which we obtain

$$\text{required } \ell = 1.95 \text{ in.} \quad (\text{use } \ell = 2 \text{ in.})$$

11.   The maximum length of slot is

$$\ell = 10(0.673) = 6.73 \text{ in.} > 2 \text{ in.} \qquad\qquad \textbf{O.K.}$$

**Use a 2 in. $\times$ $1\frac{1}{16}$ slot weld.**

---

**8-5**

## END-PLATE SHEAR CONNECTIONS

The end-plate shear connection is discussed in Section 7-9 of this text. It consists of a rectangular plate welded to the end of a beam web and bolted to the supporting member. Example 7-10 furnishes the design for the bolted portion of the connection. The design of the welded portion of the connection is treated here.

The end-plate length $L$ is always made less than the beam depth so that all the welding will be on the beam web. The plate is welded to the beam web with fillet welds, as shown in Figure 8-19. The welds should not be returned across the web at the top or bottom of the end plates, and the effective weld length should be taken as equal to the plate length minus twice the weld size. The weld size to the beam web should be such that the weld shear capacity per linear inch does not

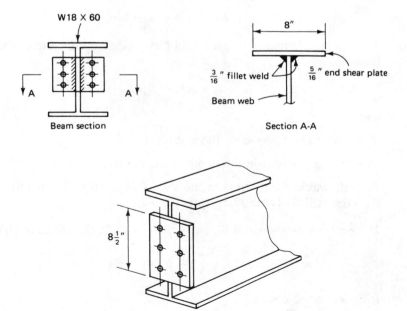

**FIGURE 8-19**   Weld for end-plate shear connection.

exceed the beam web shear capacity per linear inch. This portion of the design assumes no eccentricity.

### Example 8-7 _____

See Example 7-10 for all data. Design the welded portion of the connection using an E70 electrode (SMAW).

**Solution:**

1.  The plate length $L$ was established as $8\frac{1}{2}$ in.
2.  Assuming a $\frac{3}{16}$-in. weld on each side of the beam web, the available effective length of one weld is

$$8\frac{1}{2} - 2\left(\frac{3}{16}\right) = 8.13 \text{ in.}$$

3.  The total weld capacity is

$$2.78(8.13)(2) = 45.2 \text{ kips} > 40.0 \text{ kips (end reaction)} \qquad \textbf{O.K.}$$

4.  Check to ensure that the shear capacity of the welds (per linear inch) does not exceed the shear capacity of the web (per linear inch). The shear capacity of the welds is

$$2(2.78) = 5.56 \text{ kips/in.}$$

The shear capacity of the web (A36 steel) is

$$F_v(t_w) = 0.40F_y t_w$$
$$= 0.40(36)(0.415) = 5.98 \text{ kips/in.}$$
$$5.56 \text{ kips/in.} < 5.98 \text{ kips/in.} \qquad \textbf{O.K.}$$

Use a $\frac{3}{16}$-in. fillet weld.

## 8-6
▬▬▬▬

## ECCENTRICALLY LOADED WELDED CONNECTIONS

In Section 7-11 of the text, eccentrically loaded bolted connections are analyzed and designed. The analysis and design of an eccentrically loaded welded connection is approached in a similar way. Note in Figure 8-20 that the eccentrically applied load $P$ lies in the plane of the connection. $P$ may be resolved into a concentric load-moment combination. The concentric load acts through the centroid of the weld configuration, and the torsional moment ($M = Pe$) is with respect to the same centroid as a center of rotation. Therefore, the forces acting on the welds will be made up of two components: $P_v$ due to the axial effect of the eccentric load, and $P_m$ due to the torsional moment effect, as shown in Figure 8-21. The axial effect

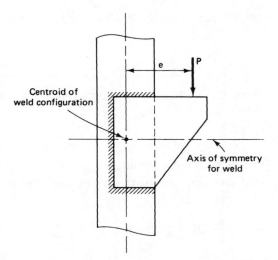

**FIGURE 8-20**   Eccentrically loaded welded connection.

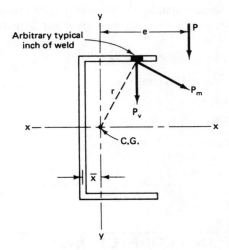

**FIGURE 8-21**   Forces in welds.

produces a load of $P/\ell$ per inch of weld, where $\ell$ is the total length of weld. This load acts in a direction parallel to $P$ and will be the same for each linear inch of weld.

$P_m$ will vary with the distance $r$ from the centroid of the weld configuration to that element of weld being considered and will act in a direction normal to the line that connects the centroid with that weld element. Therefore, the connection must be designed or analyzed so that the resultant of these two components acting at any point of the weld does not exceed the weld capacity.

The torsional load $P_m$ may be determined by applying the classic torsional stress formula to the weld configuration:

$$f_v = \frac{Mr}{J}$$

where

$f_v$ = unit shearing stress in the weld

$M$ = torsional moment $(Pe)$

$r$ = radial distance from center of gravity of the weld configuration to any point of the weld being considered

$J$ = polar moment of inertia of the weld

For purposes of design, each weld element may be assumed to be a line coincident with the root (see Figure 8-5) of the fillet weld. Hence the weld may be considered to have location and length only. Therefore, the computed unit stress in the torsional stress formula becomes a force per unit length (kips/in.) rather than a force per unit area (kips/in.²). This force per linear inch may be designated $P_m$. It should also be noted that, to make units and assumptions consistent, the polar moment of inertia is given in units of in.³ rather than in.⁴. This practice is based on the fact that the weld has only length, thereby removing one dimension from $I_x$ and $I_y$. Moment-of-inertia formulas for a length of weld are shown in Figure 8-22. For this type of problem, it is probably easier to use the polar moment of inertia in the form

$$J = I_x + I_y$$

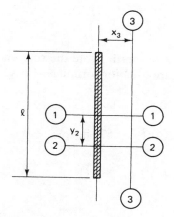

| Moment of inertia | Formula |
|---|---|
| $I_{1-1}$ | $\frac{1}{12}\ell^3$ |
| $I_{2-2}$ | $\frac{1}{12}\ell^3 + \ell y_2{}^2$ |
| $I_{3-3}$ | $*\ell x_3{}^2$ |

* Neglects moment of inertia about its own centroidal axis (which = 0)

**FIGURE 8-22**  Moment-of-inertia formulas.

**Example 8-8**

Determine the size of the fillet weld required to resist a load of 20 kips on the bracket shown in Figure 8-23. The steel is A36, and the welding is to be performed using E70 electrodes (SMAW).

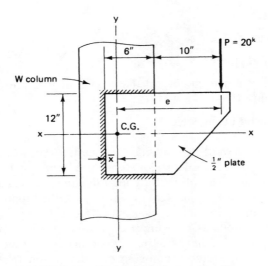

**FIGURE 8-23**   Welded bracket connection.

**Solution:**

1.  Assuming the weld to be a line, the center of gravity of the weld configuration may be obtained by taking a summation of moments of lengths of weld about the 12-in. side:

$$2(6)(3) + 12(0) = (6 + 6 + 12)(\bar{x})$$
$$\bar{x} = 1.50 \text{ in.}$$

2.  For calculation of the polar moment of inertia, note the reference axes through the weld centroid. See Figure 8-22 for formulas. Thus

$$J = I_x + I_y$$

$$I_x = \left(\frac{1}{12}\right)(12)^3 + 2(6)(6)^2 = 576 \text{ in.}^3$$

$$I_y = 2\left(\frac{1}{12}\right)(6)^3 + 2(6)(1.5)^2 + 12(1.5)^2$$

$$= 90 \text{ in.}^3$$

$$J = 576 + 90 = 666 \text{ in.}^3$$

3.     The torsional moment $M$ is

$$M = Pe = 20(14.5) = 290 \text{ in.-kips}$$

4.     The force on the weld due to the torsional moment is

$$P_m = \frac{Mr}{J}$$

Since the most stressed parts of the weld are those that are the greatest distance from the weld center of gravity, the largest $r$ value should be used. In this example, with reference to Figure 8-24,

$$\text{maximum } r = \sqrt{6^2 + 4.5^2} = 7.5 \text{ in.}$$

Therefore,

$$P_m = \frac{290(7.5)}{666} = 3.266 \text{ kips/in.}$$

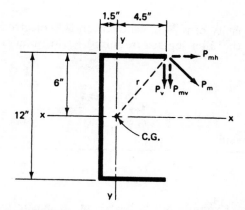

**FIGURE 8-24** Weld geometry and forces.

5.     Calculating the vertical and horizontal components of $P_m$, we have

$$P_{mh} = \frac{6.0}{7.5}(3.266) = 2.61 \text{ kips/in.}$$

$$P_{mv} = \frac{4.5}{7.5}(3.266) = 1.96 \text{ kips/in.}$$

6.     The force on the weld due to axial effect of the eccentric load is

$$P_v = \frac{P}{\ell} = \frac{20}{24} = 0.83 \text{ kip/in.}$$

7.   Adding the forces vectorially and determining the resultant force (see Figure 8-25) gives us

$$R = \sqrt{2.61^2 + (1.96 + 0.83)^2}$$

$$= 3.82 \text{ kips/in.}$$

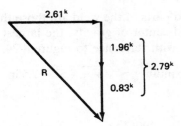

**FIGURE 8-25**   Force resolution.

8.   Since the capacity of a $\frac{1}{16}$-in. weld for an E70 electrode is 0.925 kip/in. (Table 8-1), the fillet leg size required is (in number of sixteenths of an inch)

$$\frac{3.82}{0.925} = 4.13$$

**Use a $\frac{5}{16}$-in. fillet weld.**

---

The solution used in Example 8-8 is based on an elastic method designated Alternate Method 1—Elastic in the ASDM, Part 4. In this method, the assumption is made that each element of weld will resist an equal share of the components of the load and a proportional share of the eccentric moment portion of the load (dependent on the element's distance from the center of gravity of the weld configuration). This method, although providing a simplified and conservative approach, does not result in a consistent factor of safety. As a result, an ultimate strength method is provided in the ASDM, 9th edition, and is the recommended approach. The elastic method is still acceptable for conditions where the ASDM tables do not apply.

As with eccentrically loaded bolted connections, the ASDM furnishes a new method, designated Alternate Method 2, that permits extension of the ASDM weld tables to eccentric loads that are inclined at an angle $\Theta$ from the vertical. Discussion of this new method is provided in Part 4 of the ASDM.

As a means of comparison, Example 8-8 is recalculated using the ASDM ultimate strength method.

### Example 8-9

Rework Example 8-8 using the ASDM ultimate strength method. This is simplified through the use of the ASDM, Part 4, Table XXIII. Referring to ASDM nomenclature for Tables XIX–XXVI (Part 4), we find

$$\ell = 12 \text{ in.} = \text{length of vertical weld}$$

$$k\ell = 6 \text{ in.} = \text{length of horizontal weld}$$

$$A = 16 \text{ in.} = \text{distance from vertical weld to } P$$

**Solution:**

$$k = \frac{k\ell}{\ell} = \frac{6}{12} = 0.5 \qquad A = 16 \text{ in.}$$

Enter Table XXIII with $k = 0.5$ and obtain $x = 0.125$. Thus

$$x\ell = 0.125(12) = 1.50 \text{ in.}$$

Check with Example 8-8. This is the distance from the vertical weld to the center of gravity of the weld group. Thus

$$a\ell = A - x\ell = 16 - 1.50 = 14.5 \text{ in.}$$

$$a = \frac{a\ell}{\ell} = \frac{14.5}{12} = 1.21$$

Interpolating between $a = 1.20$ and $a = 1.40$ for $k = 0.5$ yields

$$C = 0.532$$

$$C_1 = 1.0 \text{ for E70 electrode}$$

The required minimum size of weld, in sixteenths of an inch, is

$$D = \frac{P}{CC_1\ell} = \frac{20}{0.532(1.0)(12)}$$

$$= 3.13 \text{ (sixteenths)}$$

**Use a $\frac{1}{4}$-in. fillet weld.**

**8-7**

## UNSTIFFENED WELDED SEATED BEAM CONNECTIONS

Unstiffened bolted seated beam connections were discussed in Section 7-8 of this text. Bolts were used to connect the vertical leg of the seat angle to the supporting member. We now discuss the use of welds for the connection of the vertical leg of the seat angle.

The required bearing length for the beam on the horizontal leg of the seat angle, the length of the angle, and the thickness of the angle are all determined in the same manner as that used for the unstiffened bolted seated beam connection. The attachment of the seat angle to the supporting member is achieved by welds along the vertical edges of the angle. The welds are considered eccentrically loaded, with the load applied outside the plane of the welds.

The eccentric load is resolved into an axial load and a moment. The axial load is assumed to be uniformly distributed over the total length of weld. The weld stress (shear stress) resulting from the axial load may be computed from

$$f_v = \frac{P}{\text{total weld length}}$$

where

$P$ = beam reaction

$f_v$ = actual shear stress

The effect of the moment is to create a bending stress distribution varying linearly from zero at the weld neutral axis, which is assumed to occur at the middepth of the weld. The maximum bending stress (tension) in the weld is assumed to occur at the top end of the vertical weld.

The combined effect of the shear stress (a result of the axial load) and the bending stress (a result of the moment) will be the vectorial sum of the two. Since the stresses act at right angles to each other, the resultant stress may be computed from

$$f_r = \sqrt{f_h^2 + f_v^2}$$

where

$f_r$ = resultant stress

$f_h$ = maximum bending stress at top edge of weld

$f_v$ = actual shear stress

### Example 8-10

Calculate the allowable load for the seat angle-beam combination shown in Figure 8-26. All steel is A36. Assume an E70 electrode (SMAW).

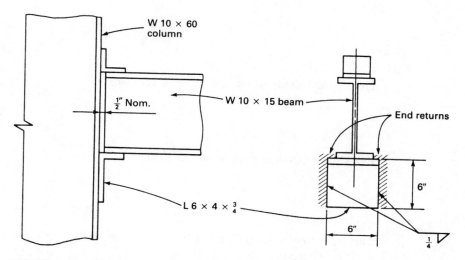

**FIGURE 8-26**   Welded seated beam connection.

**Solution:**

See Example 7-9 for the determination of the allowable load based on web yielding and bending of the seat angle. The analysis is the same for this example; the welded connection of the seat angle to the column flange must be checked, however.

The welds are subjected to both bending moment and direct (axial) load. The bending moment with respect to the face of the column flange is

$$M = R\left(0.75 + \frac{b}{2}\right)$$

$$= 20.7\left(0.75 + \frac{2.08}{2}\right) = 37.1 \text{ in.-kips}$$

(Note that bearing length $b$ is obtained from Example 7-9.)

Since there are two vertical lines of weld, the bending moment per line is

$$\frac{37.1}{2} = 18.6 \text{ in.-kips}$$

The bending stress distribution in these vertical welds is based on the assumption that the neutral axis occurs at the middepth of the weld. For each line of weld, the horizontal stress due to the bending moment is based on the 6-in. vertical leg. Therefore, the length of weld is 6 in. (Review Section 8-6 for stress per linear inch determination.)

$$f_h = \frac{Mc}{I} = \frac{18.6(3)}{\left(\dfrac{6^3}{12}\right)} = 3.1 \text{ kips/in.}$$

The vertical shear stress due to the direct load is

$$f_v = \frac{20.7}{12} = 1.73 \text{ kips/in.}$$

The resultant stress may be obtained from

$$f_r = \sqrt{f_h^2 + f_v^2}$$
$$= \sqrt{3.1^2 + 1.73^2} = 3.55 \text{ kips/in.}$$

The weld that is furnished is $\frac{1}{4}$ in. (see Figure 8-26), with a capacity of 3.70 kips/in. Therefore, this weld is satisfactory, and the allowable load for this connection is 20.7 kips (92.1 kN).

## 8-8

## WELDED FRAMED BEAM CONNECTIONS

Bolted framed beam connections are discussed in Chapter 7 of this text. In this section fillet welds, rather than high-strength bolts, are used as the fastening technique. As in the bolted connection, angles are used on the beam web to transmit the end reaction. The connection may be categorized as a simple beam connection.

The angles may be shop-welded to the beam web and subsequently field-welded to the supporting member, as shown in Figure 8-27. A combination-type connection

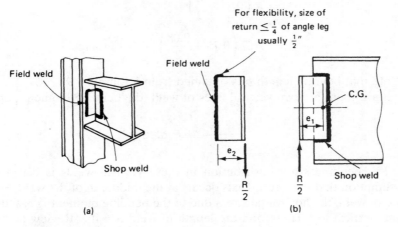

**FIGURE 8-27**  Welded framed beam connection.

**PHOTO 8-1**   Typical framed connection between a beam and the flange of a column. The angles have been shop-welded to the beam web and field-bolted to the column. Note the bolted connection for the open web steel joist (see Chapter 9).

is often used, in which the angles are shop-welded to the beam web but field-connected to the supporting member with high-strength bolts.

The bolted framed beam connection is designed to resist shear only, and all eccentricity is neglected. In the welded framed beam connection, however, the eccentricity of the end reaction with respect to the welds *is* taken into consideration.

Each angle is subject to a vertical shear equal to $R/2$, where $R$ is the beam reaction. In Figure 8-28a and b, this force may be observed as eccentric to both the shop and field welds, thus subjecting the welds to both an axial load and a moment. The moment may be expressed as

$$\frac{R}{2}(e)$$

where $e$ is the eccentricity, as shown in Figure 8-28, and may be either $e_1$ or $e_2$.

The design or analysis of the shop weld is basically the same as for the eccentrically loaded welded connection described in Section 8-6 of this text. The field welds, however, are subject to a rotational effect that forces the top portion of the web angles against the beam web, and the bottom portion of the angles is pushed apart, as indicated in Figure 8-28b. Hence the resistance to this rotation is the bearing of the angles on the beam web together with a horizontal shear stress in the field weld. It is commonly assumed that the neutral axis is located at one-sixth of the

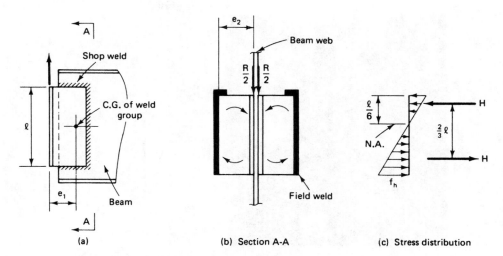

**FIGURE 8-28**  Connection behavior.

length of the angles from the top and that a triangular stress distribution exists, as shown in Figure 8-28c. The resisting moment becomes $H(\frac{2}{3}\ell)$, which must be equal to the applied moment $(R/2)(e_2)$. Equating the two in terms of the horizontal shear stress gives us

$$\frac{R}{2}(e_2) = \frac{1}{2}(f_h)\left(\frac{5}{6}\ell\right)\left(\frac{2}{3}\ell\right)$$

from which we obtain

$$f_h = \frac{R(e_2)}{0.56(\ell^2)}$$

The vertical shear $f_v$ in the field weld is equal to $R/2$ divided by the length of the weld. The vertical shear stress and horizontal shear stress may then be combined vectorially to determine the maximum shear stress:

$$f_r = \sqrt{f_h^2 + f_v^2}$$

**Example 8-11** _____

Design the shop and field welds for the W21 × 73 framed beam connection shown in Figure 8-29. Use an E70 electrode (SMAW). The web angles are L3 × 3 × $\frac{3}{8}$ × 10 in. The beam reaction is 50 kips. All structural steel is A36.

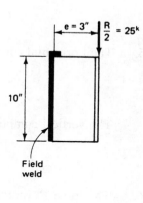

**FIGURE 8-29**   Weld configuration.

## Solution:

*Design of shop weld to beam web:*

1.  Locate the center of gravity (find $\bar{x}$) of the weld by summation of moments of lengths of weld about the 10-in. side:

$$\bar{x} = \frac{\Sigma \ell x}{\Sigma x} = \frac{2(2.5)(1.25) + 10(0)}{10 + 2.5 + 2.5}$$

$$\bar{x} = 0.42 \text{ in.}$$

2.  Determine the polar moment of inertia:

$$J = I_x + I_y$$

$$I_x = \left(\frac{1}{12}\right)(10)^3 + 2(2.5)(5)^2 = 208.3 \text{ in.}^3$$

$$I_y = (2)\left(\frac{1}{12}\right)(2.5)^3 + (2)(2.5)\left(\frac{2.5}{2} - 0.42\right)^2 + 10(0.42)^2$$

$$= 7.8 \text{ in.}^3$$

$$J = 208.3 + 7.8 = 216.1 \text{ in.}^3$$

3.  The torsional moment is

$$M = Pe$$

$$= \frac{R}{2}(e)$$

$$= 25(3.00 - 0.42) = 64.5 \text{ in.-kips}$$

4.   $r = \sqrt{(2.5 - 0.42)^2 + 5^2} = 5.42$ in. The force on the weld due to the torsional moment (at point $A$) is

$$P_m = \frac{Mr}{J} = \frac{64.5(5.42)}{216.1} = 1.62 \text{ kips/in.}$$

5.   The horizontal component of $P_m$ is

$$\frac{5}{5.42}(1.62) = 1.49 \text{ kips/in.}$$

The vertical component of $P_m$ is

$$\frac{2.08}{5.42}(1.62) = 0.62 \text{ kip/in.}$$

6.   The force $P_v$ on the weld due to the axial effect of the eccentric load where $P = R/2$ is

$$P_v = \frac{P}{\ell} = \frac{25}{15} = 1.67 \text{ kips/in.}$$

7.   Adding the forces vectorially and determining the resultant force $F$ gives

$$F^2 = 1.49^2 + (0.62 + 1.67)^2$$
$$F = 2.73 \text{ kips/in.}$$

8.   The fillet weld leg size required (number of sixteenths) is

$$D = \frac{2.73}{0.925} = 2.95$$

**Use a $\frac{3}{16}$-in. fillet weld.**

9.   Check to ensure that the shear capacity of the $\frac{3}{16}$-in. fillet weld (on each side of the beam web) does not exceed the shear capacity of the web. The shear capacity of the welds (per linear inch, with reference to Table 8-1 of this chapter) is

$$2(2.78) = 5.56 \text{ kips/in.}$$

The shear capacity of the web (A36 steel) is

$$F_v(t_w) = 0.40F_y t_w$$
$$= 0.40(36)(0.455) = 6.55 \text{ kips/in.}$$

$$5.56 \text{ kips/in.} < 6.55 \text{ kips/in.} \qquad\qquad \textbf{O.K.}$$

The $\frac{3}{16}$-in. fillet weld is satisfactory.

*Design of field weld to the supporting member (end return is neglected):*

1.  The horizontal shear due to the rotational effect, as shown in Figure 8-28b, is

$$f_h = \frac{Re}{0.56\ell^2} = \frac{50(3)}{0.56(10)^2} = 2.68 \text{ kips/in.}$$

2.  The vertical shear is

$$f_v = \frac{50}{2(10)} = 2.5 \text{ kips/in.}$$

3.  The resultant shear is

$$f_r = \sqrt{f_h^2 + f_v^2}$$
$$= \sqrt{2.68^2 + 2.5^2} = 3.67 \text{ kips/in.}$$

4.  The fillet weld size required (number of sixteenths) is

$$D = \frac{3.67}{0.925} = 3.97$$

**Use a $\frac{1}{4}$-in. fillet weld.**

Example 8-11 illustrates an elastic design method, which is based on the assumption that each unit of weld supports an equal share of the vertical load and a proportional share (dependent on the unit element's distance from the centroid of the group) of the eccentric moment portion of the load. This method is simple and conservative. The ASDM, Part 4, Tables XIX–XXVI, provide for an inelastic (ultimate strength) method. Reference 2 contains discussion of the design of connection angle welds using inelastic design.

**8-9**

## WELDING SYMBOLS

The use of welding symbols as a means of communication has been standardized by the American Welding Society. To prepare drawings properly, the steel detailer must have a means of accurately conveying complete information about the welding to the shop and erection personnel so that this information conforms to the intent of the designer. To avoid misunderstanding and confusion, it is important that the same standard system of weld symbols be used by everyone involved. The reader is referred to the ASDM, Part 4, Welded Joints, Standard Symbols.

Since fillet welds and simple butt welds compose 95% of most structural steel welding, a few of the more common symbols for fillet welds and groove welds are shown in Figure 8-30.

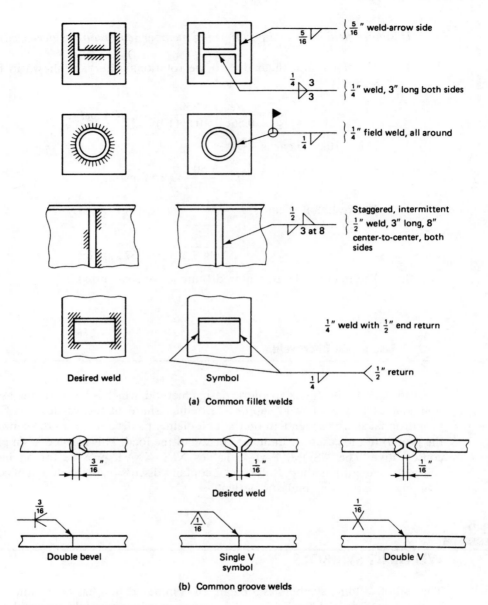

$\frac{5}{16}''$ weld-arrow side

$\frac{1}{4}''$ weld, 3" long both sides

$\frac{1}{4}''$ field weld, all around

Staggered, intermittent $\frac{1}{2}''$ weld, 3" long, 8" center-to-center, both sides

$\frac{1}{4}''$ weld with $\frac{1}{2}''$ end return

$\frac{1}{2}''$ return

Desired weld        Symbol

(a)  Common fillet welds

Double bevel        Single V symbol        Double V

Desired weld

(b)  Common groove welds

**FIGURE 8-30**  Weld symbols.

## 8-10
███

## WELDING INSPECTION

In the inspection phase, one is concerned primarily with the soundness and quality of a welded joint or weldment. Inspection should begin prior to the actual welding and should continue during welding as well as after the welding is completed. All personnel engaged in inspection operations should be familiar with their company inspection methods as well as all governing codes or standards. The service conditions to which the weldment might be subjected must be known and carefully evaluated before an inspection method can be specified.

The inspection process is only as good as the quality of the inspectors. Employment of competent inspectors is only one aspect of assuring weld quality. In addition, good welding procedures and the use of qualified and certified welders contribute to an acceptable weld. The weld testing methods generally used for structures may be categorized as nondestructive and include visual, magnetic particle, radiographic, liquid penetrant, and ultrasonic methods.

*Visual inspection* is probably the most widely used of all inspection methods. It is simple and inexpensive, and the only equipment commonly used is a magnifying glass. Although many factors are beyond the scope of visual examination, it must be regarded as one of the most important methods for determining weld quality. Visual inspection should begin before the first arc is struck. The materials should be examined to see whether they meet specifications for quality, type, size, cleanliness, and freedom from defects. Foreign matter, such as grease, paint, oil, oxide film, and heavy scale, which could be detrimental to the weld, should be removed. The pieces to be joined should be checked for straightness, flatness, and dimensions. Warped, bent, improperly cut, or damaged pieces should be repaired or rejected. Alignment, fit-up of parts, and joint preparation should be checked. Inspection prior to welding also includes verification that the correct process and procedures are to be employed and that the electrode type and size are as specified.

Inspection during welding may detect errors and defects that can easily be remedied. It prevents minor defects from piling up into major defects and leading to ultimate rejection. When more than one layer of filler metal is to be deposited, it may be necessary to inspect each layer before a subsequent layer is deposited. The greater the degree of supervision and inspection during welding, the greater the probability of the joint being satisfactory and efficient in service.

Visual inspection after the weldment has been completed is also useful in evaluating quality even if ultrasonic, radiographic, or other methods are to be employed. The following quality factors can usually be determined by visual means: dimensional accuracy of the weldment, conformity to specification requirements, weld appearance, and surface flaws such as cracks and porosity. With only surface defects visible to the eye, however, additional nondestructive methods may be necessary and specified.

*Magnetic particle inspection* is a nondestructive method used to detect the presence of cracks and seams in magnetic materials. It is not applicable to nonmagnetic materials. This method will detect surface discontinuities that are too fine to be seen with the naked eye, those that lie slightly below the surface, and when special equipment is used, the more deeply seated discontinuities. The basic principle involved in magnetic particle inspection is that when a magnetic field is established in a piece of ferromagnetic material that contains one or more discontinuities in the path of the magnetic flux, minute poles are set up at the discontinuities. These poles have a stronger attraction for the magnetic particles than the surrounding surface of the material. The particles form a pattern or indication on the surface that assumes the approximate shape of the discontinuity. Magnetic particle inspection is a relatively low-cost method of inspection and is considered outstanding for detecting surface cracks. It is also used to advantage on heavy weldments and assemblies.

*Radiographic inspection* is one of the most widely used techniques for showing the presence and nature of macroscopic defects and other discontinuities in the interior of welds. This test method is based on the ability of X-rays and gamma rays to penetrate metal and other opaque materials and produce an image on sensitized film or a fluorescent screen. It is a nondestructive test method and offers a permanent record when recorded on film. It is a relatively expensive type of inspection, and due to the radiation hazard, requires extensive safety precautions. Considerable skill is required in choosing angles of exposure, operating the equipment, and interpreting the results.

*Liquid penetrant inspection* is a nondestructive method for locating surface cracks and pinholes that are not visible to the naked eye. It is a favored technique for locating leaks in welds, and it can be applied where magnetic particle inspection cannot be used, such as with nonferrous metals. Fluorescent or dye penetrating substances may be used for liquid penetrant inspection.

Fluorescent penetrant inspection makes use of a highly fluorescent liquid with unusual penetrating qualities. It is applied to the surface of the part to be inspected and is drawn into extremely small surface openings by capillary action. The excess liquid is then removed from the part, a "developer" is used to draw the penetrant to the surface, and the resulting indication is viewed by ultraviolet (black) light. The high contrast between the fluorescent material and the background makes possible the detection of minute traces of penetrant.

Dye penetrant inspection is similar to fluorescent penetrant inspection except that dyes visible under ordinary light are used. By eliminating the need for ultraviolet light, greater portability in equipment is achieved.

*Ultrasonic inspection* is a rapid and efficient nondestructive method of detecting, locating, and measuring both surface and subsurface defects in the weldment and/or base materials. Flaws that cannot be discovered by the other methods, and even cracks small enough to be termed microseparations, may be detected. Ultrasonic testing makes use of an electrically timed wave of the same nature as a sound wave, but of a higher pitch or frequency. The frequencies used are far above those heard by the human ear, hence the name ultrasonic. The sound waves or vibrations are propagated in the metal that is being inspected until a discontinuity or change of density is reached. At these points, some of the vibrational energy is reflected back and indicated on a cathode-ray tube. The pattern on the face of the tube is thus a representation of the reflected signal and of the defect. The ultrasonic method requires special commercial equipment, and a high degree of skill is required in interpreting the cathode-ray tube patterns.

Those who require more detailed information on inspection and all phases of welding should obtain Reference 3.

## REFERENCES

[1]  *Structural Welding Code—Steel,* AWS D1.1-88, American Welding Society, 2501 N.W. Seventh Street, Miami, FL 33125.

[2]  K. M. Loomis, et al., "A Design Aid for Connection Angle Welds Subjected to Combined Shear and Axial Loads," *AISC Engineering Journal,* 4th Qtr., 1985.

[3]  *Welding Handbook,* latest edition, American Welding Society, 2501 N.W. Seventh Street, Miami, FL 33125. A multivolume series covering practically all aspects of welding.

## PROBLEMS

**8-1.**  Determine the allowable tensile load that may be applied to the connection shown. The steel is A36, and the electrode used was an E70 (SMAW). The weld is a $\frac{3}{8}$-in. fillet weld.

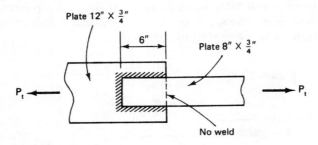

Plate 12" $\times \frac{3}{4}$"

6"

Plate 8" $\times \frac{3}{4}$"

$P_t$

$P_t$

No weld

**PROBLEM 8-1**

**8-2.**  Determine the allowable tensile load that may be applied to the connection shown. The steel is A36, and the electrode used was an E70 (SMAW). The weld is a $\frac{1}{4}$-in. fillet weld.

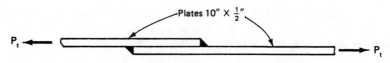

**PROBLEM 8-2**

**8-3.** Determine the maximum tensile load that can be applied to the wide-flange tension member shown. The welds are $\frac{1}{4}$-in. fillet welds. The steel is A36, and the electrode used was an E70.

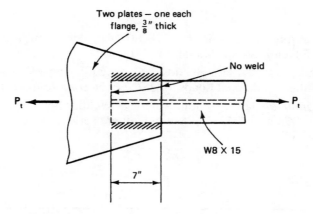

**PROBLEM 8-3**

**8-4.** A fillet weld made between two steel pieces intersecting at right angles was determined to have one $\frac{1}{4}$-in. leg and one $\frac{9}{16}$-in. leg. The electrode was an E80 (SMAW). Determine the strength of this weld in kips/in.

**8-5.** Design longitudinal fillet welds to develop the full tensile capacity of the 8-in. $\times$ $\frac{1}{2}$-in. plate shown. The steel is A36, and the electrode is an E70 (SMAW). Standard end returns are used.

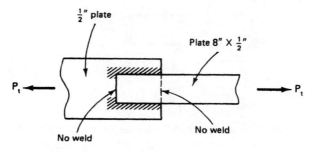

**PROBLEM 8-5**

**8-6.** Design an end connection using longitudinal welds and an end transverse weld to develop the full tensile capacity of the angle shown. Use A36 steel and E70 electrodes (SMAW).

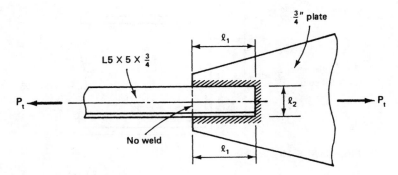

**PROBLEM 8-6**

**8-7.** Design an end connection using only longitudinal welds to develop the full tensile capacity of the angle described. The member is subjected to repeated stress variations. Use A36 steel and an E70 electrode (SMAW). The angle is an L7 × 4 × $\frac{3}{4}$ with the long leg connected to a $\frac{9}{16}$-in. gusset plate. Use the maximum allowable weld size.

**8-8.** Rework Problem 8-7 using a transverse weld at the end of the angle in addition to the longitudinal welds.

**8-9.** Design the connection of a C15 × 40 to a $\frac{3}{8}$-in. gusset plate as shown to develop the full tensile capacity of the channel. Welding is not permitted across the back of the channel. All structural steel is A36. Use an E70 electrode (SMAW). The maximum length of lap on the gusset plate is 6 in. due to space limitations.

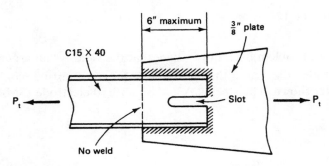

**PROBLEM 8-9**

**8-10.** Compute the maximum allowable beam reaction for the end-plate shear connection shown. All structural steel is A36. The electrode is an E70 (SMAW).

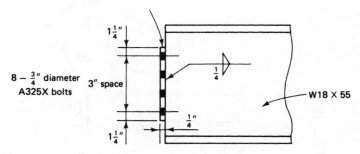

$8 - \frac{3}{4}''$ diameter A325X bolts

$1\frac{1}{4}''$   $3''$ space   $1\frac{1}{4}''$   $\frac{1}{4}$   $\frac{1}{4}''$   W18 × 55

**PROBLEM 8-10**

**8-11.** With reference to Problem 7-22, design the welded portion of the end-plate shear connection using an E70 electrode (SMAW).

**8-12.** Calculate the maximum resultant force (kips/in.) to be resisted by the welds shown.

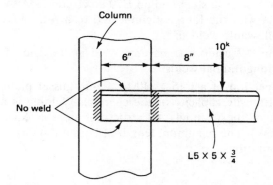

Column   6"   8"   $10^k$   No weld   L5 × 5 × $\frac{3}{4}$

**PROBLEM 8-12**

**8-13.** Rework Problem 8-12, but assume that the weld is "all around."

**8-14.** Determine the size of the fillet weld required to resist a 65-kip load on the bracket shown. The steel is A36. Use an E70 electrode (SMAW).

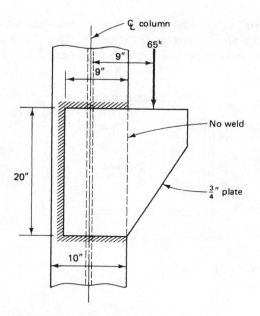

**PROBLEM 8-14**

**8-15.** Rework Problem 8-14 using the ASDM ultimate strength method (using ASDM tables).

**8-16.** An unstiffened seated beam connection for a beam-to-column-web connection is to be composed of an L8 × 4 × 1 (6 in. long), with the long leg of the seat angle shop-welded to the column web with $\frac{5}{16}$-in. fillet welds. The welds are placed as shown in Figure 8-26. The beam is a W16 × 77, and the column is a W10 × 49. All steel is A36. Assume an E70 electrode (SMAW). Determine the allowable load for the connection.

**8-17.** Compute the maximum allowable beam reaction for the welded framed beam connection shown. All structural steel is A36, and the electrode used was an E70 (SMAW).

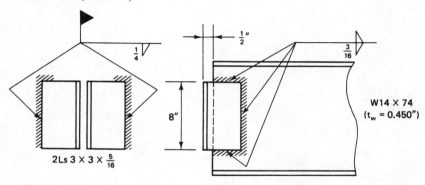

**PROBLEM 8-17**

**8-18.** Design the shop and field welds for the framed connection for the W18 × 71 beam shown. Use an E70 electrode (SMAW). All structural steel is A36. The beam reaction is 40 kips, and the web angles are L3 × 3 × ¼ × 10 in.

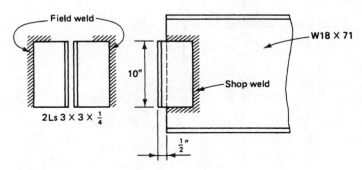

**PROBLEM 8-18**

# CHAPTER 9

# Open Web Steel Joists and Metal Deck

9-1   INTRODUCTION TO STEEL JOISTS

9-2   OPEN WEB STEEL JOISTS, K-SERIES

9-3   FLOOR VIBRATIONS

9-4   CORRUGATED STEEL DECK

## 9-1

### INTRODUCTION TO STEEL JOISTS

Steel joists are standardized prefabricated trusses generally used for the direct support of floor decks and/or roof decks in buildings. They are commonly used in combination with a corrugated steel metal deck and generally provide an efficient and economical floor or roof system in lightly loaded buildings. A typical floor or roof system includes other suspended or supported materials (in addition to the joists and metal deck), such as suspended ceilings and roofing materials. Discussion of these other materials is not included in this text. The reader is referred to texts on construction materials and appropriate manufacturers' literature (see the references at the end of this chapter). There are three categories of steel joists available: (1) Open Web Steel Joists, K-Series; (2) Longspan Steel Joists, LH-Series; and (3) Deep Longspan Steel Joists, DLH-Series. Primary structural members called *Joist Girders* are also available.

*Open Web Steel Joists, K-Series* are members that have the appearance of shallow trusses with parallel chords. They are completely standardized as to length, depth, and load-carrying capacities. They are suitable for use in both floor and roof applications. The standard depths for the K-Series joists are 8 in. to 30 in., varying by 2-in. increments. Span lengths of up to 60 ft are common for the K-Series joists.

In the United States, the design, fabrication, and erection of steel joists are generally accomplished in accordance with the requirements of the specifications published by the Steel Joist Institute (SJI) [1]. The SJI publication is a valuable reference source and should be obtained by the reader who desires more in-depth information. The institute is a nonprofit organization of steel joist manufacturers with the primary purpose of promoting the use of steel joists. The *Standard Specifications* first appeared in 1928 and has been modified over the years to reflect progress in research, manufacturing, materials, and welding techniques.

*Longspan Steel Joists, LH-Series* and *Deep Longspan Steel Joists, DLH-Series* are similar in general appearance to the K-series joists, but are deeper and span greater distances. Longspan Series (LH) joists have been standardized in depths from 18 in. to 48 in. for clear spans to 96 ft and are generally used in both floor and roof applications. Deep Longspan Series (DLH) joists have been standardized in depths from 52 in. to 72 in. for clear spans up to 144 ft and are generally used for roof applications. The specifications applying to the LH-Series and the DLH-Series joists are found in Reference 1. Both series can be furnished either with parallel chords or with single- or double-pitched top chords to provide sufficient slope for roof drainage.

The design of the LH- and the DLH-Series is based on a steel yield strength of at least 36,000 psi but not greater than 50,000 psi. The standard designation for LH-Series joists—for example, 28LH09—furnishes first the depth of the member, 28 in. The designation LH represents the series. The final number, 05 through 13 (for the 28-in.-deep members), denotes the relative size of the chords, the size increasing with the number. Other depth LH-Series joists may have the last two digits different from those of the 28-in.-deep series. The deep longspan series designation is similar, but the designation DLH replaces LH.

The design of all standardized joists has become the responsibility of the joist manufacturers. The selection of which joists to use, irrespective of series, involves the use of standard load tables as furnished by the SJI and, commonly, by the joist manufacturer as well. Hence, when the designer of a building decides on the use of steel joists as part of the floor or roof system, the designer does not *design* the joists, but rather *selects* the proper joists from the load tables based on span length, loading, and joist spacing. The K-Series (excluding the new KCS grouping within the K-Series, discussed shortly), LH-Series, and DLH-Series joists are all designed as simply supported *uniformly* loaded trusses supporting a floor or roof deck so that the top chords of the joists are adequately braced against lateral buckling. Where joists are used under conditions different from those for which they were originally designed, they must be investigated and modified as necessary since the load tables are no longer applicable.

The SJI tables used for the selection of K-Series joists are contained in Appendix A (Tables A-1 and A2) in the back of this text. Two load-carrying capacities are tabulated in the body of the tables for each joist-span combination. The upper number represents the *total* load-carrying-capacity of the joist. The *dead* load, including the weight of the joist, should be deducted to determine the *live*-load-carrying capacity of the joist. The lower number (normally printed in color in the SJI tables) represents the load per linear foot of joist that will produce a deflection of span/360. Normally, it is only the live load that is significant in this consideration. Loads that will produce a deflection of span/240 may be obtained by multiplying the lower number by 1.5. Example problems will demonstrate the use of these tables. Tables for the selection of LH-Series and DLH-Series joists are found in reference 1.

The Steel Joist Institute defines *span* for K-Series joists as shown in Figure 9-1. Note that for joists supported on rolled steel shapes (and on joist girders), the span

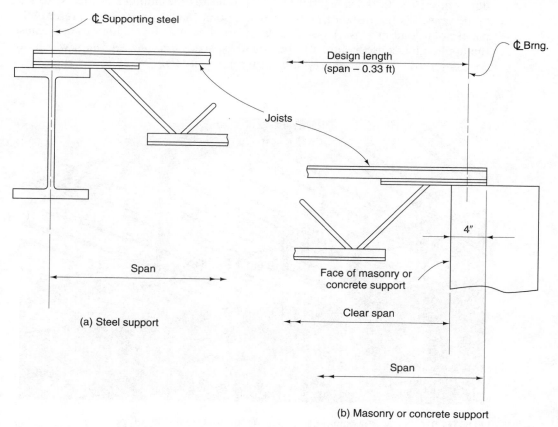

**FIGURE 9-1**   Definition of span for typical end connections for K-Series joists.

is defined as the distance between centerlines of supports. This is the length on which the selection of the joist is based. For joists on masonry or concrete supports, the span is measured to the end of the joist. Assuming a 4-in. (0.33-ft) bearing length, the selection of the joist is based on the center-of-bearing to center-of-bearing length, which is called the *design length*. In the case of masonry or concrete supports, then, the design length can be calculated as the span length minus 0.33 ft (or the clear span plus 0.33 ft).

Another type of open web steel prefabricated truss is the joist girder. This product is used as a primary framing member and is designed as a simple span supporting panel point concentrated loads from a floor or roof system. It may be observed in Photo 9-1. These members have been standardized for depths from 20 in. to 72 in. and span lengths to 60 ft (although some manufacturers furnish joist girders up to 96 in. in depth and with a span length of 100 ft).

Standard specifications for joist girders, as well as standard load tables titled Design Guide Weight Tables for Joist Girders, are furnished in SJI *Standard Specifications*. A typical standard designation is **48G8N8.8K,** where 48 indicates the girder depth in inches, G indicates "girder," 8N indicates the number of joist spaces (a "joist space" is the distance between panel point loads), and 8.8K represents the panel point load in kips. A detailed designation such as this is important because the applied loads are assumed to be equally spaced *concentrated loads,* which are considered to act at the *panel points* of the joist girders.

**PHOTO 9-1**   Typical roof and deck construction for a two-story office building. Open web steel joists for the roof are being welded to supporting joist girders.

Our discussion in this chapter is based on the U.S. Customary System as the primary unit system. The SJI, in its most recent (40th edition) publication of specifications and load tables for joists and joist girders, has included metric nomenclature and equivalents (see Reference 1). In the load tables, depths are tabulated in millimeters (mm), approximate mass is given in both kilograms per meter (kg/m) and kilonewtons per meter (kN/m), spans are in millimeters (mm), and safe uniformly distributed loads are given in kilonewtons per meter (kN/m).

Joists are typically designed for in-plane (vertical) load and have very little resistance to out-of-plane (lateral) load. Because of this, the time period from the initial placement of the joists at their final locations in the structure until all required bridging has been installed is extremely critical. The applicable SJI K-Series specifications covering application and erection stability and handling are summarized in the following section. The reader is encouraged to refer to the latest SJI publications for complete specifications for all steel joists as well as the recommended code of standard practice for steel joists and joist girders.

## 9-2

## OPEN WEB STEEL JOISTS, K-SERIES

The design of the K-Series joist *chord* is based on a steel minimum yield strength of 50,000 psi. The design of the *web members* may be based on a steel minimum yield strength of 36,000 psi or 50,000 psi.

An example of the standard designation for K-Series joists is 22K7. The depth of this joist is 22 in. K represents the series, and the number 7 denotes the relative size of the chords of the joist. Chord sizes are designated by the numbers 3 through 12, the size increasing with increasing number. The chord and web members may vary in shape and makeup from manufacturer to manufacturer, but the design and the capacity of the joists must conform to the SJI specifications and to the standardized load tables. The K-Series standard load table is applicable where the joists are installed up to a maximum slope of $\frac{1}{2}$ in. per foot.

The use of open web steel joists in any given application must be based on SJI requirements as furnished in its standard specifications. These requirements for the K-Series joists are summarized as follows:

1.   The span of a joist must not exceed 24 times its depth.
2.   K-Series joists supported by masonry or concrete are to bear on steel bearing plates. The ends of the joists shall extend a distance of not less than 4 in. over the masonry or concrete support and be anchored to the steel bearing plate. The plate shall be located not more than $\frac{1}{2}$ in. from the face of the wall and shall be not less than 6 in. wide perpendicular to the length of the joist. The steel bearing plate must be designed in accordance with the latest ASDM and shall be furnished by other than the joist manufacturer.

For the case where 4-in. bearing cannot be furnished, a special design of the steel bearing plate is necessary. The joist must then bear a minimum of $2\frac{1}{2}$ in. on the plate, however.

The ends of K-series joists must extend a distance of not less than $2\frac{1}{2}$ in. over structural steel supports.

3.  In construction that uses joists, bridging and bridging anchors are required for the primary purpose of furnishing lateral stability for the joists, particularly during the construction phase. The bridging spans between and perpendicular to the steel joists.

    It is required that one end of all joists be attached to their supports before allowing the weight of an erector on the joists. When bolted connections are used, the bolts must be snug tightened. All bridging must be completely installed and the joists permanently fastened into place before the application of any construction loads. Even under the weight of an erector, the joists may exhibit some degree of lateral instability until the bridging is installed. The bridging also serves the purpose of holding the steel joists in position as shown on the plans. The minimum number of rows of bridging is a function of the joist chord size and span length. A table is furnished in the standard specifications that establishes the required number of rows of bridging. Spacing of bridging rows should be approximately equal. Reference 2 also contains pertinent information. Two permissible types of bridging may be observed in Figure 9-2. *Horizontal bridging* (Figure 9-2a) consists of two continuous horizontal steel members, one attached to the top chord and the other attached to the bottom chord by means of welding or mechanical fasteners. The attachment must be capable of resisting a horizontal force of not less than 700 lb. If the bridging member is a round bar, the diameter must be at least $\frac{1}{2}$ in. The maximum slenderness ratio ($\ell/r$) of the bridging member cannot exceed 300, where $\ell$ is the distance between bridging attachments and $r$ is the least radius of gyration of the bridging member. The bridging member shall be designed for a compressive force of 0.24 times the area of the top chord. *Diagonal bridging* (Figure 9-2b) consists of cross-bracing with a maximum $\ell/r$ of 200, with $\ell$ and $r$ as defined previously. Where the cross-bracing members connect at their intersection, $\ell$ is the distance between the intersection attachment and chord attachment. The ends of all bridging lines terminating at walls or beams must be properly anchored. A typical detail may be observed in Figure 9-2b.

    Where the span of the joist exceeds the erection stability span, as indicated by the shaded areas in the load tables, the row of bridging nearest the midspan of the joist shall be diagonal bridging (cross-bracing) with bolted connections at the chords of the joists and the intersections of the cross-braces. Furthermore, the hoisting cables must not be released until this bolted diagonal bridging is completely installed.

4.  Positive end anchorage of the joists must be provided in addition to the required end bearing length. Ends of K-Series joists resting on structural steel supports or on steel bearing plates on masonry or concrete shall be attached

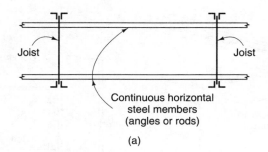

(a)

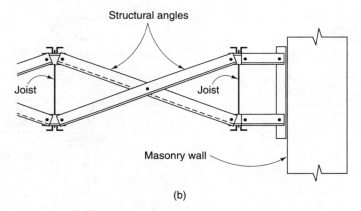

(b)

**FIGURE 9-2**  Typical bridging.

thereto with a minimum of two $\frac{1}{8}$-in. fillet welds 1 in. long (as shown in Figure 9-3a), with two $\frac{1}{2}$-in.-diameter bolts, or with a combination of one $\frac{1}{8}$-in. fillet weld 1 in. long and one $\frac{1}{2}$-in.-diameter bolt. Where structural steel columns are not framed in at least two directions with structural steel members, the joists at column lines must be field-bolted at the columns to assure some measure of lateral stability during construction (see Figure 9-3b).

5.  Joist extensions are frequently used with K-Series joists to support a variety of overhang conditions. Two types are shown in Figures 9-3c and 9-3d. The first is the *Top Chord Extension (S Type)*, which has only the top chord angles extended. The second is the *Extended End (R Type)*, in which the standard $2\frac{1}{2}$-in. end bearing depth is maintained over the entire length of the extension. The R Type (*reinforced*) involves reinforcing the top chord. The S Type (*simple*) is more economical and should be specified whenever possible.

Load tables for K-Series Top Chord Extension and Extended Ends are furnished by the SJI. Specific designs and load tables, however, are generally furnished by the various joist manufacturers and can be used to advantage.

6.  Ceiling extensions (Figure 9-3b) in the form of an extended bottom chord element or a loose unit, whichever is standard with the joist manufacturer,

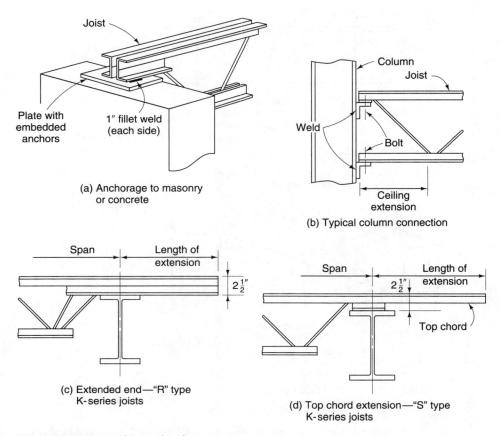

(a) Anchorage to masonry or concrete

(b) Typical column connection

(c) Extended end—"R" type K-series joists

(d) Top chord extension—"S" type K-series joists

**FIGURE 9-3**   Typical joist details.

are frequently used to support ceilings that are to be attached directly to the bottom of the joists. They are not furnished for the support of suspended ceilings.

7.  When joists are used in conjunction with a corrugated metal deck and concrete slab, the cast-in-place slab should not be less than 2 in. thick.

The typical standard K-Series joist is designed for a simple span subjected to uniformly distributed load for its full span length, resulting in a linear shear distribution (maximum at the supports and zero at midspan) and a parabolic moment distribution (zero at the supports and maximum at midspan). The KCS joist is a new type of K-Series joist developed to overcome some of the limitations of the standard K-Series joist. The KCS joist may be used for special design applications requiring a joist capable of supporting nonuniform loads, concentrated loads, or combinations thereof in addition to or independent of the normal uniform load.

The KCS joists are designed in accordance with the SJI *Standard Specifications* for K-Series joists and range in depth from 10 in. to 30 in. Load tables furnished by the SJI provide the shear and moment capacity of each joist. The designer must calculate the maximum moment and shear imposed and then select the appropriate KCS joist.

**Example 9-1**

Select open web steel K-Series joists for a floor system of a typical interior bay of a commercial building, as shown in Figure 9-4. The joists are to span in the direction indicated. The span length is 26 ft. The allowable live load deflection is span/360. The floor loadings are 40 psf superimposed dead load (DL) and 100 psf live load (LL). Use a joist spacing of 2 ft–0 in. on center.

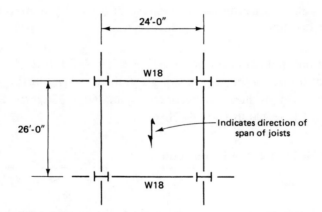

**FIGURE 9-4** Framing plan; typical interior bay.

**Solution:**

Calculate the load per linear foot on a joist:

$$DL = \ \ 40(2) = \ \ 80 \, \text{lb/ft}$$
$$LL = 100(2) = \underline{200 \, \text{lb/ft}}$$
$$\text{Total} = 280 \, \text{lb/ft}$$

The minimum joist depth, as per the SJI *Standard Specifications*, is span/24. Thus

$$\frac{26(12)}{24} = 13 \, \text{in.}$$

Refer to the K-Series Standard Load Table (Table A-1 in the appendices) and select the following joists as possible solutions. Note that we select only joists that have total load capacities greater than 280 lb/ft (plus the joist weight) and tabulated load to cause span/360 deflection greater than 200 lb/ft.

| Joist | Weight (lb/ft) | Tabulated load-carrying capacity (lb/ft) |
|-------|----------------|-------------------------------------------|
| 18K4  | 7.2            | 328/222                                   |
| 20K3  | 6.7            | 304/236                                   |

The 20K3 is the most economical (lightest, at 6.7 lb/ft); therefore, this joist will be checked first.

1.  Superimposed load capacity = 304 − 6.7 = 297.3 lb/ft
2.  Actual total superimposed load = 280 lb/ft < 297.3 lb/ft          **O.K.**
3.  LL that will produce a deflection of span/360 = 236 lb/ft
4.  Actual LL = 200 lb/ft < 236 lb/ft                                **O.K.**

**Use the 20K3 at 2 ft–0 in. on center.**

Note that the selection of the most economical joist may be simplified by using the K-Series Economy Table (Table A-2 in the appendices). In this table, the K-Series joists are arranged in order of increasing weight per foot. To use the table, determine the span (ft) and the load (lb/ft). Enter the table with the span and read across until a joist is found that satisfies the load requirement and the depth requirement. Example 9-2 uses this table.

**Example 9-2**

Select open web steel joists for a roof system of a typical interior bay of a commercial building, as shown in Figure 9-5. Assume a joist span length of 48 ft. The roof loadings are 25 psf superimposed dead load (DL) and 45 psf snow load (LL). Use a joist spacing of 3 ft–6 in. on center. The allowable live load deflection is span/240, and the desired maximum joist depth is 26 in.

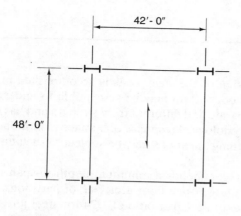

**FIGURE 9-5**  Framing plan—typical interior bay.

**Solution:**

Calculate the load per linear foot on the joist:

$$DL = 25(3.5) = \ \ 87.5 \ lb/ft$$

$$snow \ load = 45(3.5) = \underline{157.5 \ lb/ft}$$

$$total = \ \ 245 \ lb/ft$$

The minimum joist depth as per the SJI *Standard Specifications* is span/24. Thus

$$\frac{1}{24}(48)(12) = 24 \ in.$$

From Table A-2 in the appendices, the lightest K-Series joist with sufficient capacity is the 30K9, with total load capacity of 266 lb/ft and a live load deflection figure of 160 lb/ft. Its depth, however, exceeds the desired maximum depth of 26 in.

We next try a 26K10, with a total load capacity of 272 lb/ft. Adding the joist weight, we have

$$245 + 13.8 = 259 \ lb/ft < 272 \ lb/ft \qquad \textbf{O.K.}$$

The LL that will produce a deflection of span/360 is 140 lb/ft. The LL that will produce a deflection of span/240 is therefore calculated as

$$140 \times 1.5 = 210 \ lb/ft$$

The actual LL is 157.5 lb/ft:

$$157.5 \ lb/ft < 210 \ lb/ft \qquad \textbf{O.K.}$$

**Use the 26K10 at 3 ft–6 in. on center.**

**9-3**

## FLOOR VIBRATIONS

Even when the structural design of the steel joists is accomplished in accordance with design specifications, a floor system may be susceptible to undesirable vibrations. This phenomenon is separate and different from strength and has to do mainly with the psychological and physiological response of humans to motion. Large open floor areas without floor to ceiling partitions may be subject to such undesirable vibrations.

The ASDS Commentary recommends a minimum depth-to-span ratio of 1/20 for a steel beam supporting a large open floor area free of partitions. In addition, the SJI requires a minimum depth-to-span ratio of 1/24 for steel joists, although a generally accepted practice for steel joist roofs and floors is to use a minimum depth-to-span ratio of 1/20. Even if these recommendations and requirements are satisfied, a vibration analysis should be made, particularly when a floor system is composed of steel joists that support a thin concrete slab placed on steel metal deck. References 3 and 4 contain relatively brief and sufficiently accurate methods that can be used to determine (1) whether disturbing vibrations will be present in a floor system, and (2) possible design solutions for the problem.

**9-4**

## CORRUGATED STEEL DECK

Steel deck is commonly used in conjunction with steel joists in floor and roof systems, as shown in Figure 9-6. Most decking used in buildings today is designed, manufactured, and erected in accordance with the Steel Deck Institute specifications and code of recommended standard practice [5]. The *Specifications for the Design*

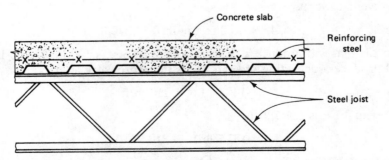

**FIGURE 9-6**  Corrugated steel deck on open web steel joists.

*of Cold-Formed Steel Structural Members* of the American Iron and Steel Institute [6] also apply.

Steel decks are cold-formed steel products with longitudinal ribs of various configurations distinctive to individual manufacturers. Originally, steel decks were used only for roof construction; as a result of testing and research, however, applications now include floor decks as well. All steel decks are cold-formed from various thicknesses of sheet steel. Whereas it has been common practice (and still is) to classify steel decks by a gage designation, the Steel Deck Institute has replaced the gage value with a design thickness as the unit of measure in references to material thickness. It should be noted that the gage designation is still used, but is now termed the "Type No." Both designations, Type No. and material thickness, are tabulated by the Steel Deck Institute.

When used as a roof deck, the decking acts as a structural supporting member. It must support the applied roof system, such as rigid insulation and built-up roofing or insulating concrete and roofing material, plus roof snow load and/or design live load. When used as the structural supporting member, the steel deck must be galvanized, aluminized, or prime painted. The primer coat is intended to protect the steel for only a short period of exposure in ordinary atmospheric conditions and is considered an impermanent and provisional coating. Field painting of prime painted deck is recommended where the deck is exposed to corrosive or high-moisture environments. Roof deck may be furnished with narrow ribs, intermediate ribs, wide ribs, or deep ribs, and in various thicknesses depending on the manufacturer. All manufacturers have load tables indicating load-carrying capacities for the different types of decks for varying span lengths.

When used as a floor deck, the steel deck may act in two different ways.

1.   As a form only, for a structural concrete slab until the concrete reaches its design strength. In this case the design load will consist of the weight of the wet concrete plus a construction live load of 20 psf (uniformly distributed load) or a concentrated load of 150 lbs acting on a section of deck that is 1 ft–0 in. wide. In this application the steel deck is generally furnished uncoated, but it may also be furnished galvanized or painted with a shop coat of primer paint (one or both sides). If uncoated or painted deck is used as the form, the weight of the concrete slab must be deducted from the allowable live load of the reinforced concrete slab. If galvanized form is used, the weight of the slab is considered to be permanently carried by the deck and need not be deducted from the live load. For this application, the steel deck is designated "Non-Composite Steel Form Deck" by the Steel Deck Institute.

2.   The steel deck may also act compositely with the concrete slab. The steel deck in effect acts as a permanent form providing the positive moment reinforcement for the concrete slab and may eliminate the need for any additional reinforcement. Composite floor decks are designed to interlock positively with the overlying concrete, resulting in unit action. The interlocking process is achieved by mechanical means, deck profile, surface bond, or a combination of these. Since the composite steel deck is the positive moment reinforcement

for the slab, it must be designed to last the life of the structure. Therefore, the minimum recommended finish is a galvanized coating. For this application, the steel deck is designated "Composite Steel Floor Deck" by the Steel Deck Institute. Welded wire fabric should be provided in all composite floor deck slabs, primarily for purposes of crack control rather than negative reinforcing.

All steel deck, whether roof, form, or floor deck, when in place and properly attached, is usually assumed to provide lateral restraint for the compression flange (assuming simple spans) of the supporting members. The steel deck offers satisfactory in-plane stiffness in a structure, assuming proper attachments to the supports. This is particularly critical where steel deck is supported by open web steel joists, since the joists have a minimum of lateral stability with their compression flange unsupported laterally. Connections to the supporting member flange are usually accomplished by mechanical fasteners such as naillike fasteners driven with either pneumatic devices or powder-actuated tools, by self-tapping screws, or by puddle welding from the top through the metal deck. For thin metal decks (less than Type No. 22) that have a minimum thickness less than 0.028 in., welding washers must be used for the welded connections (see Photo 9-2).

The spacing of the connections should not exceed 12 in. on centers, or as otherwise recommended, and the deck should be connected to all supporting members.

**PHOTO 9-2**   This metal deck has been welded to supporting open web steel joists. Square welding washers have been used to reinforce the thin decking in the weld area.

Prior to specifying any steel deck, the reader is encouraged to consult the manufacturers' literature. Design load tables should be read carefully and used with caution, since there is little consistency from one manufacturer to another with respect to table format.

Composite action for supporting steel beams can be achieved by welding shear studs through the metal deck onto the top flange of the beam, as shown in Figure 5-17. As discussed in Chapter 5, composite action depends on the steel beam-concrete interaction, and tests have demonstrated that the ribs of the steel deck do not interfere with this interaction. Limitations as to stud diameter and length may exist, however, along with other design criteria for the various steel decks.

## REFERENCES

[1] Steel Joist Institute, *Standard Specifications Load Tables and Weight Tables*, published annually by The Steel Joist Institute, 1205 48th Ave. North, Myrtle Beach, SC 29577.

[2] Steel Joist Institute, *Spacing of Bridging for Open Web Steel Joists*, Technical Digest No. 2, The Steel Joist Institute, 1205 48th Ave. North, Myrtle Beach, SC 29577.

[3] Kenneth H. Lenzen, "Vibration of Steel Joist—Concrete Floors," *AISC Engineering Journal*, July 1966.

[4] Steel Joist Institute, *Vibration of Steel Joist—Concrete Slab Floors*, Technical Digest No. 5, The Steel Joist Institute, 1205 48th Ave. North, Myrtle Beach, SC 29577.

[5] Steel Deck Institute, *Design Manual for Composite Decks, Form Decks, Roof Decks, and Cellular Metal Floor Deck with Electrical Distribution*. Publication No. 28, 1993, The Steel Deck Institute, P.O. Box 9506, Canton, OH 44711.

[6] American Iron and Steel Institute, *Specifications for the Design of Cold-Formed Steel Structural Members*, American Iron and Steel Institute, 1000 16th Street, N.W., Washington, DC 20036.

## PROBLEMS

*Note: 1. Span lengths in these problems are center-to-center of supports.*
*2. Refer to the K-Series Standard Load Table and Economy Table in Appendix A of this text.*

**9-1.** **(a)** What is the designation of the shallowest K-Series joist?

**(b)** What is the designation of the deepest, heaviest K-Series joist?

**(c)** What is the approximate weight of a 24K8?

**(d)** What is the total safe load for a 22K7 on a 38-ft span?

**(e)** For a 24K6 on a span of 34 ft, what live load will cause a deflection of span/360?

**9-2.** Calculate the permissible total live load for a 30K11 joist that has a span length of 56 feet and supports 150 lb/ft estimated dead load in addition to its own weight. Assume the maximum allowable live load deflection to be span/360.

**9-3.** Determine the total live load that will produce a deflection of span/240 for a 22K6 joist that has a span length of 40 ft.

**9-4.** A 28K6 joist spans 50 ft and supports a roof with a superimposed dead load of 20 psf excluding its own weight. Joists are spaced 4 ft on center. Determine the allowable live load (psf) if the allowable live load deflection = span/240.

**9-5.** Select steel joists for a floor system. The span length is to be 34 ft. The floor loading consists of 100 psf live load and 40 psf superimposed dead load. The allowable live load deflection is span/360. Use a joist spacing of 2 ft–8 in. on center.

**9-6.** If 24K7 joists were available for the floor in Problem 9-5, determine the maximum spacing for these joists in order that they be acceptable to carry the load.

**9-7.** Select open web steel joists (K-Series) for a floor system with a span length of 21 ft. The floor loading is live load 60 psf and superimposed dead load 30 psf. The allowable live load deflection is span/240. Use a joist spacing of 2 ft–0 in.

**9-8.** Select open web steel joists (K-Series) for a roof system with a span length of 40 ft. The roof loading is snow load 45 psf and superimposed dead load 20 psf. The allowable live load deflection (snow load) is span/360. Use a joist spacing of 5 ft–0 in. The maximum desired depth is 26 in.

**9-9.** Select open web steel joists for a floor system with a span length of 30 ft. Floor loading is live load 100 psf and superimposed dead load 40 psf. Allowable live load deflection is span/360. Use a joist spacing of 2 ft–6 in.

**9-10.** The job must go forward. Determine the maximum joist spacing to be used for the floor system of Problem 9-9 if the only joist immediately available from the local supplier is the 24K4.

# CHAPTER 10

# Continuous Construction and Plastic Design

## 10-1

### INTRODUCTION

Continuous construction may be defined as a structural system in which individual members are integrally attached so that they behave as single members. For example, if three simply supported beams were placed end to end with their ends connected by moment-resisting connections, the resulting beam would be considered a *continuous beam* spanning four supports, as shown in Figure 10-1. A continuous beam may be considered to be any beam that spans more than two supports. This type of member is frequently used in modern structures and generally offers economy with respect to the beam itself when compared with a series of simple beams over the same spans. The effect of the beam continuity is to reduce the maximum bending moment from that of the simple beam, thereby reducing the required beam size. Continuity in a structure also generally serves to increase stiffness and decrease deflections. Other factors associated with continuous beams may offset the beam savings, however, making the question of total structural economy difficult to establish.

   *Continuous frames,* which may include one-story rigid frames as well as multistory frames, are structures whose individual members are rigidly connected to each

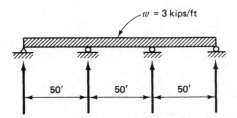

**FIGURE 10-1**   Three-span continuous beam.

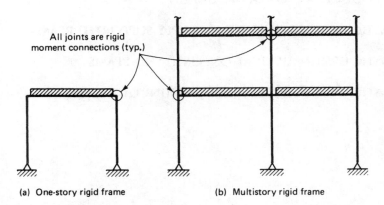

(a)  One-story rigid frame          (b)  Multistory rigid frame

**FIGURE 10-2**   Continuous frames.

other with moment-resisting connections, as shown in Figure 10-2, thereby preventing relative rotation under load.

Continuous beams and frames are categorized as statically indeterminate structures, which means that the moments, shears, and external reactions cannot be found by the condition of static equilibrium alone. It is not the intent of this text to review or introduce the many analytical techniques that may be used for the determination of the moments, shears, and reactions in indeterminate structures. Since the structural design of continuous beams is similar to the design of simple beams, and since multistory frame design is beyond the scope of this text, the coverage herein is limited to continuous beams.

**Example 10-1** _____

Compare the maximum bending moments and shears for the beams shown in Figures 10-3 and 10-4. Use the ASDM, Part 2.

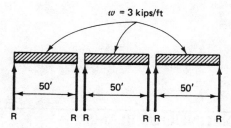

**FIGURE 10-3**    Simple beams (three spans).

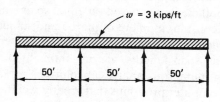

**FIGURE 10-4**    Three-span continuous beam.

**Solution:**

For the simple beam of Figure 10-3,

$$\text{maximum } M = \frac{wL^2}{8} = \frac{3(50)^2}{8} = 938 \text{ ft-kips (1272 kN·m)}$$

$$\text{maximum } V = \frac{wL}{2} = \frac{3(50)}{2} = 75 \text{ kips (334 kN)}$$

reaction at the end of each span $(R) = 75$ kips (334 kN)

The maximum moment is positive (compression in the top) and occurs at the middle of the 50-ft spans.

For the three-span continuous beam of Figure 10-4, utilizing shear and moment coefficients (discussed in Section 10-2),

$$\text{maximum } M = 0.100wL^2 = 0.100(3)(50)^2 = 750 \text{ ft-kips (1017 kN·m)}$$

$$\text{maximum } V = 0.600wL = 0.600(3)(50) = 90 \text{ kips (400 kN)}$$

$$\text{maximum reaction} = 1.10wL = 1.10(3)(50) = 165 \text{ kips(734 kN)}$$

The maximum moment is negative and occurs at the interior supports. The maximum shear occurs at the end-span side of the interior supports.

The least maximum bending moment occurs with the three-span continuous beam, hence economy will result with respect to beam size.

## 10-2

## ELASTIC DESIGN OF CONTINUOUS BEAMS

In an effort to simplify continuity and expedite the planning and design phases of a structure, the ASDM, Part 2, furnishes moment and shear coefficients for multiple-span continuous beams. These coefficients are based on equal span lengths and are applicable for steel beams with a constant moment of inertia and subject to loading conditions as shown in the diagrams.

For conditions such as unequal span lengths, varying moment of inertia, and combinations of types of loading, numerous analytical solutions are available. In addition, numerous commercial computer programs that greatly facilitate the analysis of continuous beams are available.

After determination of the continuous beam positive and negative bending moments, the ASDS, Section F1.1, permits a modification of these values. According to Section F1.1, a continuous structural steel *compact* member may be designed on the basis of nine-tenths of the maximum negative moments produced by gravity loading, provided that the maximum positive moments are increased by one-tenth of the average negative moments at the adjacent supports. The moment adjustments apply only to gravity loads and not to lateral loads (such as wind). In addition, they do not apply to steel members with yield points greater than 65 ksi, hybrid members, or to moments produced by loads on cantilevers.

**PHOTO 10-1**    Variable depth multispan, continuous, welded plate girder bridge over the Hudson River between Troy and Watervliet, New York. (Courtesy of the New York State Department of Transportation.)

### Example 10-2

Design a W-shape continuous beam for the conditions shown in Figure 10-5. Use A36 steel and ASD design approach. Assume continuous lateral support for the compression flange. (Consider moment and shear.)

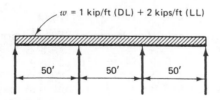

**FIGURE 10-5**    Continuous beam.

**Solution:**

1.  Refer to Case No. 36 in the Beam Diagrams and Deflections section in the ASDM, Part 2. The dead load negative moment (all spans loaded) at the interior supports is

$$M = 0.100wL^2$$

$$= 0.100(1.0)(50)^2 = 250 \text{ ft-kips}$$

The dead load positive moment in the end spans is

$$M = 0.080wL^2$$

$$= 0.080(1.0)(50)^2 = 200 \text{ ft-kips}$$

This occurs at a distance of $0.40L$ from the end support:

$$0.40(50) = 20 \text{ ft}$$

2.   The live load moments are determined with careful regard to the loading patterns that produce maximum conditions.

   (a)   Maximum positive moment in end spans—load the end spans only (Case No. 35):

$$M = 0.1013wL^2$$

$$= 0.1013(2)(50)^2 = 507 \text{ ft-kips}$$

   This occurs at a distance of $0.450L$ from the end support:

$$0.450(50) = 22.5 \text{ ft}$$

   (b)   The maximum negative moment at one interior support—load two adjacent spans with one end span unloaded (Case No. 34) is

$$M = 0.1167wL^2$$

$$= 0.1167(2)(50)^2 = 584 \text{ ft-kips}$$

   (c)   The negative moment at interior supports with only the end spans loaded (Case No. 35) is

$$M = 0.050wL^2$$

$$= 0.050(2)(50)^2 = 250 \text{ ft-kips}$$

3.   The maximum negative moment for design (ASDS, Section F1.1) is

$$M = 0.9(250 + 584) = 751 \text{ ft-kips}$$

The maximum positive moment for design (assume that maximum dead load and live load moments occur at the same location) is determined by increasing the calculated maximum positive moment by one-tenth of the average of the negative moments at each end of the span. Thus

$$M = (200 + 507) + 0.1\left(\frac{0 + 250}{2}\right)$$

$$= 707 + 13 = 720 \text{ ft-kips}$$

4.   Negative moment > positive moment; therefore (assuming that $F_b = 24$ ksi),

$$\text{required } S_x = \frac{M}{F_b} = \frac{751(12)}{24} = 376 \text{ in.}^3$$

5.   From the ASDM, Part 2, select a W33 × 130, $S_x = 406$ in.³. The section is compact; therefore, the assumed $F_b$ is satisfactory.

6.   Check the shear:

$$\text{maximum } V_{DL} = 0.6wL = 0.6(1)(50) = 30 \text{ kips}$$

$$\text{maximum } V_{LL} = 0.617wL = 0.617(2)(50) = 62 \text{ kips}$$

$$\text{total maximum } V = 92 \text{ kips}$$

The maximum shear $V$ occurs at the end span face of the interior support when the opposite end span is unloaded. From the Allowable Uniform Load Tables, ASDM, Part 2, the allowable shear for a W33 × 130 is 276 kips. Since 276 kips > 92 kips, the section is satisfactory. **Use a W33 × 130.**

Depending on the structure, a complete continuous beam design would normally involve other design considerations, such as the determination of the deflections and the need for bearing stiffeners at supports. In addition, bearing plates may be necessary, as may be a beam splice.

The designer should also be aware that in areas of negative moment in continuous beams, the bottom flange is the compression flange. If only the top flange is laterally supported, the lateral support conditions for the bottom flange must be considered. $F_b$ may be affected. Since $L_b$ is defined as the distance between lateral support points on the *compression flange*, the unsupported length may be considered to end at a point of lateral support or at a point of inflection (moment = 0). Moment diagrams should be carefully studied when considering $L_b$ for continuous beams.

## 10-3

### INTRODUCTION TO PLASTIC DESIGN

As discussed in Chapter 1, most structural steel beams are designed in accordance with the allowable stress design method, whereby an actual bending stress $f_b$ induced by applied loads may not exceed an allowable bending stress $F_b$. The ASDS, Chapter F, indicates that this allowable bending stress is always substantially less than the steel yield stress $F_y$ and, therefore, as depicted in Figure 1-4, must lie on the initial straight-line portion of the steel stress-strain curve. Hence, in this method of design, bending stresses are kept within the elastic range.

The *plastic design method* takes advantage of the substantial reserve strength of a steel beam that exists after the yield stress has been reached at some location. Extensive tests have shown that bending members can support loads in excess of those that would initially induce the yield stress. Therefore, the plastic theory utilizes the stress-strain relationships through the plastic range up to the start of strain hardening. The strain-hardening range could theoretically permit steel members to withstand additional stress; the corresponding strains and resulting structural deformations would be so large, however, that the structure would no longer be usable. The assumption is made in plastic design, therefore, that strains do not reach the strain-hardening range.

The idealized stress-strain diagram of Figure 1-4 is based on the assumption that the maximum stress through the plastic range does not exceed the yield stress $F_y$. The *stress* within the plastic range is assumed to be constant despite the fact that *strain* increases. As strains increase from the elastic range into the plastic range on a beam cross section, however, there is a distinct change in the shape of the resulting bending stress distribution diagram. The shape of the bending stress diagram that is assumed in allowable stress design is shown in Figure 10-6a. This general shape exists up to the time at which the maximum bending stress (at the extreme outside fiber of the beam) becomes $F_y$. In this range unit strain varies linearly from zero at the neutral axis to a maximum at the outer fibers, and since unit stress is proportional to unit strain (in the elastic range only), the stress variation also varies linearly from zero at the neutral axis to a maximum at the outermost fibers. When the outermost fibers *first* reach the yield stress $F_y$, and the rest of the cross section is still stressed to less than $F_y$, the resisting moment existing in the beam is

$$M_y = F_y S_x$$

and may be termed the *yield moment.* If the moment is increased beyond the yield moment, the outer fibers, which have been stressed to their yield stress, will continue to have the same stress, but at the same time additional strains will occur, and unit stress will no longer be proportional to unit strain. Any required *additional* resisting moment will then be furnished by *the fibers nearer to the neutral axis,* and the stress distribution will take the form shown in Figure 10-6b. This process will continue, with more parts of the beam cross section stressed to the yield point, as shown in Figure 10-6c, until a fully plastic rectangular stress distribution develops, as shown in Figure 10-6d. At this point, the unit strain has become so large that practically

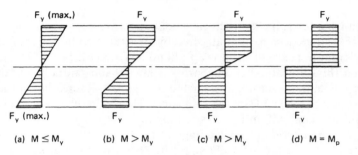

**FIGURE 10-6**  Bending stress distribution.

the entire cross section has yielded, and it is assumed that no additional moment can be resisted. The moment that exists at this point is called the *plastic moment* $M_p$. For the hypothetical cross section shown in Figure 10-7, the magnitude of $M_p$ is determined as follows ($C_f$ and $C_w$ are internal compressive forces in the flange and web, respectively):

$$M_p = C_f y_f + C_w y_w$$

$$= F_y(8)(1)(12 - 1) + F_y(0.75)(5)\left(\frac{10}{2}\right)$$

$$= 106.75 F_y$$

Assuming that $F_y = 36$ ksi,

$$M_p = 106.75(36) = 3843 \text{ in.-kips}$$

The resistance to bending at this point may also be expressed as

$$M_p = F_y Z$$

where

$M_p$ = plastic moment

$Z$ = plastic section modulus (in.$^3$)

$F_y$ = yield stress

The plastic section modulus of the cross section is equal to the numerical sum of the moments of the areas of the cross section above and below the neutral axis, taken about the neutral axis. For the cross section of Figure 10-7, $Z$ was determined

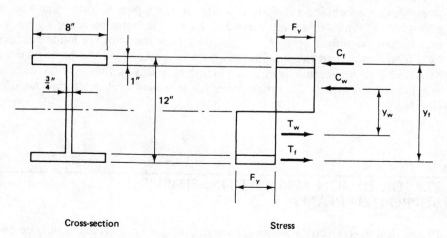

Cross-section                                    Stress

**FIGURE 10-7**   Diagram for determination of $M_p$.

as 106.75 in.[3] in the calculation of $M_p$. The plastic section modulus is tabulated for W and M shapes in the ASDM, Part 2, Plastic Design Selection Table.

The ratio $M_p/M_y$ is called the *shape factor* and may be described as a measure of the plastic moment strength in comparison to the yield moment strength. For the cross section of Figure 10-7, the shape factor is calculated as follows:

$$M_y = F_y S_x = F_y \frac{I_x}{C}$$

$I_x$ may be determined to be 548 in.[4], and, again assuming that $F_y = 36$ ksi,

$$M_y = 36\left(\frac{548}{6}\right) = 3288 \text{ in.-kips}$$

from which we obtain

$$\text{shape factor} = \frac{M_p}{M_y} = \frac{3843}{3288} = 1.17$$

For most wide-flange shapes (the cross section of Figure 10-7 is hypothetical) with bending occurring about the strong axis, the value of the shape factor lies between 1.10 and 1.23.

Assuming a progressive increase in a beam loading, the actual bending moment induced at some location would eventually reach the plastic moment strength $M_p$. When this occurs, a *plastic hinge* is said to have formed, and no additional moment can be resisted at that location. Although the effect of a plastic hinge may extend for some distance along the beam, it is assumed for analysis and design purposes to be localized in a single plane. When sufficient plastic hinges have formed so that no further loading may be supported, a *mechanism* is said to have been created. This may be defined as an arrangement of plastic hinges and/or real hinges that would permit collapse of a structural member.

All structural members are designed based on some factor of safety. In allowable stress design a bending member is designed to support working loads so that an allowable bending stress is not exceeded. Hence the factor of safety against yielding may be considered to be $F_y/F_b$. In plastic design the working loads are increased by a *load factor,* and the bending member is designed on the basis of the plastic or *collapse strength*. In essence, the factor of safety is the load factor, and as a minimum must equal 1.70 times the given live load and dead load (ASDS, Section N.1).

## 10-4

## PLASTIC DESIGN APPLICATION: SIMPLY SUPPORTED BEAMS

Plastic design is of little advantage for simply supported beams. It may be economical, however, for statically indeterminate members such as fixed end (restrained) or continuous beams. A simple beam will fail if one plastic hinge develops, since real hinges exist at each support.

Assuming a W shape subjected to a concentrated load at midspan, a plastic hinge will develop at the point of maximum moment (under the load in this case) as the loading is increased. The combination of the plastic hinge with the two real hinges at the supports creates a collapse mechanism, as shown in Figure 10-8, and failure is assumed to have occurred.

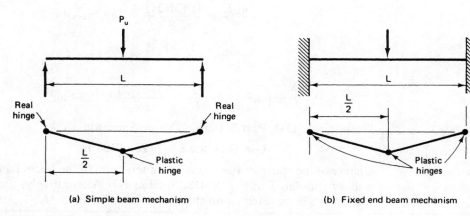

(a)  Simple beam mechanism          (b)  Fixed end beam mechanism

**FIGURE 10-8**   Collapse mechanisms.

## Example 10-3

A laterally supported ($L_b = 0$) simple-span beam with a 24 ft–0 in. span length must support a uniformly distributed working load of 3.0 kips/ft (which includes an assumed weight of beam). Use A36 steel and the ASDS, and disregard shear and deflection. Select the lightest W shape by

(a)   The elastic design method (allowable stress design)
(b)   The plastic design method

### Solution:

(a)   *Elastic design method:*

$$M = \frac{wL^2}{8} = \frac{3(24)^2}{8} = 216 \text{ ft-kips}$$

Assume a compact section and $F_b = 24$ ksi. Thus

$$\text{required } S_x = \frac{M}{F_b} = \frac{216(12)}{24} = 108 \text{ in.}^3$$

From ASDM, Part 2, Allowable Stress Design Selection Table:

**Use W24 × 55**     ($S_x = 114$ in.$^3$)

This section is compact and the assumptions are valid. Therefore, the section is O.K. since 114 in.$^3$ > 108 in.$^3$

(b)   *Plastic design method:* Knowing that the moment will reach $M_p$ at collapse and will occur at midspan, a beam must be selected that has an $M_p$ value at least equal to the moment $M_u$ created by the factored loads. Thus

$$M_u = \frac{w_u L^2}{8} = \frac{1.7(3)(24)^2}{8} = 367.2 \text{ ft-kips}$$

$$M_u = M_p = F_y Z$$

Therefore,

$$\text{required } Z_x = \frac{M_u}{F_y} = \frac{367.2(12)}{36} = 122.4 \text{ in.}^3$$

From the ASDM, Part 2, Plastic Design Selection Table:

**Use W24 × 55**     ($Z_x = 134$ in.$^3$)

Disregard the check of the width–thickness ratio of the beam flange and web as stipulated in the ASDS, Section N7. Alternatively, the beam could have been selected on the basis of its tabulated $M_p$.

## 10-5

## PLASTIC DESIGN APPLICATION: FIXED-END BEAMS

For a beam that is fixed at both ends and that supports a uniform load (ASDM, Part 2, Beam Diagrams and Formulas, Case No. 15), expressions for positive and negative moments are shown in Figure 10-9. These expressions are valid for the

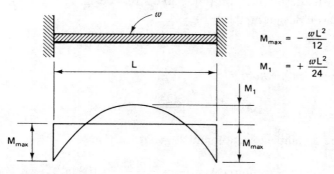

$$M_{max} = -\frac{wL^2}{12}$$

$$M_1 = +\frac{wL^2}{24}$$

**FIGURE 10-9**   Fixed-end beam.

allowable stress design method. They may also be used for the case in which the uniformly distributed load is increased to that load $w_y$ that will induce a maximum bending stress equal to the yield stress.

As the load is further increased, plastic hinges will form at the fixed ends. These are the points of maximum moment. These points will allow rotation to take place without resisting any more of the applied moment. That is, a constant moment $M_p$ will exist at the hinges. A further increase in load must then be resisted by sections of the beam that are less stressed. The load may be increased until the moment at some other point reaches the plastic moment $M_p$. In this case this will occur at midspan, and a third plastic hinge will be developed. The combination of the three plastic hinges, when formed, constitutes a collapse mechanism, and the beam is no longer capable of supporting additional load.

**Example 10-4** _____

A laterally supported ($L_b = 0$) fixed-end beam with a 20 ft–0 in. span length must support a uniformly distributed working load of 5.0 kips/ft (which includes an assumed weight of beam). Use A36 steel and the ASDS, and disregard shear and deflection. Select the lightest W shape by

(a)   The elastic design method (allowable stress design)
(b)   The plastic design method

**Solution:**

(a)   *Elastic design method:* The maximum negative moment (at supports) is

$$M = -\frac{wL^2}{12} = \frac{5.0(20)^2}{12} = 166.7 \text{ ft-kips}$$

The maximum positive moment (at midspan) is

$$M = +\frac{wL^2}{24} = \frac{5.0(20)^2}{24} = 83.3 \text{ ft-kips}$$

To modify the moments in accordance with ASDS, Section F1.1, first determine the maximum negative moment for design:

$$-M = 0.9(166.7) = 150 \text{ ft-kips}$$

The maximum positive moment for design is

$$+M = 83.3 + (0.1)(166.7)$$

$$= 100.0 \text{ ft-kips}$$

Assume a compact section and $F_b = 24$ ksi. Thus

$$\text{required } S_x = \frac{M}{F_b} = \frac{150(12)}{24} = 75 \text{ in.}^3$$

From ASDM, Part 2, Allowable Stress Design Selection Table:

$$\text{Use } \mathbf{W21 \times 44} \qquad (S_x = 81.6 \text{ in.}^3)$$

This section is compact and the assumptions are valid. Therefore, the section in O.K. since 81.6 in.$^3$ > 75 in.$^3$.

(b)  *Plastic design method:* Knowing that at the instant of collapse the moment $M_p$ will occur at supports as well as midspan (three plastic hinge mechanism), an expression for $M_p$ is developed using a free-body approach, as shown in Figure 10-10. $\Sigma\, M$ about midspan:

$$\frac{w_u L}{2}\left(\frac{L}{2}\right) - w_u\left(\frac{L}{2}\right)\left(\frac{L}{4}\right) - M_p - M_p = 0$$

$$\frac{w_u L^2}{4} - \frac{w_u L^2}{8} = 2M_p$$

$$M_p = \frac{w_u L^2}{16}$$

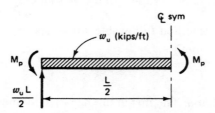

**FIGURE 10-10**   Free-body diagram.

Using a load factor of 1.7, the factored load is

$$w_u = 1.7(5) = 8.5 \text{ kips/ft}$$

$$M_p = \frac{w_u L^2}{16} = \frac{8.5(20)^2}{16} = 212.5 \text{ ft-kips}$$

$$\text{required } Z_x = \frac{M_p}{F_y} = \frac{212.5(12)}{36} = 70.8 \text{ in.}^3$$

From the ASDM, Part 2, Plastic Design Selection Table,

$$\text{Use } \mathbf{W16 \times 40} \qquad (Z_x = 72.9 \text{ in.}^3)$$

$$72.9 \text{ in.}^3 > 70.8 \text{ in.}^3 \qquad\qquad\qquad \textbf{O.K.}$$

Disregard the check of the width–thickness ratio of the beam flange and web as stipulated in the ASDS, Section N7.

## 10-6

## PLASTIC DESIGN APPLICATION: CONTINUOUS BEAMS

A more practical and realistic application of plastic design is the design of continuous beams. They are relatively common in structures compared with fixed-end beams, which are seldom encountered.

If the loads supported by a continuous beam are increased proportionately, the ultimate or maximum loading is reached when the weakest span is reduced to a mechanism by the formation of plastic hinges progressively at points of maximum moment. In the case of three or more identically loaded equal spans having simple supports at the outer ends, mechanisms will form in the end spans at a magnitude of loading less than that required to form mechanisms in the interior spans. As a means of comparison between plastic design and elastic design, the continuous beam of Example 10-2 will be redesigned using the plastic design method.

**Example 10-5** _____

Redesign the beam of Example 10-2 using the ASDS, Chapter N, Plastic Design method. The steel is A36.

**Solution:**

Using a load factor of 1.7, the total factored load on any one of the spans is

$$1.7(3.0) = 5.1 \text{ kips/ft}$$

Considering the end span first, a plastic hinge will develop initially at the interior support, since this is the location of the elastic maximum moment within that span. The ultimate or maximum load condition would be reached when another plastic hinge develops within the span. The end span would then have one real hinge and two plastic hinges, thereby forming a collapse mechanism, as shown in Figure 10-11. This second plastic hinge will occur as

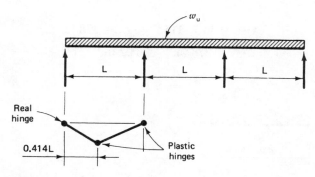

**FIGURE 10-11**   Continuous beam mechanism.

a result of some ultimate load and will form at a point $0.414L$ from the simply supported end. This location is applicable for the end spans of continuous beams of two or more equal spans having identical uniformly distributed loading. It is the location of maximum moment $M_p$ within the end span, and shear will be zero at this point.

The required plastic moment strength $M_p$ that must exist to resist the ultimate load may be obtained using a free-body approach, as shown in Figure 10-12. Since shear is zero at the right end of the free-body diagram, the left reaction must be $0.414w_uL$. Then taking $\Sigma M$ about plane P,

$$0.414w_uL(0.414L) - w_u(0.414L)\left(0.414\frac{L}{2}\right) - M_p = 0$$

$$M_p = 0.0858w_uL^2$$

If the interior span is now considered, plastic hinges will be developed simultaneously at each end, and after an increase in loading, a third plastic hinge will form at midspan in a manner similar to the single-span fixed-end beam previously discussed. The required $M_p$ to develop the plastic hinge at midspan would also be the same as that of the fixed-end single span shown previously as

$$M_p = \frac{w_uL^2}{16} = 0.0625w_uL^2$$

Since the interior and end spans are of the same length, the beam selection will be based on the largest required $M_p$.

Using

$$M_p = 0.0858w_uL^2$$

$$= 0.0858(5.1)(50)^2$$

$$= 1094 \text{ ft-kips}$$

the required plastic modulus is

$$Z_x = \frac{M_p}{F_y} = \frac{1094(12)}{36} = 364.7 \text{ in.}^3$$

**Use W30 × 116**      $(Z_x = 378 \text{ in.}^3)$

$$378 \text{ in.}^3 > 364.7 \text{ in.}^3 \qquad\qquad \textbf{O.K.}$$

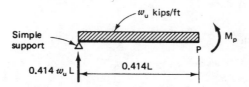

**FIGURE 10-12**  Free-body diagram.

For a complete design, shear and deflection should also be checked, as well as the width-thickness ratios of the beam flange and web in accordance with the ASDS, Section N7.

The plastic design method used in these examples is generally designated the *equilibrium method*. As beams and their spans and loadings become more complex, however, a more practical and simpler method is recommended. This alternative method is generally designated the *virtual work method* or *mechanism method*. This method is also recommended for statically indeterminate frames, which constitute another type of structure where the plastic design method is applicable.

For a more extensive coverage of plastic design, the reader is referred to References 1 through 3.

## REFERENCES

[1]  L. S. Beedle, *Plastic Design of Steel Frames* (New York: John Wiley & Sons, Inc., 1958).

[2]  *Plastic Design in Steel*, ASCE Manual of Engineering Practice, No. 41, 2nd ed., 1971.

[3]  *Plastic Design of Braced Multistory Steel Frames*, Publication M004, American Institute of Steel Construction, 1968.

## PROBLEMS

*Note: For the following problems, all steel is A36.*

**10-1.**  Design a two-span continuous beam for the conditions shown. Select the lightest W shape. Assume continuous lateral support for the compression flange. The load includes an assumed weight of beam. Use the ASD design approach and consider moment and shear. (*Hint:* Refer to Case No. 12, Beam Diagrams and Formulas, ASDM, Part 2.)

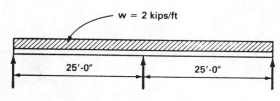

**PROBLEM 10-1**

**10-2.** Design a three-span continuous beam for the conditions shown. Select the lightest W shape. Assume continuous lateral support for the compression flange. The dead load includes an assumed weight of beam. Use the ASD design approach and consider moment and shear.

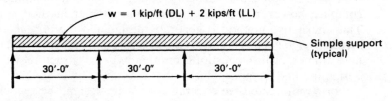

**PROBLEM 10-2**

**10-3.** Rework Problem 10-2 assuming three simply supported beams, each with a 30-ft span length.

**10-4.** Rework Problem 10-2 assuming end spans cantilevering over the interior supports as shown. Note that the center span is simply supported at the two hinges. Compare the results of Problems 10-2, 10-3, and 10-4 based on weight only.

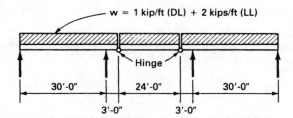

**PROBLEM 10-4**

**10-5.** Redesign the beam of Problem 10-1 using the ASDS, Chapter N, Plastic Design Method. Consider moment only.

**10-6.** Redesign the beam of Problem 10-2 using the ASDS, Chapter N, Plastic Design Method. Consider moment only.

# CHAPTER 11

# Structural Steel Detailing: Beams

11-1  INTRODUCTION

11-2  OBTAINING THE STEEL

11-3  DRAWING PREPARATION

11-4  BEAM DETAILS

## 11-1

### INTRODUCTION

The construction sequence of a steel-framed building, as discussed in Section 1-5 of this text, consists of three sequential phases that occur (generally) after the design sequence. These phases may be categorized as the detailing, fabrication, and erection phases.

In the detailing phase, information from the design drawings, which have been prepared by the architect and/or engineer, is used to develop detail drawings. These are generally called *shop drawings,* and they must convey all the information necessary for shop fabrication of the multitude of structural members and their connections in a given structure. They must also convey information relative to field erection procedures and sequences.

The actual production of the shop drawings is the job of the *detailer*. The detailer must develop the ideas conveyed by the design drawings to the point where individual members and all the many required components of the structure may be fabricated. A great deal of practical knowledge is required. The drawings and schedules generated by the detailer will be instrumental in coordination of the work in the fabrication and erection phases. Good detailers can do much to promote economy in steel construction.

## 11-2

### OBTAINING THE STEEL

Most structural steel fabricators are staffed and equipped for the detailing, fabrication, and erection portions of a project, but must purchase the steel from the rolling mills. Economy does not permit most fabricators to keep in stock a very large inventory of structural steel. In an effort to expedite the structural steel portion of a project, the fabricator will prepare an advance bill of materials based purely on the design drawings for the purpose of ordering the steel from the rolling mills. This order is placed before shop drawings are prepared so as to expedite getting some steel into the shop. In reality, the preparation of the shop drawings has started

**PHOTO 11-1**   Structural steel in the fabricator's shop. The ends of the stacked wide-flange beams on the left have been coped and punched in readiness for application of other detail material, which is placed nearby.

at the same time the advance bill of materials has been prepared; shop fabrication cannot be started until the ordered material has arrived at the fabricator's plant, however. When the order arrives, quantities and items are checked and the steel is stored, ready for fabrication. To facilitate fabrication and with efficient planning, it may be possible to order the larger and longer pieces cut to detail sizes at the mill. Because the mills assess extra charges for cutting to detail sizes, however, it is generally more economical to order the structural shapes in long lengths from which several shorter lengths can be shop-cut.

## 11-3

## DRAWING PREPARATION

The initial step in the preparation of the shop drawings is to prepare a set of erection drawings, which give all information required for the layout and installation of the structural steel. These drawings show each steel piece or subassembly of pieces with its assigned shipping or erection mark to identify and locate it in its correct position in the structure. The erection plans include an *anchor bolt plan*. This is one of the first drawings made since the anchor bolts must be set prior to the erection of the steel. This drawing locates all the anchor bolts that will be embedded in the concrete foundations and which serve to anchor the steel frame to the supporting foundation. This aspect is discussed further in Chapter 12 of this text.

Sometimes reproductions of the architect's or engineer's design drawings are used as erection plans. Erection marks and instructions are added to the plans, and the drawing is given a new number.

Erection drawings not only show the exact location of every piece with its erection mark, but also the sequence of erection when the project is large. Large areas of framing are usually divided into separate sections called *installments*. These installments permit fabricated pieces to be delivered on a detailed schedule to predetermined locations at the project site without expensive rehandling costs. This advance planning helps to establish detailing, fabrication, shipping, and erection schedules.

The main objective of the detailing phase is to prepare the structural steel detail or shop drawings. These drawings, which are prepared by detailers, are subsequently used in the shop to fabricate each individual piece of steel required in the structure. From the information furnished in the contract documents (design plans and specifications), the detailer prepares complete and explicit details for each structural member. To avoid the repetition of labeling each sketch with the same information, shop notes are placed on the shop drawings indicating bolt size, size of open holes, type of material, paint, and other pertinent information required by the shop. In addition, since the design drawings generally show only a few connections and those are seldom completely developed, it is up to the detailer to develop them fully, including modifications to facilitate fabrication and erection. The detailer must also design other typical connections as well as any special connections not shown on the design drawings. Hence the detailer must be familiar with the ASDM [1] as well as standard design and detailing practices. After the structural members are detailed and shop drawings completed, it is general practice to have more experienced personnel, called *checkers,* review and verify all sketches and dimensions on the drawings. To help the shop assemble the material required to fabricate the various shipping pieces detailed on the shop drawings, a shop bill of material is prepared as part of each shop drawing. Most fabricators provide standard shop drawing sheets with a preprinted shop bill form on the sheet. To provide space on the drawing, separate shop bill forms are sometimes used.

Before fabrication can begin, the finished shop drawings must be approved by the architect/engineer or some other owner-designated representative. This applies to all shop details and erection plans, since all the shop drawings contain additional information not specifically shown on the design drawings.

## 11-4

## BEAM DETAILS

The required details for fabrication of a beam are shown on a shop drawing. Generally, each beam in a roof or floor framing system makes a convenient shipping and erection unit. All features that have an impact on the erection of the beam must be investigated. Beam connection holes must match the location of similar holes in the supporting members. Proper erection clearances must be provided and possible interferences must be eliminated so that the beam can be swung or lowered into position for connection to its supporting members.

In detailing a beam the detailer must first design the end connections to transmit the beam load to its supporting members. The necessary information for the connection design is generally furnished on the design drawings. This should include information on the type of construction, design loads, shears, reactions, moments, and axial forces where applicable.

A partial framing plan taken from a design drawing is shown in Figure 11-1. It represents a portion of a typical interior bay of a steel-framed building floor system. Unless otherwise noted, one may assume that members shown on such a framing plan are to be

1.  Placed parallel or at right angles to one another with their webs in a vertical plane
2.  Located at some specific elevation and set level end to end

Beam designations (A3, B4, etc.) are generally not furnished on a design drawing, but are furnished here for purposes of explanation. As a means of introduction

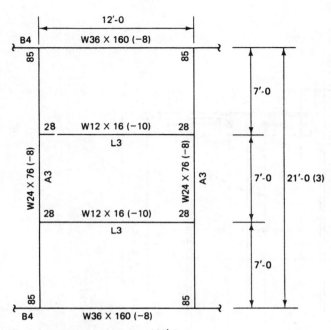

**FIGURE 11-1** Partial framing plan; first floor.

to the detailing process, Figure 11-2 depicts the detailed beam A3. The various dimensions, together with other information numbered in the sketch, are referenced sequentially in the following discussion.

1. *Beam A3* is the beam designation and is sometimes called the *shipping and erection mark.* The derivation of this designation may be according to some method, such as a capital letter followed by the number of the drawing on which the beam is detailed. Various marking systems are used throughout the industry, and the reader is referred to AISC publications (see References 2 and 3) for further discussion.

2. The *assembly piece marks* are *a* and *b*. These are used where the assembly to be shipped is composed of several pieces. In this case, there are five pieces per shipping unit—the beam itself and four connection angles. If a piece is plain (nothing attached to it), it may not have an assembly mark.

3. The 21'-0 dimension for A3 (Figure 11-1) represents the beam's *theoretical span length,* center to center of supporting members. Note that the inch symbol (") is omitted from the dimensions. This is common practice (and we will follow it for the duration of our detailing discussion).

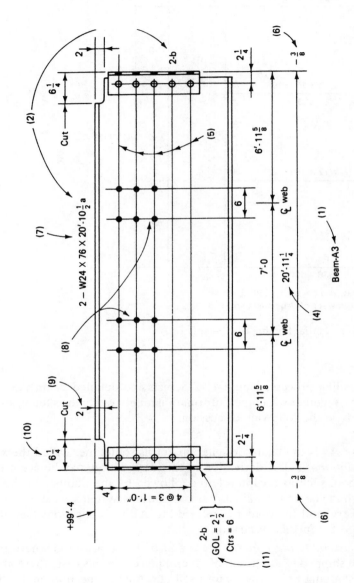

**FIGURE 11-2** Beam details.

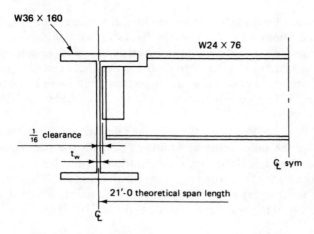

**FIGURE 11-3** Working sketch—each end of beam.

4. The $20'-11\frac{1}{4}$ dimension represents the *beam assembly unit length* back-to-back of connection angles. It is established as the theoretical span length minus one-half the web thickness of the supporting member at each end. Another $\frac{1}{16}$ in. is subtracted for each end of the beam (see Figure 11-3). The beam assembly unit length is then calculated from

$$\text{theoretical span length} - \frac{t_w}{2} - \frac{t_w}{2} - 2 \times \left(\frac{1}{16}\right)$$

The length should then be rounded to the nearest $\frac{1}{16}$ in. Calculating the beam assembly unit length, we have

$$(21'-0) - \left(\frac{5}{16} + \frac{1}{16}\right)2 = 20'-11\frac{1}{4}$$

5. These lines are the gage lines on the beam web. Preferably, all bolt holes should fall on these gage lines. In establishing the gage lines, it is necessary to consider the beam's own end connections as well as all the members that frame into the beam web.

6. The *setback distance* is $-\frac{3}{8}$. It is the distance from the centerline of the supporting beam to the back of the connection angle:

$$\text{setback} = \frac{t_w}{2} + \frac{1}{16}$$

In this case

$$\text{setback} = \frac{5}{16} + \frac{1}{16} = \frac{3}{8}$$

7. The *ordered*, or *billed*, *length* of the beam is $20'-10\frac{1}{2}$. It is sometimes shown only in the bill of materials. In this case it is shown together with the assembly

piece mark. It should be computed and specified to the nearest $\frac{1}{2}$ in. so that the beam will stop about $\frac{1}{2}$ in. short of the backs of the connection angles. This allows for inaccurate cutting of the beam length and eliminates possible recutting or trimming. The actual difference between the ordered beam length and length back-to-back of connection angles will be between $\frac{3}{8}$ and $1\frac{5}{8}$ in. based on ASDM cutting tolerances of $\frac{3}{8}$ in. over and under (ASDM, Part 1).

8.  This represents the hole locations for the connections for beams L3. The vertical location of the holes must be coordinated with the connections on the supported member L3. Note that the centerline of the L3 beam web is referenced from the back of the connection angles on beam A3. The dimension is determined by subtracting $\frac{1}{16}$ in. clearance and one-half the web thickness of B4 from the centerline-to-centerline dimension of $7'-0$. Thus

$$(7'-0) - \frac{1}{16} - \frac{5}{16} = 6'-11\frac{5}{8}$$

9.  This represents the depth of cope. Coping is required where tops of beams and supporting girders are at the same elevation or will interfere with each other. The cope depth is generally established so that the horizontal cut is at the level of the toe of fillet of the supporting member. The depth of cope dimension is usually rounded up to the next $\frac{1}{4}$ in.

10. This represents the length of cope. It is dimensioned from the back of the connection angles and should allow for $\frac{1}{2}$ to $\frac{3}{4}$ in. clearance at the edge of the flange. The dimension is usually rounded up to the next $\frac{1}{4}$ in. With reference to Figure 11-4, in this case,

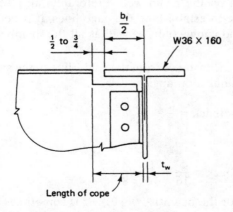

**FIGURE 11-4** Working sketch.

$$\text{length of cope} = \frac{b_f}{2} - \frac{t_w}{2} + \frac{1}{2}$$

$$= \frac{12}{2} - \frac{5}{16} + \frac{1}{2} = 6\frac{3}{16}$$

**Use $6\frac{1}{4}$ in.**

11.  GOL = $2\frac{1}{2}$ refers to the gage distance on the outstanding legs of the connection angles. Ctrs = 6 represents the transverse spacing of the gage lines for the two outstanding legs of the angles. It is generally called the *spread*, as shown in Figure 11-5. This value must be computed taking into consideration the assembling clearances for threaded fasteners as furnished in the ASDM, Part 4. A detailed example is furnished in Example 11-1. If adequate clearances cannot be furnished, the bolts through the outstanding legs of the connection angles must be staggered with respect to those through the beam web.

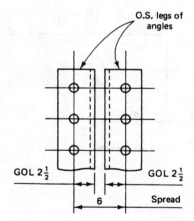

**FIGURE 11-5**  Working sketch.

**Example 11-1** _____

The partial framing plan shown in Figure 11-6 is that of a typical interior bay of a roof system in a one-story steel-framed building. All structural steel is A36. The bolts are $\frac{3}{4}$-in.-diameter A325N in standard holes (where welding is used, use E70 electrodes, shielded metal-arc welding).

Detail beams B1, B2, and G1 in accordance with the latest ASDS.

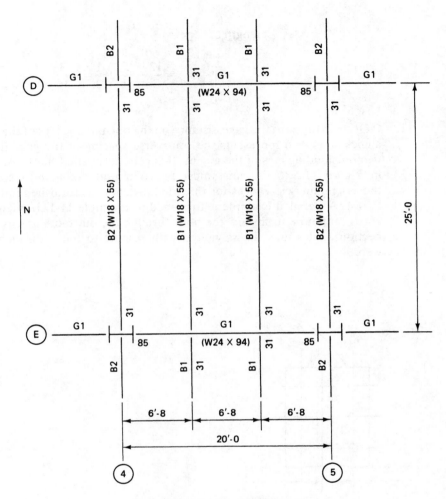

**FIGURE 11-6**   Partial framing plan.

**Solution:**

This is a lengthy detailing problem. It will be accomplished step by step in accordance with the following outline:

I.  Beam B1
    A.   Connection to B1 web
    B.   Connection to G1 web

       C.   Framing angle leg sizes

       D.   Required dimensions

  II.  Beam B2

       A.   Select seated beam connection

       B.   Required dimensions

 III.  Girder G1

       A.   Connection to G1 web

       B.   Connection to column flange

       C.   Required dimensions

All table references are to the ASDM, Part 4, unless otherwise noted.

  I.  *Beam B1:* Since the detailing process includes the design of the end connections, the initial step will be to select the most economical end connections for each end of the beam. Beam B1 is a W18 × 55 that frames into girder G1, which is a W24 × 94. High-strength bolted framed connections will be used. The end reaction of 31 kips is given on the framing plan. In some instances the end reactions are not furnished. When this occurs, and when it is obvious that the beam loading is only a uniformly distributed load, the end connection must be designed to support one-half the uniform load capacity of the beam as furnished in the ASDM, Part 2, Allowable Uniform Load Tables. In this case the reaction noted on the drawing is 31 kips, and it will be used for design.

      A.   Consider the part of the connection through the W18 × 55 web.

         1.   To provide stability during erection, it is recommended (ASDM, Part 4, Table II, discussion) that the minimum length of connection angle be at least one-half the $T$ dimension as furnished in the ASDM, Part 1. For a W18 × 55, $T$ is $15\frac{1}{2}$ in. Therefore, the minimum angle length $L$ is

$$\text{minimum } L = \frac{T}{2} = \frac{15.5}{2} = 7.75 \text{ in.}$$

         2.   Based on bolt shear, from Table II-A, select a connection with three rows ($n = 3$) of $\frac{3}{4}$-in.-diameter A325N bolts. The angles for this connection are to be $\frac{5}{16}$ in. thick, and the capacity based on bolt shear is 55.7 kips. This exceeds the 31-kip reaction. The length of the angles is $8\frac{1}{2}$ in., which is greater than the required minimum of 7.75 in.

         3.   Check the capacity of the connection based on *shear on the net area* of the connecting angles (using $F_v = 0.30 \, F_u$). Figure 11-7 shows a tentative layout of the connection with regard to bolt spacing, vertical edge distance, and horizontal edge distance. Tops of flanges are at the same elevation, and dimen-

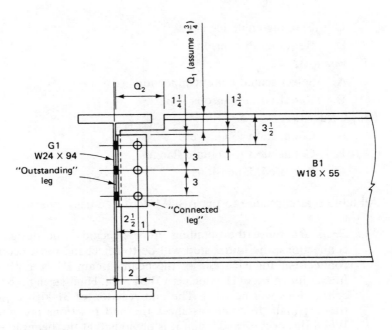

**FIGURE 11-7**   Working sketch.

sion $Q_l$ is determined so that the horizontal cut of the cope is at or below the toe of the fillet of the girder G1 ($k = 1\frac{5}{8}$ in. for the W24 × 94). This dimension is rounded to the next $\frac{1}{4}$ in. Therefore, use $Q_l$ of $1\frac{3}{4}$ in. The capacity based on shear on the net area of the angles is calculated as follows:

$$\text{hole diameter} = \frac{3}{4} + \frac{1}{8} = 0.875 \text{ in.}$$

$$A_v = 2\left(\frac{5}{16}\right)[8.5 - 3(0.875)] = 3.67 \text{ in.}^2$$

$$\text{capacity} = A_v F_v = 3.67(0.30)(58) = 63.9 \text{ kips}$$

$$63.9 \text{ kips} > 31.0 \text{ kips} \qquad \textbf{O.K.}$$

4.   The bearing on both the beam (B1) web and connection angles must be checked. Consider the capacity in bearing on the beam web ($t_w = 0.390$ in.). As shown in Figure 11-7, the bolt spacing is 3 in., which is $4.0d$, where $d$ is the bolt diameter. (This exceeds the $2.67d$ minimum spacing and the $3.0d$ desirable minimum spacing.) The edge distance parallel to the line of force (from the center of the top bolt to the cope of the beam web) is $1\frac{3}{4}$ in., which is $2.33d$. This exceeds the minimum

required edge distance along a line of transmitted force. Refer to ASDS Sections J3.9 and J3.7. Therefore, Table I-E applies. Since the web thickness (0.390 in.) of beam B1 is less than the sum of the thicknesses of the two $\frac{5}{16}$-in.-thick angles, the bearing on the beam web is the more critical of the two. The allowable load from Table I-E is

$$3(52.2)(0.309) = 61.1 \text{ kips}$$

$$61.1 \text{ kips} > 31 \text{ kips} \qquad \textbf{O.K.}$$

5.    Check the capacity of the connection based on web tear-out (block shear). The working sketch (Figure 11-7) indicates that B1 must be coped at each end.

$$\text{hole diameter} = \frac{3}{4} + \frac{1}{8} = 0.875 \text{ in.}$$

$$P_t = A_v(0.30)F_u + A_t(0.50)F_u$$

$$A_v = 0.390[7.25 - 2.5(0.875)] = 1.974 \text{ in.}^2$$

$$A_t = 0.390[2.0 - 0.5(0.875)] = 0.609 \text{ in.}^2$$

$$P_t = 1.974(0.30)(58) + 0.609(0.50)(58)$$

$$= 52.0 \text{ kips}$$

$$52.0 \text{ kips} > 31.0 \text{ kips} \qquad \textbf{O.K.}$$

The connection to the web of B1 as sketched in Figure 11-7 is adequate.

B.    Consider the part of the connection through the web of the supporting member G1, which is a W24 × 94. The six high-strength bolts must support a 31-kip reaction *from each side* of G1. Therefore, the total reaction to G1 is 62 kips.

1.    Check the capacity of the connection in double shear. From Table II-A, the capacity of three $\frac{3}{4}$-in.-diameter A325N bolts in double shear is 55.7 kips. For six bolts, the capacity is

$$55.7(2) = 111.4 \text{ kips}$$

$$111.4 \text{ kips} > 62 \text{ kips} \qquad \textbf{O.K.}$$

2.    Check the capacity of the connection in bearing on the web of girder G1. The bolt spacing is 3 in., which is 4.0$d$. This exceeds the desirable minimum of 3.0$d$. The vertical edge distance is not applicable for this check (it is in excess of 1.5$d$). Therefore, Table I-E is applicable. The web thickness

of G1 is 0.515 in. The allowable load based on bearing on the girder web is

$$6(52.2)(0.515) = 161 \text{ kips}$$

$$161 \text{ kips} > 62 \text{ kips} \qquad \textbf{O.K.}$$

3.   Check the capacity of the connection in bearing on the $\frac{5}{16}$-in. angle thickness. Consider the two angles that connect the end of beam B1 to the web of G1. The bolt spacing is 3 in., which is $4.0d$ ($> 3.0d$, O.K.). The vertical edge distance is $1\frac{1}{4}$ in., which is $1.67d$ ($> 1.5d$, O.K.). Therefore, Table I-E is applicable. Since there are two connection angles with three bolts through each one, the allowable load based on bearing on the angle thickness is

$$6(16.3) = 97.8 \text{ kips}$$

$$97.8 \text{ kips} > 31 \text{ kips} \qquad \textbf{O.K.}$$

4.   Shear on the net area of the connection angles is the same as for the other part of the connection (63.9 kips) and therefore is O.K.

5.   Web tear-out (block shear) is not applicable for the girder web part of the connection since the girder is not coped at this location. The six-bolt connection to the web of G1 is adequate. (The angle thickness is $\frac{5}{16}$ in., and the length of angle is $8\frac{1}{2}$ in.)

C.   Establish the leg sizes of the connection angles.

1.   With reference to Figure 11-7, the length of the connected leg (so called because it is the leg of the angle connected to the beam when the beam assembly is shipped) may be determined by summing the clearance distance and the edge distances on the beam web and the angle:

$$\frac{1}{2} + 2 + 1 = 3\frac{1}{2} \text{ in.}$$

Use a $3\frac{1}{2}$-in. connected leg with a gage as shown.

2.   The length required for the outstanding leg of the connection angles is a function of the required assembling clearances.

The angles will be shop-bolted to the web of B1 and field-bolted to the web of G1. Therefore, adequate wrench tightening clearance must be provided for this field installation.

Pertinent data for these calculations are found in the ASDM, Part 4, Assembling Clearances for Threaded Fasten-

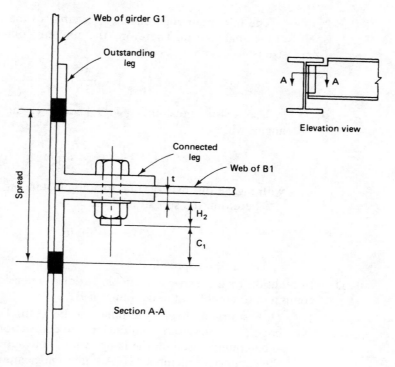

**FIGURE 11-8**   Working sketch.

ers. With reference to Figure 11-8, the following dimensions may be established:

$$C_1 = \text{clearance for tightening} = 1\frac{1}{4}\text{ in.}$$

$$H_2 = \text{shank extension} = 1\frac{3}{8}\text{ in.}$$

(includes flat washer, nut, and projection)

$$t = \text{angle thickness} = \frac{5}{16}\text{ in.}$$

The sum of the foregoing represents the minimum gage distance for the angle outstanding leg:

$$C_1 + H_2 + t = 1\frac{1}{4} + 1\frac{3}{8} + \frac{5}{16} = 2\frac{15}{16}\text{ in.}$$

The minimum edge distance required to the rolled edge of the angle is 1 in. Therefore, the minimum outstanding leg size is

$$2\frac{15}{16} + 1 = 3\frac{15}{16}\text{ in.}$$

Use a 4-in. outstanding leg. Therefore, the connection angles will be

$$2\text{ L4} \times 3\frac{1}{2} \times \frac{5}{16} \times 8\frac{1}{2}\text{ in.}$$

with a gage distance of 3 in. for the outstanding leg.
  Therefore, the spread must be

$$2(3) + \frac{3}{8} = 6\frac{3}{8}\text{ in.}$$

Use a $6\frac{3}{8}$-in. spread.

D.   In addition to the end connections, various dimensions must be computed to complete the detailing of B1.

1.   $Q_1$ has already been determined to be $1\frac{3}{4}$ in. The length of cope $Q_2$ is measured from the back of connection angles and is computed based on the flange width and web thickness of the supporting member G1. A minimum clearance of $\frac{1}{2}$ in. is commonly provided. Thus

$$\text{required } Q_2 = \frac{b_f}{2} - \frac{t_w}{2} + \frac{1}{2}$$

$$= \frac{9.125}{2} - \frac{1}{4} + \frac{1}{2}$$

$$= 4.81\text{ in.}$$

$Q_2$ should be rounded up to the next $\frac{1}{4}$ in. Therefore, use $Q_2$ of 5 in.

  The vertical and horizontal cuts ($Q_1$ and $Q_2$) should be connected and shaped notch free to a radius of $\frac{1}{2}$ in. (see ASDM, Part 4, Fabricating Practices).

2.   The setback distance required is the distance from the centerline of the supporting beam to the back of the connection angle:

$$\text{required setback} = \frac{t_w}{2} + \frac{1}{16}$$

$$= \frac{1}{4} + \frac{1}{16} = \frac{5}{16}\text{ in.}$$

3. The beam assembly unit length is the distance back-to-back of connection angles and is calculated from

$$\text{theoretical span length} - \left(\frac{t_w}{2} + \frac{1}{16}\right)(2) = (25'{-}0) - \left(\frac{5}{16}\right)(2)$$

$$= 24'{-}11\frac{3}{8}$$

4. The ordered, or billed, length of the beam should be established so that the ends stop approximately $\frac{1}{2}$ in. short of the backs of the connection angles. This dimension is rounded to the nearest $\frac{1}{2}$ in.:

$$\left(24'{-}11\frac{3}{8}\right) - \frac{1}{2} - \frac{1}{2} = 24'{-}10\frac{3}{8}$$

The ordered length is $24'{-}10\frac{1}{2}$. The complete detail of beam B1 is shown in Figure 11-9.

II. *Beam B2:* With reference to Figure 11-6, it may be observed that beam B2 is supported at each end by columns and must fit in between the column flanges. Beam B2 is a W18 × 55, and the column is a W10 × 60. An unstiffened seated beam connection will be used at each end, with the beam seat shop-bolted to the column and field-bolted to beam B2. The beam reaction is 31 kips. Use $\frac{3}{4}$-in.-diameter A325N bolts in standard holes.

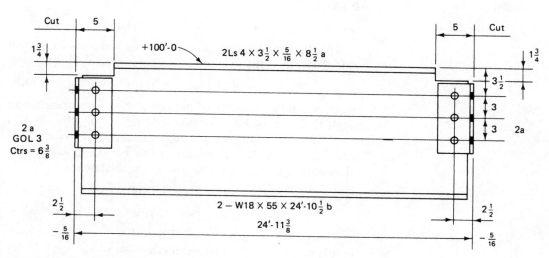

**FIGURE 11-9**  Beam B1.

A.   Select the bolted seated beam connection.

1.   Select the angle thickness based on an angle length of 6 in. determined as follows. The angle length cannot exceed the $T$ dimension of the supporting column (W10 × 60) since the angle will be bolted to the column web. The $T$ dimension is $7\frac{5}{8}$ in. Therefore, use an angle length of 6 in.

Enter Table V-A, ASDM, Part 4. With a beam (B2) web thickness of $\frac{3}{8}$ in., the required angle thickness is 1 in. The outstanding leg capacity (based on an outstanding leg of 4 in.) is 41.1 kips.

$$41.1 \text{ kips} > 31.0 \text{ kips} \qquad \textbf{O.K.}$$

2.   Enter Table V-C to select the type of seat angle. With $\frac{3}{4}$-in.-diameter A325N high-strength bolts, a Type B connection must be used. Fastener shear capacity is 37.1 kips.

$$37.1 \text{ kips} > 31.0 \text{ kips} \qquad \textbf{O.K.}$$

3.   Enter Table V-D to select the seat angle. With a Type B connection, an 8 × 4 angle is available in 1 in. thickness.

4.   Check the capacity of the connection in bearing on the column web.

The Type B connection has four bolts on two rows with two bolts per row in the line of force. Bolt spacing is 3 in., which is $4.0d$ ($> 3.0d$, O.K.). The vertical edge distance for the column web is appreciably in excess of $1.5d$ and is not considered. Therefore, Table I-E is applicable.

The allowable load in bearing on the column web ($t_w = 0.420$ in.) is

$$4(52.2)(0.420) = 87.7 \text{ kips}$$

Two B2 beams frame into the column web, one from each side. Therefore, the reaction to the column is $2(31) = 62$ kips:

$$87.7 \text{ kips} > 62 \text{ kips} \qquad \textbf{O.K.}$$

5.   Next, check the capacity of the connection in bearing on an angle thickness of 1 in. The bolt spacing is 3 in., which is $4.0d$ ($> 3.0d$, O.K.). The vertical edge distance is 2 in., which is $2.67d$ ($> 1.5d$, O.K.). Therefore, Table I-E is applicable. The allowable load in bearing on the angle thickness is

$$4(52.2) = 209 \text{ kips}$$

$$209 \text{ kips} > 31 \text{ kips} \qquad \textbf{O.K.}$$

Therefore, use a seat angle L8 × 4 × 1 × 6 with four $\frac{3}{4}$-in.-diameter A325 N high-strength bolts in the vertical leg (Type B pattern) at each end of beam B2.

In addition, select a top angle (discussed in Section 7-7 of this text). Since this is a roof framing system with the top of the column at approximately the same elevation as the tops of the beams and girders, the top angle cannot be located on the top flange, but must be placed at the optional location as indicated in the ASDM, Part 4. Use an angle L4 × 3 × $\frac{1}{4}$ × $5\frac{1}{2}$ connected to the beam web and column web with a total of four $\frac{3}{4}$-in.-diameter A325 high-strength bolts.

6. Check web crippling for beam B2. Since web yielding has been incorporated into the values of Table V, the beam must be checked for web crippling. From the Allowable Uniform Load Tables, Part 2 of the ASDM,

$$R_3 = 39.4 \text{ kips}$$

$$R_4 = 3.18 \text{ kips/in}$$

The allowable end reaction may be computed from

$$R = R_3 + NR_4$$

$$= 39.4 + \left(4 - \frac{3}{4}\right)(3.18)$$

$$= 49.7 \text{ kips}$$

$$49.7 \text{ kips} > 31 \text{ kips} \qquad \textbf{O.K.}$$

B. In addition to the selection of the seat angle, various dimensions must be computed to complete the detailing of B2. This is best accomplished with the use of working sketches, as shown in Figure 11-10.

1. The setback distance required is the distance from the centerline of the column to the end of the beam. Approximately $\frac{1}{2}$-in. erection clearance should be used between the end of the beam and the face of the column web:

$$\frac{t_w}{2} + \frac{1}{2} = \frac{1}{4} + \frac{1}{2} = \frac{3}{4} \text{ in.}$$

2. The beam overall length, or ordered length, is the theoretical span length minus the setback:

$$(25'-0) - \frac{3}{4} - \frac{3}{4} = 24'-10\frac{1}{2}$$

3. As shown in the ASDM, Part I, Standard Mill Practice, the cutting tolerance for this beam is $\frac{3}{8}$ in. over and $\frac{3}{8}$ in. under.

   Depending on the final length of the beam, adjustments by the shop may be necessary for the location of holes and

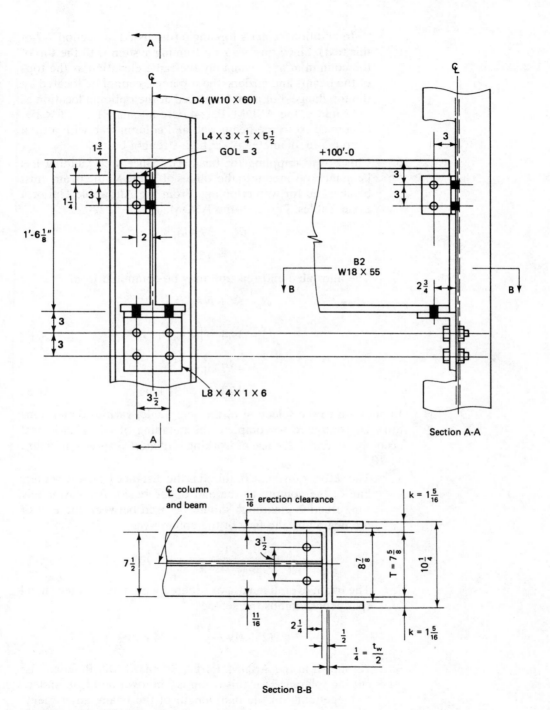

**FIGURE 11-10**  Working sketch.

edge distances. The distance between the holes at the ends of the bottom flange, however, is fixed at 24'–6, as shown in Figure 11-11.

4. It is not necessary to cut the top and bottom beam flanges to fit between column flanges, since an $\frac{11}{16}$-in. erection clearance exists. An erection clearance of at least $\frac{1}{2}$ in. must be furnished.

5. The top angle, which connects the beam web to the column web is located below the toe of the fillet of beam B2 (defined by the $k$ dimension). If a vertical edge distance of $1\frac{1}{4}$ in. is provided for the top angle and the angle is located $1\frac{3}{4}$ in. down from the top of B2, the upper gage line will be conveniently located 3 in. from the top of B2. This angle will be bolted hand-tight and shipped with the column. It will be removed in the field and then replaced properly after the placement of beam B2 (see Chapter 12).

6. The distance between gage lines on the column web may be set at $3\frac{1}{2}$ in., which conforms to the angle length of 6 in. as shown in the ASDM, Part 4, Seated Beam Connections.

7. The distance from the end of beam B2 to the web holes for the top angle is the angle gage distance of 3 in. minus the erection clearance of $\frac{1}{2}$ in. from the end of the beam to the face of the column web. The $2\frac{1}{2}$-in. edge distance is shown in Figure 11-11. Note that the detail of beam B2 in Figure 11-11 does not include the seat angle or the top angle. Since they will be shipped with the column, they are detailed with the column (see Chapter 12).

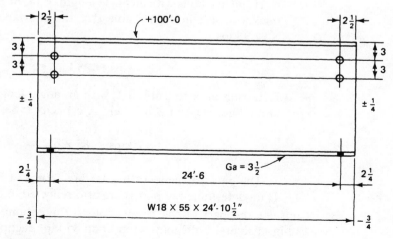

**FIGURE 11-11**   Beam B2.

III.   *Girder G1:* G1 is a W24 × 94 supporting the end reactions of four B1 beams and in turn is supported by W10 × 60 columns (which will be designated D4). The girder end connections will be framed connections shop-welded to the girder web and field-bolted to the column flange. Recall that the beam-to-girder connections were detailed with (and will be shipped with) beams B1. High-strength bolts (A325N) $\frac{3}{4}$ in. in diameter in standard holes and E70 electrodes will be used. The end reaction as furnished on the framing plan (Figure 11-6) is 85 kips.

A.   Consider the part of the connection to the W24 × 94 web.

1.   The framing angle length will again be taken as at least one-half of the *T* dimension, which for the W24 × 94 is 21 in. Therefore, use an angle length of at least $10\frac{1}{2}$ in. Enter Table III, ASDM, Part 4, under weld A, and find the closest allowable load greater than 85 kips.

Select a weld size of $\frac{1}{4}$ in., with a capacity of 97.0 kips and a required length of angle of $11\frac{1}{2}$ in. This requires a $\frac{5}{16}$-in.-thick angle to meet the weld requirement of the ASDS, Section J2.2. The 0.515-in. web thickness of the W24 × 94 is greater than the minimum web thickness tabulated in Table III. Therefore, no capacity reduction is required.

2.   Note in Table IIA, ASDM, Part 4, that the $11\frac{1}{2}$-in. angle length provides for four $\frac{3}{4}$-in.-diameter A325N high-strength bolts, with a capacity of 74.2 kips. This represents the capacity of the eight bolts through the outstanding legs of the connection angles (in single shear) to the column flange. This is less than the beam reaction of 85 kips; therefore, the angle length must be increased.

Using five bolts with an angle length of $14\frac{1}{2}$ in. and an angle thickness of $\frac{5}{16}$ in. from Table IIA, the allowable bolt shear is 92.8 kips:

$$92.8 \text{ kips} > 85 \text{ kips} \qquad \textbf{O.K.}$$

3.   Referring back to Table III, with an angle length of $14\frac{1}{2}$ in. and a fillet weld of $\frac{3}{16}$ in., weld A capacity is 88.7 kips:

$$88.7 \text{ kips} > 85 \text{ kips} \qquad \textbf{O.K.}$$

The data determined thus far are shown in Figure 11-12. Data determined in part B are also reflected in Figure 11-12.

B.   Consider the part of the connection to the column flange. The 10 $\frac{3}{4}$-in.-diameter bolts must support an 85-kip reaction. The shear capacity of the bolts has been checked in part A.

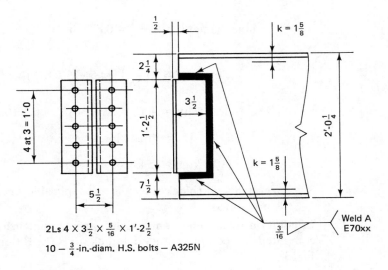

**FIGURE 11-12**  Working sketch.

1.   Check the adequacy in bearing on the column flange and on the outstanding legs of the angles. For the column, $t_f = 0.680$ in. The angles are $\frac{5}{16}$ in. (0.313 in.) thick and are, therefore, more critical (they are thinner than the web). Bolt spacing is 3 in., which is $4.0d$ ($> 3.0d$, O.K.). The vertical edge distance for the angles is $1\frac{1}{4}$ in. (vertical edge distance is not a consideration for the column flange), which is $1.67d$ ($> 1.5d$, O.K.). Therefore, Table I-E is applicable. The capacity in bearing on the angle thickness is calculated as

$$10(16.3) = 163 \text{ kips}$$

$$163 \text{ kips} > 85 \text{ kips} \qquad \textbf{O.K.}$$

The 10-bolt connection to the column flange is adequate.

To accommodate usual gages, the angle leg widths are generally taken as $4 \times 3\frac{1}{2}$ with the 4-in. leg outstanding. Therefore, the connection angles for this connection are

$$2L4 \times 3\frac{1}{2} \times \frac{5}{16} \times 1'{-}2\frac{1}{2}$$

2.   Check the capacity of the connection based on shear on the net area of the connecting angles. Refer to Figure 11-12. Consider shear on the net area taken through the line of five bolt holes in each angle.

$$\text{hole diameter} = \frac{3}{4} + \frac{1}{8} = 0.875 \text{ in.}$$

$$A_v = 2\left(\frac{5}{16}\right)[14.5 - 5(0.875)] = 6.33 \text{ in.}^2$$

$$\text{capacity} = A_v F_v = A_v(0.30)F_u = 6.33(0.30)(58)$$

$$= 110.4 \text{ kips}$$

$$110.4 \text{ kips} > 85 \text{ kips} \qquad\qquad \textbf{O.K.}$$

3.   Web tear-out (block shear) is not applicable since the girder is not coped.

C.   In addition to the end connections, various dimensions must be computed to complete the detailing of G1. Refer to Figure 11-13.

1.   The setback distance required is the distance from the centerline of the supporting column to the back of the connection angle:

$$\frac{d}{2} + \frac{1}{16} = \frac{10\frac{1}{4}}{2} + \frac{1}{16} = 5\frac{3}{16} \text{ in.}$$

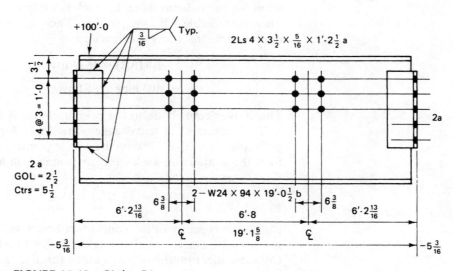

**FIGURE 11-13**   Girder G1.

2. The beam assembly unit length is the distance back-to-back of angles and is calculated as

$$\text{theoretical span length} - \left(\frac{d}{2} + \frac{1}{16}\right) 2$$

$$= (20'{-}0) - \left(\frac{10\frac{1}{4}}{2} + \frac{1}{16}\right) 2$$

$$= 19'{-}1\frac{5}{8}$$

3. The ordered or billed length of the beam is established so that the ends are $\frac{1}{2}$ in. short of the backs of the connection angles. The ordered length is rounded to the nearest $\frac{1}{2}$ in.

$$\left(19'{-}1\frac{5}{8}\right) - \frac{1}{2} - \frac{1}{2} = 19'{-}0\frac{5}{8}$$

Use an ordered length of $19'{-}0\frac{1}{2}$.

4. The usual gage for the 4-in. outstanding leg of the framing angle is $2\frac{1}{2}$ in. The maximum spread is calculated as

$$2 \times 2\frac{1}{2} + t_w = 5 + \frac{1}{2} = 5\frac{1}{2}\,\text{in.}$$

5. The dimension from the backs of the connection angles to the centerline of the connections for the B1 beams is determined by subtracting the setback distance from the centerline of column to the centerline of B1 dimension:

$$(6'{-}8) - 5\frac{3}{16} = 6'{-}2\frac{13}{16}$$

## REFERENCES

[1] *Manual of Steel Construction, Allowable Stress Design,* 9th ed., American Institute of Steel Construction, 1 East Wacker Drive, Suite 3100, Chicago, IL 60601.

[2] *Detailing for Steel Construction* (AISC Publication No. M013), American Institute of Steel Construction, 1 East Wacker Drive, Suite 3100, Chicago, IL 60601 (1983).

[3] *Engineering for Steel Construction* (AISC Publication No. M014), American Institute of Steel Construction, 1 East Wacker Drive, Chicago, IL 60601 (1984).

## PROBLEMS

*Note: In the following problems use $F_u = 58$ ksi for A36 steel and assume a class A contact surface for all slip-critical connections. All bolts are placed in standard holes.*

**11-1.**  A W16 × 36 beam is to connect to the flange of a W12 × 53 column. The beam reaction is 40 kips. All structural steel is A36. Select a suitable framed connection using $\frac{3}{4}$-in.-diameter A325SC high-strength bolts. Include a working sketch of the connection and complete data for the angles to be used.

**11-2.**  A W24 × 68 beam is to connect to the web of a W33 × 118 girder. The elevation of the top of the beam is to be 3 in. below the top of the girder. The beam reaction is 90 kips. Select a suitable framed connection using $\frac{3}{4}$-in.-diameter A325X high-strength bolts. Include a working sketch of the connection and complete data for the angles to be used. All structural steel is A36.

**11-3.**  Rework Problem 11-2, except that the beam reaction is 75 kips, and the tops of the beam and girder are to be at the same elevation.

**11-4.**  Select a suitable framed beam connection for the beam-to-girder connection shown. All structural steel is A36. Use $\frac{7}{8}$-in.-diameter A325N high-strength bolts. Include a working sketch and complete data for the angles to be used.

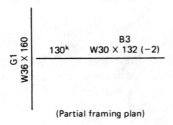

(Partial framing plan)

**PROBLEM 11-4**

**11-5.**  Select framed connections for beam B4 shown. Completely detail the beam. All structural steel is A36. Beam reactions are not given on the design drawings. Use $\frac{7}{8}$-in.-diameter A325N high-strength bolts for the connections.

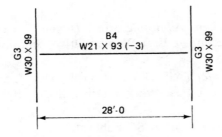

**PROBLEM 11-5**

**11-6.** Rework Problem 11-5 but use E70 electrodes and weld the framing angles to the web of B4. Use $\frac{7}{8}$-in.-diameter A325N bolts for the girder web connection.

**11-7.** Refer to Figure 7-24a in Chapter 7. Design the seated beam connection assuming that the beam is a W16 × 77 and the column is a W8 × 35. All structural steel is A36. Use $\frac{7}{8}$-in.-diameter A325X bolts. The beam reaction is 35 kips.

**11-8.** Rework Problem 11-7 except that the supporting column is a W12 × 65 and the reaction is 40 kips.

**11-9.** Select unstiffened seated beam connections for the beam shown. Completely detail the beam. All structural steel is A36. Use $\frac{3}{4}$-in.-diameter A325 N high-strength bolts for the connections. Note that the columns are not similar.

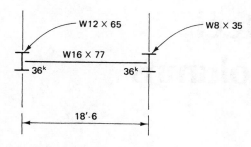

**PROBLEM 11-9**

# CHAPTER 12

# Structural Steel Detailing: Columns

12-1   INTRODUCTION

12-2   COLUMN BASE DETAILS

12-3   COLUMN DETAILS

12-4   SHOP DRAWINGS OF COLUMNS

## 12-1

### INTRODUCTION

Columns and their associated base plates are the first structural steel members to be erected, and they must be among the first to be fabricated. Therefore, shop drawings of the columns are generally prepared before the shop drawings of the other structural members.

The information needed for detailing columns is furnished on the structural drawings of the contract documents. The structural drawings (often termed *engineering drawings* or *design drawings*) depict the floor and roof framing plans. These drawings also locate the column centerlines and orient the columns with respect to direction of flanges and webs. Special enlarged details clarify special framing conditions, such as off-center beams for spandrel framing or for framing around stairwells.

The structural drawings also normally contain a column schedule, which provides the fabricator with information on the size and length of the columns as well as splice location and column base plate information. Some form of grid system is used throughout the entire set of contract documents (structural, architectural, mechanical, etc.) to establish a consistent means of identifying all columns. The grid system usually reflects, in some way, the general shape of the building. The simplest is the rectangular grid, although grid shapes may be radial for circular buildings, repetitive triangles for sprawling buildings, or irregular patterns to meet some other shape. For the common rectangular grid, one normally uses a simple numerical sequence, beginning with number 1, in one direction and a letter sequence, beginning with A, in the other direction. Thus a column at the intersection of lines E and 5 would be uniquely identified by designation E5.

A sample column schedule is shown in Figure 12-1. As may be observed, the column schedule furnishes column loads, various building elevations, and all column base plate information. Various forms of column schedules are used by the many design offices; they all convey the same important information, however.

## 12-2

## COLUMN BASE DETAILS

Typical column bases and column base plate design are discussed in Chapter 3 of this text. Typical column base details for any given building are generally furnished on the structural drawings. Two such typical details are shown in Figure 12-2. All column base details require a base plate and anchor bolts as well as other detail items as necessary. The anchor bolts fix the column base to the foundation. The construction of the foundation is usually the responsibility of the general contractor, not the steel fabricator/erector. The anchor bolts are generally set in place by the masonry contractor, but detailed and furnished by the fabricator. Since foundation construction always precedes steel erection, it is necessary for the fabricator to prepare an anchor bolt plan as soon as possible. This plan, which may be similar in appearance to the foundation plan, gives complete information for field placement. It furnishes anchor bolt sizes, lengths, exact locations, erection marks, elevations at top of base plates, grout thicknesses, and the length of anchor bolts above the top of concrete. In addition, base plate erection marks and location, as well as top of concrete elevations, are furnished. Not only does the masonry contractor use the anchor bolt plan to set the bolts, but the steel erector also uses the plan to set the base plates. On occasion, loose base plates may be small enough to be set manually and may be placed by the masonry contractor. Heavy base plates (which may weigh several tons) are generally set by the erector with the use of a crane.

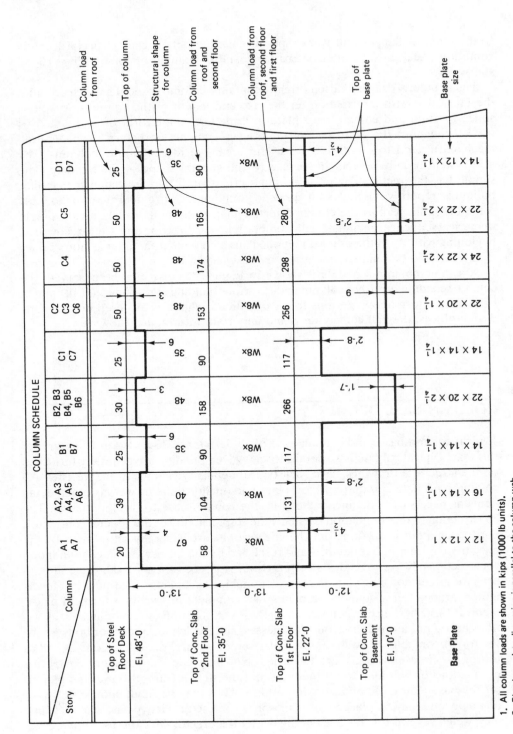

## COLUMN SCHEDULE

| Story | Column | A1<br>A7 | A2, A3<br>A4, A5<br>A6 | B1<br>B7 | B2, B3<br>B4, B5<br>B6 | C1<br>C7 | C2<br>C3<br>C6 | C4 | C5 | D1<br>D7 |
|---|---|---|---|---|---|---|---|---|---|---|
| Top of Steel<br>Roof Deck<br>El. 48'-0 | | 20 | 39 | 25 | 30 | 25 | 50 | 50 | 50 | 25 |
| | 13'-0 | 4 | | 6 | 3 | 6 | | | 48 | 6 |
| | | 67 | 40 | 35 | 48 | 35 | 48 | 48 | 48 | 35 |
| Top of Conc. Slab<br>2nd Floor<br>El. 35'-0 | | W8× | W8× | W8× | W8× | W8× | W8× | W8× | W8× | W8× |
| | 13'-0 | 58 | 104 | 90 | 158 | 90 | 153 | 174 | 165 | 90 |
| Top of Conc. Slab<br>1st Floor<br>El. 22'-0 | | 131 | 131 | 117 | 266 | 117 | 256 | 298 | 280 | 90 |
| | 12'-0 | 4½ | 2'-8 | | 1'-7 | 2'-8 | 6 | | 2'-5 | 4½ |
| Top of Conc. Slab<br>Basement<br>El. 10'-0 | | | | | | | | | | |
| **Base Plate** | | 12 × 12 × 1 | 16 × 14 × 1¼ | 14 × 14 × 1¼ | 22 × 20 × 2¼ | 14 × 14 × 1¼ | 22 × 20 × 1¼ | 24 × 22 × 2¼ | 22 × 22 × 2¼ | 14 × 12 × 1¼ |

Column load from roof

Top of column

Structural shape for column

Column load from roof and second floor

Column load from roof, second floor and first floor

Top of base plate

Base plate size

1. All column loads are shown in kips (1000 lb units).
2. First base plate dimension is parallel to the column web.

**FIGURE 12-1** Typical column schedule.

424

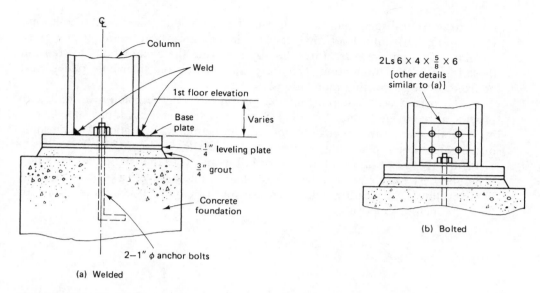

**FIGURE 12-2**   Typical column base details (as shown on design drawings).

Small base plates are often attached to the columns in the shop. When this is the case, $\frac{1}{4}$-in.-thick steel leveling plates are normally installed on the foundation to provide a smooth bearing area. The leveling plates are easy to handle and are conveniently set level to the prescribed elevation prior to the erection of the columns. Leveling plates are generally furnished by the fabricator and placed by the masonry contractor.

Large base plates are set to elevation and leveled using shims of various thicknesses or by leveling screws with nuts welded to the edges of the base plate. The top of the rough masonry foundation is usually set approximately 1 in. below the bottom of the base plate to provide for adjustment and subsequent grouting. After a plate has been carefully leveled up to the proper elevation (through the use of the shims or leveling screws), cement grout is worked under the plate to build up the foundation and provide contact bearing over the full area of the underside of the plate. The design may have specified one or more large holes near the center of the plate through which the grout may be poured. The object is to ensure an even distribution across the entire undersurface. Such holes need not be drilled but may be burned with an acetylene torch.

Anchor bolt holes in base plates and in any of the other fittings at the bases of columns are generally made $\frac{5}{16}$ in. to 1 in. larger than the diameter of the bolts to allow for inaccuracies in the setting of the bolts. The anchor bolts serve additional purposes during erection of a frame. They aid in centering the column and also serve to prevent displacement or collapse of columns due to accidental collisions. In addition, horizontal loads may tend to induce a vertical uplift on any given column of a completed structure, thereby requiring the anchor bolts to transmit these forces to the foundation. Therefore, all columns and their base plates must be fixed to the supporting foundations.

In the past it was required that the bottom end of the column be milled to provide a true overall contact bearing. Some saws used in present-day shops, however, produce end surfaces that do not require the milling operation. The finishing of the column base plate must conform to the ASDS, Section M2.8.

Figure 12-3 is a portion of an anchor bolt plan. Note that the plan shows bolt locations, base plate mark numbers, base plate locations, leveling plate mark numbers, leveling plate locations, elevations for tops of base plates, and elevations for tops of concrete supports.

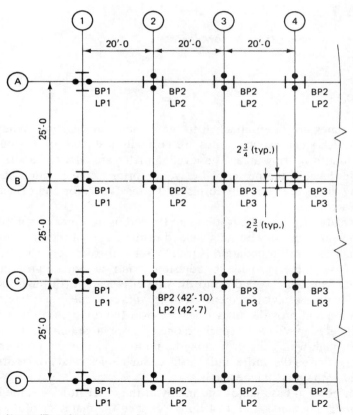

Notes: (Top of base plate elev.)
(Top of concrete elev.)

**FIGURE 12-3** Typical anchor bolt plan.

## 12-3

### COLUMN DETAILS

In multistory structures, the elevations at which the column sizes change provide a convenient means of dividing the framing vertically into *tiers* for ease in handling and erection. This requires the use of field splices to hold the column sections together and transfer the column loads. Generally, columns are spliced at two-story intervals. The splices are located far enough above the floor lines so that the connections will not interfere with beam framing details.

The transfer of load in multistory building columns is generally achieved entirely by direct bearing through finished contact surfaces. The splice material, which usually consists of plates and fasteners, serves principally to hold all parts securely in place. Suggested details for column splices are furnished in the ASDM, Part 4. One- or two-story buildings are usually designed with a constant column section for the full height of the column. Therefore, no splices are needed. Pinholes are occasionally furnished in the splice plates for erection purposes, as shown in the ASDM, Part 4.

**PHOTO 12-1**   Structural steel in the fabricator's shop. These wide-flange columns have been prepared for bolted splices. The lower two members have shim plates tack-welded in place. Note that the direction that the flange is to face (west or east) has been marked on the columns.

## 12-4

### SHOP DRAWINGS OF COLUMNS

It is traditional in the preparation of shop drawings of columns to detail the columns either in the vertical position with the base of the column at the bottom of the sheet or in the horizontal position with the column base to the left. If the column is simple with little detail material, it is detailed upright. A complex column with complicated beam bracing and, perhaps, truss connections is detailed in the horizontal position.

It is fairly common that columns have some form of connection to both flanges and to both sides of the web. As a result, it is standard practice to assign a letter to each of the four faces of the column. This identification by letter is helpful to the shop worker in laying out the work and reduces the probability of shop errors. Usually, the detailer will select the *flange face* that contains the most detail (fittings and fabrication) and label that face "A." Then, looking down on top of the column, the lettering will continue alphabetically in a counterclockwise direction around the section.

As shown in Figure 12-4, for a W shape, faces A and C are always flange faces and faces B and D are always web faces. The lettering could also be arranged so that the web with the most detail material is assigned the letter B. It is seldom necessary to show a separate view of face D (a web face) for W shapes, as any fittings on face D that differ from those on face B can be shown by dashed (invisible) lines. The detail materials to be placed on face B are shown by visible lines and may be noted N.S. (near side). Face D detail material may be noted F.S. (far side). Faces that require no detail fittings or fabrication of any kind need not be shown. A note such as "Face C Plain" is advisable, however. Where detail material and fabrication on face C is identical with that on face A, a note such as "Face C Same as Face A" may be added.

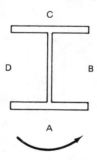

**FIGURE 12-4**   Column marking (W shape).

Where a transverse section through a column is needed to describe detail material, it should be taken from the face B view and shown looking down toward the column base. Columns that are alike except for minor differences may be detailed on the same sketch and the differences defined with notes. Combining details for too many columns on one sketch, however, can result in a complicated drawing and cause shop errors. If the columns cannot be detailed with simple notes describing the exceptions, separate sketches should be used. Shop errors are expensive to correct in the field.

Columns, like beams, must be given shipping marks. The mark shown on the drawing is painted on one of the flanges and near the base of the column. This is done in the shop when the detail material is assembled. This mark identifies the column during shop fabrication, shipping, and field erection.

It is generally necessary to provide a compass or direction mark on one of the column flanges so that the column can be oriented correctly in the field. A note such as "Face A North" instructs the shop to paint "North" on face A. In the field the erector will then turn the column so that the word "North" is facing north when the column is erected and in place.

**PHOTO 12-2**  Structural steel in the fabricator's shop. This wide-flange column is being prepared in the shop by welding on various brackets and seats. Accurate placement of such detail material is essential for rapid field erection.

**Example 12-1** _____

Refer to Example 11-1 and Figure 11-6 in Chapter 11 of this text. Using all the previously established design and detail data, detail column E4 in accordance with the latest ASDM. The column is a W10 × 60. The top of base plate elevation for this column is $81'-10\frac{1}{2}$. The top of steel roof deck is at elevation 100'-3.

**Solution:**

It is shown in Figure 11-6 that girder (beam) G1 frames into each flange of column E4 and that beam B2 frames into each side of the web of column E4. The connections have been selected in Example 11-1. Typical column base details are shown in Figure 12-2. The design drawings would indicate the desired type of column base detail. The base detail shown in Figure 12-5 is of the common type shown in Figure 12-2b. Explanations of the various dimensions in Figure 12-5 follow.

1. The top of steel roof deck elevation of 100′–3 and top of base plate (bottom of column) elevation of 81′–10½ would be furnished by the design drawings. The top of steel elevation of 100′–0 is shown 3 in. below the top of steel roof deck. The top of columns is set 1½ in. below the top of steel.

2. In the detailing of columns, overall dimensions and floor and roof elevations are placed prominently farthest away from the views.

3. Detail dimensions showing hole spacing and detail location are placed closest to the related view.

4. Extension dimensions, used by the shop to establish and check the location of open holes or the tops of seat angles, are measured from the finished bottom of the column shaft.

5. All dimensions relating to a connection and holes should be tied to a floor or roof level at which the connected beam exists.

6. Depths of beams framing into the column are given so that clearances and connection angles can be checked.

7. The number and location of the open holes on face A were established in Chapter 11 and shown in Figure 11-13 (girder G1).

8. The seated connection material and details on face B were selected in Chapter 11 and shown in Figure 11-10 (beam B2).

9. Girder G1 may be swung into place from either side to meet the flange holes on face A.

10. Beam B2 must be placed from above by moving the beam downward between the flanges of column E4. Therefore, top angle b must be shop-bolted hand-tight to the column for shipment, then removed and replaced in the field after placement of B2.

11. Based on the north arrow shown in Figure 11-6 and the orientation of the columns, the shop will mark "West" on face A.

12. The bottom of the column (shaft) must be milled (M1E: "Mill One End") for full and level bearing on the base plate.

13. The total assembly unit will include one 18′–0 length of a W10 × 60 with assembly mark "a" plus two angles marked "b," two angles marked "c," and two angles marked "d." All pieces will be assembled as shown in Figure 12-5 and the assembly marked E4.

Loose base plates, since they constitute separate shipping pieces, are not detailed with the columns, but rather on separate drawings, and are given individual shipping marks. Typical shop details for such loose base plates together with leveling plates are usually shown by a simple plan view. An

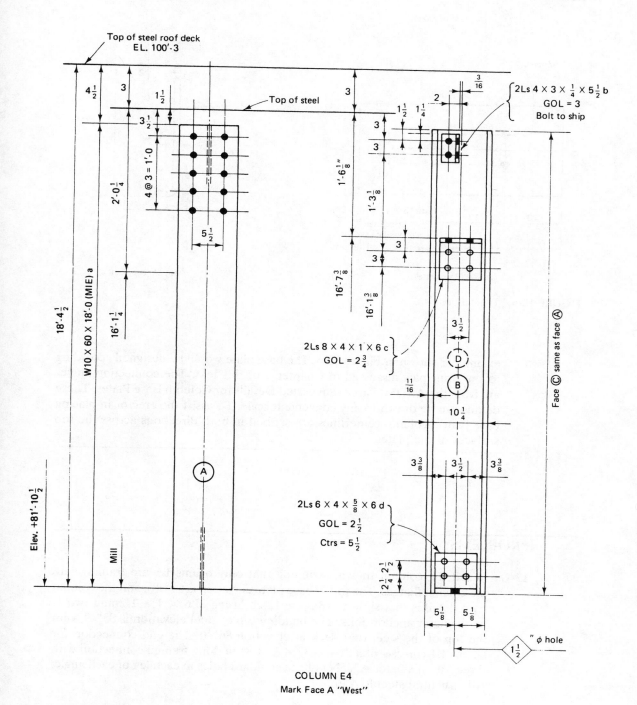

COLUMN E4

Mark Face A "West"

**FIGURE 12-5**  Typical column detail.

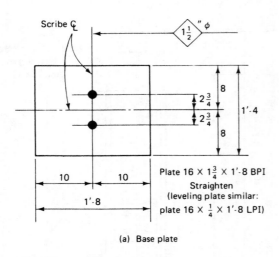

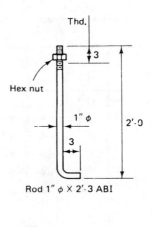

(a)  Base plate

(b)  Anchor bolt

**FIGURE 12-6**   Typical details.

example is shown in Figure 12-6. The base plate would be designed according to the principles discussed in Chapter 3 of this text. The connecting angles are from the ASDM, Part 4, Suggested Details for Column Base Plates. These details can be drawn to any convenient scale. To assist the erector in placing base plates, column centerlines are scribed in both directions across the top surfaces of the plates.

## PROBLEMS

**12-1.**   Detail the column shown, assuming that only beams B1 are framing into the column flanges. The top of base plate elevation is $63'-8\frac{1}{2}$, and the column base detail is that shown in Figure 12-2b, using 2L6 × 4 × 1/2 and two 1-in.-diameter anchor bolts. The building top of steel elevation is $85'-9$, with the top of the steel roof deck at elevation $86'-0$. The end connection for beams B1 consists of a 2L4 × $3\frac{1}{2}$ × $\frac{5}{16}$ × $8\frac{1}{2}$-in.-long framed connection with three $\frac{7}{8}$-in.-diameter A325N bolts in standard holes in each leg of each angle. All structural steel is A36.

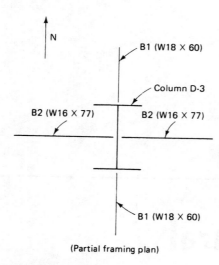

(Partial framing plan)

**PROBLEM 12-1**

**12-2.** Detail the column shown for Problem 12-1 assuming that beams B1 and B2 are framing into the column. All data from Problem 12-1 are applicable. The end connection for beams B2 is an unstiffened seated beam connection consisting of an L8 × 4 × ¾ × 6 in. seat angle with six ⅞-in.-diameter A325N bolts in standard holes.

# CHAPTER 13

## LRFD: Structural Members

### 13-1

#### INTRODUCTION

All the design procedures used in the previous chapters of this text (with the exception of Chapter 10, Sections 10-3 through 10-6) have reflected a design method based on allowable stresses. *Allowable stress design* (ASD) historically has been the method used for structural steel design and at this writing is still the preferred and most commonly used method.

As shown previously, in ASD the stress induced by applied *nominal loads* (sometimes called *service loads*) cannot exceed some code (or specification) designated allowable stress. In effect, this allowable stress represents a limit of structural usefulness and provides some assurance of structural safety. An allowable stress may be defined as

$$F_{ALL} = \frac{F_{LIM}}{F.S.}$$

where

$F_{ALL}$ = allowable stress

$F_{LIM}$ = some limiting stress value such as the yield stress $F_y$ or the tensile stress $F_u$ at which a material fractures

F.S. = factor of safety

In Chapter 10, an alternate design method designated *plastic design* was introduced. Criteria and specifications for this method were developed in the 1950s, but the method failed to gain wide acceptance due to its many limitations. In plastic design the applied nominal loads are multiplied by a load factor and members are then designed on the basis of their maximum strength. In essence, the load factor is a factor of safety.

The recent and modern approach to structural steel design is designated *load and resistance factor design* (LRFD). This method was first introduced and developed in the early 1970s. In 1986, the AISC published the first LRFD Specification. A second edition was subsequently published in 1994 and serves as the basis for this chapter and Chapter 14. The LRFD method uses a series of factors of safety called *load factors* when applied to loads and *resistance factors* when applied to member strength or resistance. Each factor is the result of a statistical study of the variability of the particular quantity and reflects the probability that the specific load or resistance is incorrect.

The method closely parallels the strength design method used for reinforced concrete. In that design method strength reduction factors are used to predict a practical strength and load factors are used to modify service loads and establish design loads for use in design calculations.

The LRFD method is considered to be the more rational method because it recognizes that some loads and material or member strengths are known with a reasonable amount of certainty whereas others are not. Based on the relative degrees of uncertainty, load and resistance factors have been developed using probability theory in combination with statistical methods. This in effect permits the setting of factors of safety in a less arbitrary fashion than is the case with ASD and ensures a more consistent margin of safety.

There is no single overwhelming advantage of LRFD over ASD. There do seem to be many minor advantages and no great disadvantages, however. According to the LRFD Specification Commentary, Section A1:

The LRFD method was devised to offer the designer greater flexibility, more rationality, and possible overall economy. . . . Also, it provides a rational methodology for transference of test results into design provisions. A more rational design procedure leading to more uniform reliability is the practical result.

In addition, the use of LRFD rather than ASD requires the designer to have a better understanding of structural behavior, categories of loadings, where loads come from, and expected load variability. It is predicted that LRFD will eventually become the predominant method for the design and analysis of structural steel.

At present, which is seen as a period of transition, there are separate manuals for ASD and LRFD. For purposes of this chapter, references will be made to the *AISC Manual of Steel Construction Load and Resistance Factor Design,* 2nd edition (see Reference 1). This manual is referenced hereafter as the *LRFD Manual,* or the *LRFDM.* We similarly refer to the AISC *Load and Resistance Factor Design Specification for Structural Steel Buildings* as the *LRFDS.*

The format of the LRFDM is, in some respects, similar to that of the *Manual of Steel Construction, Allowable Stress Design,* 9th edition, that has been referred to as the ASDM in the preceding chapters. The LRFDM 2nd edition is also divided into two volumes. Volume I contains the LRFD Specification and Commentary, tables, and design information for structural members as well as a section entitled Essentials of LRFD. Volume II contains information on connections. If the reader is beginning a formal study of the LRFD method, the *LRFD Manual* must be obtained for ready reference of all the design tables and design aids. For the purposes of our brief introduction, however, a few of the necessary tables and figures have been reproduced in the text for the convenience of the reader.

Although treatment of the load and resistance factor design method will be brief, an increasing number of texts are available on the subject. The reader is referred to References 2 through 6.

## 13-2

## BASIS FOR LRFD

The basis for LRFD may be simply stated as follows:

$$\text{design load or load effects} \leq \text{applicable design strength}$$

All members, connections, and assemblies must be proportioned to satisfy this criterion. The LRFDS expresses this in terms of *limit states* philosophy, a limit state being a condition at which a member, a connection, or the entire structure ceases to fulfill the intended function. There are two kinds of limit states: strength and serviceability. *Strength limit states* concern safety and relate to maximum load-carrying capacity (e.g., yielding or fracture of a tension member or lateral-torsional buckling of a beam.) *Serviceability limit states* refer to performance under normal service conditions (e.g., control of deflections and vibrations).

To evaluate all limit states, the variability and uncertainties associated with applied loads, material strengths, member dimensions, and workmanship must be introduced. Various factors are used for this purpose. *Resistance factors* are designated $\phi$ and will always be less than, or equal to, unity (1.0). *Load factors* are designated $\gamma$ and have a magnitude that is a function of (a) the type of load as well as (b) the combination of loads acting simultaneously.

The LRFD method, as applied to each limit state, may be expressed in *generic form* by the equation

$$\Sigma\gamma_i Q_i \leq \phi R_n \qquad \text{(LRFDM, Part 2, Eqn. 2-2)}$$

where

$i$ = a subscript that indicates the type of load (D, L, wind, etc.)

$Q_i$ = nominal (service) load

$\gamma_i$ = load factor corresponding to $Q_i$

$R_n$ = nominal resistance or strength of member

$\phi$ = resistance factor corresponding to $R_n$

Note that the left side of this expression refers to the factored load effects and essentially represents a *required* resistance or strength. The right side of the expression represents a factored resistance strength or capacity of a member. In essence, the right side represents the strength or capacity of the member or structure when it is at a limit state and is designated the *design strength*.

Load factors $\gamma$ and load combinations are furnished in the LRFDS. They are arranged so that only one of the loads will take on its maximum lifetime value in any combination. The other loads in the combination will assume some *arbitrary point-in-time* value (i.e., a value that can be expected to be on the structure at any time). This reflects the probability, for instance, that the maximum wind load will not occur with the maximum rain/ice load and at the same time that the roof is supporting the maximum snow load and live load.

Section A4 of the LRFDS directs that the following combinations shall be investigated:

| Load combination | LRFDS eqn. |
|---|---|
| $1.4D$ | (A4-1) |
| $1.2D + 1.6L + 0.5(L_r \text{ or } S \text{ or } R)$ | (A4-2) |
| $1.2D + 1.6(L_r \text{ or } S \text{ or } R) + (0.5L \text{ or } 0.8W)$ | (A4-3) |
| $1.2D + 1.3W + 0.5L + 0.5(L_r \text{ or } S \text{ or } R)$ | (A4-4) |
| $1.2D \pm 1.0E + 0.5L + 0.2S$ | (A4-5) |
| $0.9D \pm (1.3W \text{ or } 1.0E)$ | (A4-6) |

where

$D$ = dead load due to the weight of the structural elements and the permanent features on the structure

$L$ = live load due to occupancy and moveable equipment

$L_r$ = roof live load

$S$ = snow load

$R$ = load due to initial rainwater or ice exclusive of the ponding contribution

$W$ = wind load

$E$ = earthquake load determined in accordance with Part 1 of the AISC, Seismic Provisions for Structural Steel Buildings

The preceding loads represent either the loads themselves or the load effects, such as shears and moments.

Resistance factors ($\phi$) attempt to reflect the uncertainties in the nominal, or theoretical, strength (resistance) of the member. The design strength for a given limit state is the product of the nominal strength (resistance) $R_n$ and its resistance factor $\phi$. In the way of introduction, some of the resistance factors are:

|                                                                          | **LRFDS section** |
| ------------------------------------------------------------------------ | ----------------- |
| $\phi_b$ = resistance factor for flexure (0.90)                          | [F1]              |
| $\phi_c$ = resistance factor for compression (0.85)                      | [E2]              |
| $\phi_t$ = resistance factor for tension (0.90 or 0.75)                  | [D1]              |
| $\phi_v$ = resistance factor for shear (0.90)                            | [F2]              |

## 13-3

### TENSION MEMBERS

The LRFD method of analysis and design of tension members uses many of the concepts previously discussed for the ASD method of analysis and design of tension members. The assumption is made that the members are axially loaded and that a uniform tensile stress develops on the cross section. The design strength or capacity of the member previously designated as $\phi R_n$ in its generic form may be expressed as $\phi_t P_n$ for tension members.

The LRFDS, Chapter D, states that the design strength $\phi_t P_n$ of a tension member shall be the lower value obtained based on the limit states of (a) yielding in the gross section and (b) fracture in the net section. Additionally, block shear rupture strength provisions of the LRFDS, Section J4.3, apply to tension members. For the first two, the respective expressions for nominal axial strength $P_n$ are as follows:

(a)   For yielding in the gross section ($\phi_t = 0.90$),

$$P_n = F_y A_g \qquad \text{LRFDS Eqn. (D1-1)}$$

(b)   For fracture in the net section ($\phi_t = 0.75$),

$$P_n = F_u A_e \qquad \text{LRFDS Eqn. (D1-2)}$$

where

$P_n$ = nominal or theoretical axial strength (kips)

$A_e$ = effective net area (in.²) (see Chapter 2)

$A_g$ = gross area of the member (in.²)

$F_y$ = specified minimum yield stress (ksi)

$F_u$ = specified minimum tensile strength (ksi)

Note that when $P_n$ is multiplied by its respective $\phi_t$, the product represents the design tensile strength $\phi_t P_n$.

The criteria for calculation of the effective net area $A_e$ were discussed in Chapter 2 of this text. The block shear failure mode was also discussed in Chapter 2. Recall that this failure may occur along a path involving tension on one plane and shear on a perpendicular plane.

The LRFDS has adopted a more conservative approach than the ASD Specification (ASDS) for the prediction of block shear strength. In LRFD two possible block shear strengths are calculated. This approach is based on the assumption that one of the two failure planes fractures and the other yields. That is, fracture on the shear plane is accompanied by yielding on the tension plane, or fracture on the tension plane is accompanied by yielding on the shear plane. The *gross* area of the plane is used for the limit state of yielding, whether on the tensile plane or the shear plane. The *net* area of the plane is used for the limit state of fracture, whether on the tensile plane or the shear plane. Both planes contribute to the total block shear rupture strength, with the total resistance being equal to the sum of the strengths of the two planes.

The block shear rupture design strength $\phi R_n$ is calculated as follows:

(a)   When the tensile fracture value is equal to or exceeds the shear fracture value, expressed as

$$F_u A_{nt} \geq 0.6 F_u A_{nv}$$

use

$$\phi R_n = \phi[0.6 F_y A_{gv} + F_u A_{nt}] \qquad \text{LRFDS Eqn. (J4-3a)}$$

(b)    When the shear fracture value exceeds the tensile fracture value, expressed as

$$0.6F_u A_{nv} > F_u A_{nt}$$

use

$$\phi R_n = \phi[0.6F_u A_{nv} + F_y A_{gt}] \qquad \text{LRFDS Eqn. (J4-3b)}$$

where

$\phi R_n$ = block shear rupture design strength (kips)

$\phi$ = resistance factor (0.75)

$A_{gv}$ = gross area subjected to shear (in.$^2$)

$A_{gt}$ = gross area subjected to tension (in.$^2$)

$A_{nt}$ = net area subjected to tension (in.$^2$)

$A_{nv}$ = net area subjected to shear (in.$^2$)

In effect, the LRFDS states that the governing equation is the one that has the larger fracture value.

Finally, note that the governing, or controlling, design tensile strength is the *smaller* value obtained from the three cases discussed, namely (a) yielding in the gross section ($\phi_t P_n$, where $\phi_t = 0.90$); (b) tensile fracture in the net section ($\phi_t P_n$, where $\phi_t = 0.75$); and (c) block shear rupture strength ($\phi R_n$, where $\phi = 0.75$). The governing value of $\phi_t P_n$ and $\phi R_n$ must not be less than the required axial strength $P_u$.

It should be noted that in tension members with multiple gage lines of bolts, the block shear rupture strength should be checked along all possible failure paths. As an example, Figure 13-1a and b shows two different possible failure paths for the block shear check for a tension member. When investigating block shear, it is important to select possible failure paths that result in a complete rupture of the connected part. In addition, where staggered bolt patterns are used as shown in Figure 13-1c, the path for the block shear check could include a sloping line. The length and subsequent area of this sloping line may be treated using the procedure for critical net area determination discussed in Chapter 2.

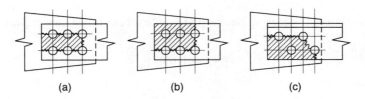

(a)            (b)            (c)

**FIGURE 13-1**   Block shear failure paths.

## Example 13-1

A tension member in a roof truss is composed of 2L5 × 3½ × ½. Assume A36 steel. The angles are connected with two gage lines of ¾-in.-diameter high-strength bolts to a ⅜-in.-thick gusset plate as shown in Figure 13-2. Compute the design tensile strength $\phi_t P_n$ for the member. Assume that the gusset plate and the connection itself are satisfactory.

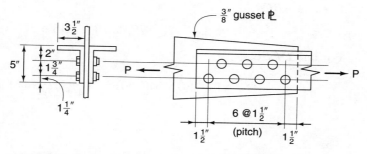

**FIGURE 13-2**   Double-angle truss tension member.

### Solution:

Note that for the two angles, $A_g = 8.0$ in.²

1.  Based on yielding in the gross section,

$$\phi_t P_n = \phi_t F_y A_g \qquad \text{LRFDS Eqn. (D1-1)}$$
$$= 0.90(36)(8.0) = 259 \text{ kips}$$

2.  Based on fracture in the net section (refer to Figure 13-3),

$$\phi_t P_n = \phi_t F_u A_e \qquad \text{LRFDS Eqn. (D1-2)}$$

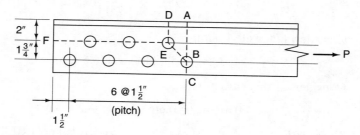

**FIGURE 13-3**   Failure lines.

where

$\phi_t = 0.75$

$F_u$ = specified minimum tensile strength, 58 ksi

$A_e$ = effective net area $(UA_n)$

$U$ = reduction coefficient (0.85) from the LRFDS, Commentary, Section B3.

For line $ABC$,

$$A_n = 8.0 - 2(0.875)(0.50) = 7.13 \text{ in.}^2$$

For line $DEBC$,

$$A_n = 8.0 - 4(0.875)(0.50) + 2\left[\frac{1.5^2}{4(1.75)}\right](0.5)$$

$$= 6.57 \text{ in.}^2$$

Therefore,

$$A_e = UA_n = 0.85(6.57) = 5.58 \text{ in.}^2$$

and

$$\phi_t P_n = 0.75(58)(5.58) = 243 \text{ kips}$$

3.  For the block shear rupture strength ($\phi = 0.75$), note that for the net area calculation, the hole diameter is taken as the fastener diameter plus $\frac{1}{8}$ in.: ($\frac{3}{4} + \frac{1}{8} = 0.875$ in.). Assuming failure path $FEBC$ as shown in Figure 13.3,

$$A_{gv} = 9.0(0.50)(2) = 9.0 \text{ in.}^2$$

$$A_{gt} = \left[3.0 + \frac{1.5^2}{4(1.75)}\right](0.50)(2) = 3.32 \text{ in.}^2$$

$$A_{nv} = [9.0 - 2.5(0.875)](0.50)(2) = 6.81 \text{ in.}^2$$

$$A_{nt} = \left[3.0 - 1.5(0.875) + \frac{1.5^2}{4(1.75)}\right](0.50)(2) = 2.01 \text{ in.}^2$$

Considering tensile fracture,

$$F_u A_{nt} = 58(2.01) = 116.6 \text{ kips}$$

Considering shear fracture,

$$0.6F_u A_{nv} = 0.6(58)(6.81) = 237 \text{ kips}$$

Since $0.6F_u A_{nv} > F_u A_{nt}$, shear fracture controls and LRFDS Equation (J4-3b) is applicable:

$$\phi R_n = \phi(0.6F_u A_{nv} + F_y A_{gt})$$
$$= 0.75[237 + 36(3.32)]$$
$$= 267 \text{ kips}$$

The design tensile strength of the member is the smallest value obtained from the three preceding considerations. Therefore, $\phi_t P_n = 243$ kips based on fracture in the net section.

## Example 13-2

A W8 × 24 of A36 steel is used as a tension member in a truss. It is connected with two lines of $\frac{3}{4}$-in.-diameter bolts in each flange as shown in Figure 13-4. There are three bolts per line and no stagger. The pitch is 3 in., and the edge distance is $1\frac{1}{2}$ in. The applied nominal loads are 60 kips dead load ($D$) and 88 kips live load ($L$). Determine whether the member is satisfactory. Use LRFDS. Assume that the gusset plates and the connection itself are satisfactory.

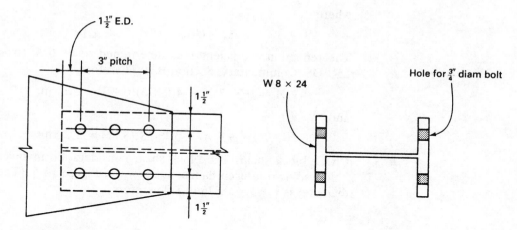

**FIGURE 13-4**  Truss tension member.

### Solution:

We will check to ensure that the required design strength is less than (or equal to) the design strength. The properties of the W8 × 24 are

$$A_g = 7.08 \text{ in.}^2$$
$$d = 7.93 \text{ in.}$$
$$b_f = 6.495 \text{ in.}$$
$$t_f = 0.40 \text{ in.}$$

I.   Calculate the required design strength $P_u$. Only two loads ($D$ and $L$) are given. LRFDS Equations (A4-1) and (A4-2) apply. The load combinations are

$$1.4D = 1.4(60) = 84 \text{ kips} \qquad \text{(A4-1)}$$

$$1.2D + 1.6L = 1.2(60) + 1.6(88) = 213 \text{ kips} \qquad \text{(A4-2)}$$

The latter combination governs. When only dead load and live load are present, Equation (A4-2) will control unless the dead load is more than eight times the live load.

II.  Calculate the design strengths.

1.   For yielding in the gross section,

$$\phi_t P_n = \phi_t F_y A_g$$

$$= 0.90(36)(7.08) = 229 \text{ kips}$$

2.   For fracture in the net section ($\phi_t = 0.75$ and hole diameter = $\frac{3}{4} + \frac{1}{8} = 0.875$ in.),

$$\phi_t P_n = \phi_t F_u A_e$$

where

$$A_e = U A_n = U(A_g - A_{\text{holes}})$$

The reduction coefficient $U$ is determined to be 0.90 (from the LRFDS, Commentary, Section B3). Therefore,

$$A_e = 0.90(7.08 - 4(0.875)(0.40)) = 5.11 \text{ in.}^2$$

and

$$\phi_t P_n = \phi_t F_u A_e = 0.75(58)(5.11) = 222 \text{ kips}$$

3.   Check block shear. The block shear consideration involves four "blocks," two in each flange, as shown in Figure 13-5. (The hole diameter is $\frac{3}{4} + \frac{1}{8} = 0.875$ in.). Therefore,

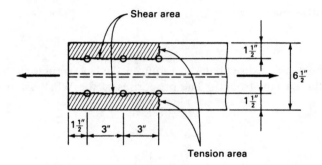

**FIGURE 13-5**   Block shear (each flange).

$$A_{gv} = 4(0.40)(7.5) = 12.0 \text{ in.}^2$$

$$A_{gt} = 4(0.40)(1.5) = 2.40 \text{ in.}^2$$

$$A_{nv} = 4(0.40)[7.50 - 2.5(0.875)] = 8.50 \text{ in.}^2$$

$$A_{nt} = 4(0.40)\left(1.5 - \frac{0.875}{2}\right) = 1.70 \text{ in.}^2$$

Considering tensile fracture,

$$F_u A_{nt} = 58(1.70) = 98.6 \text{ kips}$$

Considering shear fracture,

$$0.6 F_u A_{nt} = 0.6(58)(8.50) = 296 \text{ kips}$$

Since $0.6 F_u A_{nv} > F_u A_{nt}$, shear fracture controls and LRFDS Equation (J4-3b) is applicable:

$$\phi R_n = \phi[0.6 F_u A_{nv} + F_y A_{gt}]$$
$$= 0.75[296 + 36(2.4)]$$
$$= 287 \text{ kips}$$

The design tensile strength of the member is the smallest value obtained from the three computed design strengths. Therefore,

$$\phi_t P_n = 222 \text{ kips}$$

based on fracture in the net section. Thus the member is satisfactory, since the required design strength (213 kips) is less than the computed design strength (222 kips).

---

As discussed in Chapter 2 of this text, rods of circular cross section are commonly used for tension members. The connection of the rods to other structural members is often accomplished by threading the end of the rod and installing a nut. The design of such a member is included in Chapter 2 using the ASD approach.

Using the LRFD approach (LRFDS, Section D1), the design strength or load-carrying capacity of a threaded rod tension member may be expressed as $\phi_t P_n$. Substituting for $P_n$ as per LRFDS, Section J3.6,

$$\phi_t P_n = \phi_t F_t A_b$$

where

$P_n$ = nominal axial strength (kips)

$\phi_t$ = resistance factor (0.75) from the LRFDS, Table J3.2

$F_t$ = nominal tensile strength from the LRFDS, Table J3.2 (ksi)

$A_b$ = nominal unthreaded body area of the rod (gross bolt area from LRFDM, Volume II, Table 8-7) (in.²)

Using $F_t$ from the referenced table, the equation may be rewritten as

$$\phi_t P_n = \phi_t(0.75F_u)A_b$$

where

$F_u$ = the specified minimum tensile strength (ksi)

Note that in the LRFD design of threaded rods (with the exception of lacing, sag rods, or girts), the connection must be designed for a factored load of not less than 10 kips (LRFDS, Section J1.7). Additionally, it is generally recommended that the rod diameter be $\frac{5}{8}$ in. or greater and that the length-to-diameter ratio ($L/d$) of the rod be 500 or less. These rules of thumb are used to ensure that the rod has a minimum amount of rigidity so that it is less subject to damage during handling.

### Example 13-3

An overhead storage balcony is suspended by circular threaded rods 10 ft–0 in. on center. Each rod is subjected to floor service loads of 5 kips dead load ($D$) and 8 kips live load ($L$). Using A36 steel, determine the required rod diameter and specify the required threads.

**Solution:**

The factored load (required design strength) is calculated from

$$P_u = 1.2(5) + 1.6(8) = 18.8 \text{ kips}$$

As a limit,

$$P_u = \phi_t P_n$$
$$= \phi_t(0.75F_u)A_b$$

Therefore,

$$\text{required } A_b = \frac{P_u}{\phi_t(0.75F_u)} = \frac{18.8}{0.75(0.75)(58)} = 0.576 \text{ in.}^2$$

Using tabulated data for gross bolt area in the LRFDM, Vol. II, Part 8, Table 8-7, select a $\frac{7}{8}$-in.-diameter threaded rod ($A_b = 0.601$ in.$^2$). Using the standard thread designation, the required thread will be $\frac{7}{8}$-9UNC2A.

## 13-4

# AXIALLY LOADED COLUMNS AND OTHER COMPRESSION MEMBERS

LRFD analysis and design of axially loaded compression members use most of the concepts previously discussed in Chapter 3 of this text for ASD compression members. Here our discussion is limited to rolled W shapes whose elements (flange and web) have width–thickness ratios less than $\lambda_r$, as stipulated in the LRFDS, Section B5.1. Nearly all building columns are in this category. Such members are capable of developing their full compressive strength without localized buckling; therefore, their strength is limited by overall column buckling.

The following requirement furnishes the basis for LRFD axially loaded compression member design and analysis:

$$P_u \leq \phi_c P_n$$

where

$P_u$ = required axial compressive strength based on factored nominal loads

$\phi_c$ = resistance factor for compression member (0.85)

$P_n$ = nominal axial compressive strength of a member

The right side of the equation ($\phi_c P_n$) represents the design strength of the compression member.

$P_n$ may also be expressed as the product of maximum, or critical, stress and area:

$$P_n = A_g F_{cr} \qquad \text{LRFDS Eqn. (E2-1)}$$

where

$A_g$ = gross cross-sectional area

$F_{cr}$ = critical compressive stress

Chapter E of the LRFDS provides the necessary equations to determine the design strength of the column $\phi_c P_n$ using a slenderness parameter $\lambda_c$ instead of the traditional slenderness ratio $K\ell/r$ (which is used in the ASD method), where

$$\lambda_c = \frac{K\ell}{r\pi} \sqrt{\frac{F_y}{E}} \qquad \text{LRFDS Eqn. (E2-4)}$$

When $\lambda_c \leq 1.5$, column buckling will occur when stressed in the inelastic range:

$$F_{cr} = (0.658^{\lambda_c^2}) F_y \qquad \text{LRFDS Eqn. (E2-2)}$$

When $\lambda_c > 1.5$, column buckling will occur when stressed in the elastic range:

$$F_{cr} = \left[ \frac{0.877}{\lambda_c^2} \right] F_y \qquad \text{LRFDS Eqn. (E2-3)}$$

where

$\lambda_c$ = slenderness parameter

$K$ = effective length factor

$\ell$ = laterally unbraced length of member (in.)

$r$ = governing radius of gyration about the axis of buckling (in.)

$F_y$ = specified minimum yield stress (ksi)

$E$ = modulus of elasticity (29,000 ksi)

$F_{cr}$ = critical compressive stress (ksi)

Inspection of LRFDS Equation (E2-3) reveals that it is the Euler critical stress equation (as discussed in Chapter 3 of this text) multiplied by 0.877 to account for an initial out-of-straightness of actual columns.

LRFDS Equation (E2-2) is an empirical equation for those columns with a slenderness parameter ranging from 0 to 1.5. Most columns in buildings fall within this range. As a means of comparison, for A36 steel, $\lambda_c = 1.5$ corresponds to a slenderness ratio $K\ell/r$ of approximately 134.

It should be noted that the LRFDS still refers to the traditional slenderness ratio ($K\ell/r$) despite the introduction of the slenderness parameter $\lambda_c$. Section B7 of the LRFDS states that the slenderness ratio $K\ell/r$ for compression members *preferably* should not exceed 200. This is similar to the current ASD Specification but a change from previous ASD Specifications that did not allow $K\ell/r$ to exceed 200 for compression members.

Fortunately for the designer, many design aids are available, and the calculations for column design and analysis are much shorter and simpler than would be indicated by the preceding discussion. Note that the effective length factors ($K$) remain the same for both LRFD and ASD. Therefore, the discussions in Chapters 3 and 6 of this text are applicable.

For the following examples, Table 3-36, Design Stress for Compression Members of 36 ksi Specified Yield Stress Steel, $\phi_c = 0.85$, is shown in Figure 13-6. This table is reproduced from the Numerical Values section of the LRFDS.

The following sequence of steps is recommended for the computation of the design compressive strength of a given column (LRFD analysis):

1. Compute the slenderness ratio $K\ell/r$.
2. Using the largest $K\ell/r$, obtain the design compressive stress $\phi_c F_{cr}$ from Figure 13-6.
3. Compute the design compressive strength:

$$\phi_c P_n = \phi_c F_{cr} A_g$$

Table 3-36

**Design Stress for Compression Members of 36 ksi Specified Yield Stress Steel, $\phi_c = 0.85$[a]**

| $\frac{Kl}{r}$ | $\phi_c F_{cr}$ (ksi) | $\frac{Kl}{r}$ | $\phi_c F_{cr}$ (ksi) | $\frac{Kl}{r}$ | $\phi_c F_{cr}$ (ksi) | $\frac{Kl}{r}$ | $\phi_c F_{cr}$ (ksi) | $\frac{Kl}{r}$ | $\phi_c F_{cr}$ (ksi) |
|---|---|---|---|---|---|---|---|---|---|
| 1 | 30.60 | 41 | 28.01 | 81 | 21.66 | 121 | 14.16 | 161 | 8.23 |
| 2 | 30.59 | 42 | 27.89 | 82 | 21.48 | 122 | 13.98 | 162 | 8.13 |
| 3 | 30.59 | 43 | 27.76 | 83 | 21.29 | 123 | 13.80 | 163 | 8.03 |
| 4 | 30.57 | 44 | 27.64 | 84 | 21.11 | 124 | 13.62 | 164 | 7.93 |
| 5 | 30.56 | 45 | 27.51 | 85 | 20.92 | 125 | 13.44 | 165 | 7.84 |
| 6 | 30.54 | 46 | 27.37 | 86 | 20.73 | 126 | 13.27 | 166 | 7.74 |
| 7 | 30.52 | 47 | 27.24 | 87 | 20.54 | 127 | 13.09 | 167 | 7.65 |
| 8 | 30.50 | 48 | 27.11 | 88 | 20.36 | 128 | 12.92 | 168 | 7.56 |
| 9 | 30.47 | 49 | 26.97 | 89 | 20.17 | 129 | 12.74 | 169 | 7.47 |
| 10 | 30.44 | 50 | 26.83 | 90 | 19.98 | 130 | 12.57 | 170 | 7.38 |
| 11 | 30.41 | 51 | 26.68 | 91 | 19.79 | 131 | 12.40 | 171 | 7.30 |
| 12 | 30.37 | 52 | 26.54 | 92 | 19.60 | 132 | 12.23 | 172 | 7.21 |
| 13 | 30.33 | 53 | 26.39 | 93 | 19.41 | 133 | 12.06 | 173 | 7.13 |
| 14 | 30.29 | 54 | 26.25 | 94 | 19.22 | 134 | 11.88 | 174 | 7.05 |
| 15 | 30.24 | 55 | 26.10 | 95 | 19.03 | 135 | 11.71 | 175 | 6.97 |
| 16 | 30.19 | 56 | 25.94 | 96 | 18.84 | 136 | 11.54 | 176 | 6.89 |
| 17 | 30.14 | 57 | 25.79 | 97 | 18.65 | 137 | 11.37 | 177 | 6.81 |
| 18 | 30.08 | 58 | 25.63 | 98 | 18.46 | 138 | 11.20 | 178 | 6.73 |
| 19 | 30.02 | 59 | 25.48 | 99 | 18.27 | 139 | 11.04 | 179 | 6.66 |
| 20 | 29.96 | 60 | 25.32 | 100 | 18.08 | 140 | 10.89 | 180 | 6.59 |
| 21 | 29.90 | 61 | 25.16 | 101 | 17.89 | 141 | 10.73 | 181 | 6.51 |
| 22 | 29.83 | 62 | 24.99 | 102 | 17.70 | 142 | 10.58 | 182 | 6.44 |
| 23 | 29.76 | 63 | 24.83 | 103 | 17.51 | 143 | 10.43 | 183 | 6.37 |
| 24 | 29.69 | 64 | 24.67 | 104 | 17.32 | 144 | 10.29 | 184 | 6.30 |
| 25 | 29.61 | 65 | 24.50 | 105 | 17.13 | 145 | 10.15 | 185 | 6.23 |
| 26 | 29.53 | 66 | 24.33 | 106 | 16.94 | 146 | 10.01 | 186 | 6.17 |
| 27 | 29.45 | 67 | 24.16 | 107 | 16.75 | 147 | 9.87 | 187 | 6.10 |
| 28 | 29.36 | 68 | 23.99 | 108 | 16.56 | 148 | 9.74 | 188 | 6.04 |
| 29 | 29.28 | 69 | 23.82 | 109 | 16.37 | 149 | 9.61 | 189 | 5.97 |
| 30 | 29.18 | 70 | 23.64 | 110 | 16.19 | 150 | 9.48 | 190 | 5.91 |
| 31 | 29.09 | 71 | 23.47 | 111 | 16.00 | 151 | 9.36 | 191 | 5.85 |
| 32 | 28.99 | 72 | 23.29 | 112 | 15.81 | 152 | 9.23 | 192 | 5.79 |
| 33 | 28.90 | 73 | 23.12 | 113 | 15.63 | 153 | 9.11 | 193 | 5.73 |
| 34 | 28.79 | 74 | 22.94 | 114 | 15.44 | 154 | 9.00 | 194 | 5.67 |
| 35 | 28.69 | 75 | 22.76 | 115 | 15.26 | 155 | 8.88 | 195 | 5.61 |
| 36 | 28.58 | 76 | 22.58 | 116 | 15.07 | 156 | 8.77 | 196 | 5.55 |
| 37 | 28.47 | 77 | 22.40 | 117 | 14.89 | 157 | 8.66 | 197 | 5.50 |
| 38 | 28.36 | 78 | 22.22 | 118 | 14.70 | 158 | 8.55 | 198 | 5.44 |
| 39 | 28.25 | 79 | 22.03 | 119 | 14.52 | 159 | 8.44 | 199 | 5.39 |
| 40 | 28.13 | 80 | 21.85 | 120 | 14.34 | 160 | 8.33 | 200 | 5.33 |

[a]When element width-to-thickness ratio exceeds $\lambda_r$, see Appendix B5.3.

**FIGURE 13-6**   (Courtesy of the American Institute of Steel Construction, Inc.)

**Example 13-4** _____

Compute the design compressive strength $\phi_c P_n$ for a W12 $\times$ 120 column that has an unbraced length of 16 ft and has pin-connected ends. Use A36 steel.

**Solution:**

From the LRFDM, Volume I, or the ASDM, Part 1, W shapes, Properties and Dimensions, obtain the following:

$$r_x = 5.51 \text{ in.} \qquad r_y = 3.13 \text{ in.} \qquad A_g = 35.3 \text{ in.}^2$$

Since the member is pin-connected, use $K = 1.0$ (see Chapter 3 of this text).

1.  Compute the slenderness ratio:

$$\frac{K\ell}{r} = \frac{1.0(16)(12)}{3.13} = 61$$

2.  From Figure 13-6, $\phi_c F_{cr} = 25.16$ ksi. Note that a rounded value of $K\ell/r$ is used for the convenience of table usage.
3.  Compute the design compressive strength:

$$\phi_c P_n = \phi_c F_{cr} A_g$$
$$= 25.16(35.3) = 888 \text{ kips}$$

An easier, alternate solution to Example 13-4 is to use the column load tables in Part 3 of the LRFDM, a portion of which is reproduced in Figure 13-7. One enters the table with the known column (W12 $\times$ 120), type of steel (A36), and an effective length $KL_y$ (16 ft). $\phi_c P_n$ is then read directly without any calculations. The tabular value is 896 kips. This is approximately equal to our calculated value of 888 kips.

The column load tables may also be used for design problems as well. The design procedure using these tables has not changed from that used in ASD problems.

**Example 13-5** _____

Select the lightest W12 section for a column that will support an axial nominal load of 700 kips (350 kips D and 350 kips L). The length of the column is 20 ft, and the ends may be assumed to be pinned. Use A36 steel.

**Solution:**

Calculate the required axial compressive strength:

$$P_u = 1.2(350) + 1.6(350) = 980 \text{ kips}$$

As a limit,

# COLUMNS
## W shapes
### Design axial strength in kips ($\phi = 0.85$)

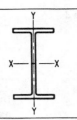

| Designation | | W12 | | | | | | | | | | | |
| --- | --- | --- | --- | --- | --- | --- | --- | --- | --- | --- | --- | --- | --- |
| **Wt./ft** | | 210 | | 190 | | 170 | | 152 | | 136 | | 120 | |
| $F_y$ | | 36 | 50 | 36 | 50 | 36 | 50 | 36 | 50 | 36 | 50 | 36 | 50 |
| | 0 | 1890 | 2630 | 1710 | 2370 | 1530 | 2120 | 1370 | 1900 | 1220 | 1700 | 1080 | 1500 |
| | 6 | 1840 | 2540 | 1660 | 2290 | 1490 | 2050 | 1330 | 1830 | 1190 | 1630 | 1050 | 1440 |
| | 7 | 1830 | 2500 | 1650 | 2260 | 1480 | 2020 | 1320 | 1810 | 1180 | 1610 | 1040 | 1420 |
| | 8 | 1810 | 2470 | 1630 | 2220 | 1460 | 1990 | 1300 | 1780 | 1160 | 1590 | 1030 | 1400 |
| | 9 | 1790 | 2430 | 1610 | 2190 | 1440 | 1960 | 1290 | 1750 | 1150 | 1560 | 1010 | 1380 |
| | 10 | 1760 | 2380 | 1590 | 2150 | 1420 | 1920 | 1270 | 1710 | 1130 | 1530 | 1000 | 1350 |
| | 11 | 1740 | 2330 | 1570 | 2100 | 1400 | 1880 | 1250 | 1680 | 1110 | 1490 | 984 | 1320 |
| | 12 | 1710 | 2280 | 1540 | 2050 | 1380 | 1840 | 1230 | 1640 | 1090 | 1460 | 966 | 1290 |
| | 13 | 1680 | 2230 | 1510 | 2000 | 1350 | 1790 | 1210 | 1590 | 1070 | 1420 | 948 | 1250 |
| | 14 | 1650 | 2170 | 1480 | 1950 | 1330 | 1740 | 1180 | 1550 | 1050 | 1380 | 928 | 1220 |
| | 15 | 1610 | 2110 | 1450 | 1900 | 1300 | 1690 | 1160 | 1510 | 1030 | 1340 | 908 | 1180 |
| | 16 | 1580 | 2040 | 1420 | 1840 | 1270 | 1640 | 1130 | 1460 | 1010 | 1290 | 886 | 1140 |
| | 17 | 1540 | 1980 | 1390 | 1780 | 1240 | 1580 | 1100 | 1410 | 980 | 1250 | 864 | 1100 |
| | 18 | 1510 | 1910 | 1350 | 1720 | 1210 | 1530 | 1070 | 1360 | 955 | 1210 | 841 | 1060 |
| | 19 | 1470 | 1840 | 1320 | 1650 | 1180 | 1470 | 1050 | 1310 | 928 | 1160 | 817 | 1020 |
| | 20 | 1430 | 1780 | 1280 | 1590 | 1140 | 1420 | 1020 | 1260 | 901 | 1110 | 793 | 976 |
| | 22 | 1340 | 1640 | 1210 | 1460 | 1070 | 1300 | 954 | 1150 | 846 | 1020 | 743 | 892 |
| | 24 | 1260 | 1490 | 1130 | 1340 | 1000 | 1180 | 891 | 1050 | 788 | 924 | 692 | 808 |
| | 26 | 1170 | 1360 | 1050 | 1210 | 933 | 1070 | 827 | 944 | 731 | 831 | 640 | 726 |
| | 28 | 1090 | 1220 | 973 | 1090 | 862 | 959 | 763 | 844 | 673 | 742 | 589 | 646 |
| | 30 | 1000 | 1090 | 895 | 967 | 792 | 852 | 700 | 749 | 617 | 656 | 538 | 569 |
| | 32 | 919 | 962 | 819 | 853 | 724 | 750 | 638 | 658 | 561 | 577 | 489 | 500 |
| | 34 | 837 | 852 | 745 | 755 | 657 | 664 | 578 | 583 | 508 | 511 | 442 | 443 |
| | 36 | 759 | 760 | 674 | 674 | 593 | 593 | 520 | 520 | 456 | 456 | 395 | 395 |
| | 38 | 682 | 682 | 605 | 605 | 532 | 532 | 467 | 467 | 409 | 409 | 355 | 355 |
| | 40 | 616 | 616 | 546 | 546 | 480 | 480 | 421 | 421 | 369 | 369 | 320 | 320 |

*Effective length KL (ft) with respect to least radius of gyration $r_y$*

| Properties | | | | | | | | | | | | | |
| --- | --- | --- | --- | --- | --- | --- | --- | --- | --- | --- | --- | --- | --- |
| $u$ | | 2.16 | 2.13 | 2.14 | 2.11 | 2.14 | 2.11 | 2.15 | 2.11 | 2.13 | 2.09 | 2.12 | 2.07 |
| $P_{wo}$ (kips) | | 558 | 774 | 465 | 646 | 389 | 540 | 333 | 462 | 276 | 383 | 232 | 322 |
| $P_{wi}$ (kips/in.) | | 42 | 59 | 38 | 53 | 35 | 48 | 31 | 44 | 28 | 40 | 26 | 36 |
| $P_{wb}$ (kips) | | 3760 | 4430 | 2700 | 3190 | 2020 | 2380 | 1500 | 1760 | 1120 | 1320 | 815 | 960 |
| $P_{fb}$ (kips) | | 731 | 1020 | 610 | 847 | 493 | 684 | 397 | 551 | 316 | 439 | 247 | 343 |
| $L_p$ (ft) | | 13.7 | 11.6 | 13.5 | 11.5 | 13.4 | 11.4 | 13.3 | 11.3 | 13.2 | 11.2 | 13.0 | 11.1 |
| $L_r$ (ft) | | 129 | 84.2 | 117 | 76.6 | 105 | 68.9 | 94.7 | 62.1 | 84.6 | 55.7 | 75.5 | 50.0 |
| $A$ (in.$^2$) | | 61.8 | | 55.8 | | 50.0 | | 44.7 | | 39.9 | | 35.3 | |
| $I_x$ (in.$^4$) | | 2140 | | 1890 | | 1650 | | 1430 | | 1240 | | 1070 | |
| $I_y$ (in.$^4$) | | 664 | | 589 | | 517 | | 454 | | 398 | | 345 | |
| $r_y$ (in.) | | 3.28 | | 3.25 | | 3.22 | | 3.19 | | 3.16 | | 3.13 | |
| Ratio $r_x / r_y$ | | 1.80 | | 1.79 | | 1.78 | | 1.77 | | 1.77 | | 1.76 | |
| $P_{ex} (KL)^2 / 10^4$ | | 61400 | | 54100 | | 47100 | | 41000 | | 35600 | | 30700 | |
| $P_{ey} (KL)^2 / 10^4$ | | 19000 | | 16900 | | 14800 | | 13000 | | 11400 | | 9900 | |

**FIGURE 13-7** Column load table.

$$P_u = \phi_c P_n = 980 \text{ kips}$$

Therefore, enter the column load table (Figure 13-7) with $KL_y$ of 20 ft and the required axial strength of 980 kips. Select a W12 × 152 that has a design axial strength of 1020 kips. Since 1020 kips > 980 kips, the W12 × 152 is satisfactory.

---

**Example 13-6** _____

Compute the design compressive strength $\phi_c P_n$ for an axial-loaded W12 × 170 column of A36 steel with an unbraced length of 12 feet about the $y$-$y$ axis and 24 feet about the $x$-$x$ axis. Assume that the member is pin-connected at the top and fixed at the bottom. (Assume a pin connection at midheight.)

**Solution:**

The properties of the W12 × 170 are

$$r_x = 5.74 \text{ in.} \qquad r_y = 3.22 \text{ in.} \qquad A_g = 50.0 \text{ in.}^2$$

1.  Compute the slenderness ratio with respect to each axis:

$$\frac{K\ell_y}{r_y} = \frac{1.0(12)(12)}{3.22} = 44.7 \qquad \text{(top part of column)}$$

$$\frac{K\ell_y}{r_y} = \frac{0.8(12)(12)}{3.22} = 35.8 \qquad \text{(bottom part of column)}$$

$$\frac{K\ell_x}{r_x} = \frac{0.8(24)(12)}{5.74} = 40.1$$

The weak axis governs, and the controlling slenderness ratio is 44.7 (use 45).

2.  From Figure 13-6, $\phi_c F_{cr} = 27.51$ ksi.
3.  Compute the design compressive strength:

$$\phi_c P_n = \phi_c F_{cr} A_g$$

$$= 27.51(50) = 1376 \text{ kips}$$

Note that since the weak axis controls, this calculation can be checked using the column load table of Figure 13-7 with a $KL$ of 12 ft.

---

# 13-5

## BENDING MEMBERS

The LRFD method for the design and analysis of bending members involves the following initial steps:

1. The calculation of the design load and load combinations
2. The calculation of the *required* design strength expressed in terms of flexural strength $M_u$ and shear strength $V_u$
3. The calculation of the *furnished* design strength expressed in terms of flexural strength $\phi_b M_n$ and shear strength $\phi_v V_n$

Since the first two steps have been discussed previously in this text, our primary concern in this section, with respect to the LRFD approach, is to establish the procedure for determining a member's design flexural strength. We also discuss the design shear strength, deflections, and block shear considerations.

## Design Flexural Strength

The design flexural strength criterion may be expressed as follows:

$$M_u \leq \phi_b M_n$$

where

$M_u$ = required flexural strength based on factored loads

$\phi_b$ = resistance factor for flexure (0.90)

$M_n$ = nominal flexural strength

The product $\phi_b M_n$ represents the furnished or design flexural strength and is limited by

1. Lateral-torsional buckling of the member (as discussed in Chapter 4 of this text)
2. Local buckling of one or more compressive elements of the member (e.g., the flange or web)
3. The formation of a plastic hinge at a specific location (as discussed in Chapter 10 of this text)

In this introductory section, the design strengths are determined for *compact* shapes only. A section is compact if its flanges are continuously connected to the web and the width–thickness ratios of its compressive elements do not exceed the limiting width–thickness ratios $\lambda_p$ as furnished in Table B5.1 of the LRFDS. This is similar to the provision of the AISC ASD Specification. Therefore, for compact sections (which include all rolled W shapes of A36 steel with the exception of the W6 × 15), a member's flexural strength *will not* be limited by a localized buckling of its compressive elements. Hence, for our discussion, a member's flexural strength will be limited by lateral-torsional buckling or the formation of a plasic hinge. Lateral-torsional buckling reflects an overall instability condition of a bending member and is in part a function of the unbraced length of the compression flange. The plastic hinge condition is one where the entire member cross section at some location has yielded. This represents the upper limit of usefulness of the member cross section. Only compact sections with adequate lateral support of the compression flange to prevent lateral-torsional buckling can attain the upper limit of flexural strength that occurs upon formation of the plastic hinge.

It then becomes evident that the unbraced length of the compression flange $L_b$ of a bending member represents a critical factor with respect to its flexural strength. This is displayed graphically in Figure 13-8, which shows the basic relationship between the nominal flexural strength $M_n$ of a typical compact shape bending member and the unbraced length of its compression flange.

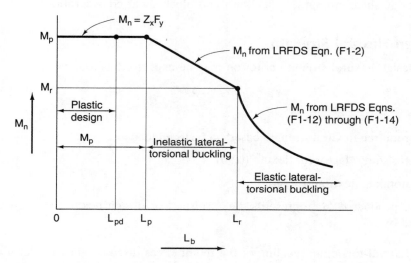

**FIGURE 13-8**   $M_n$ as a function of $L_b$ for a typical compact shape ($C_b = 1.0$).

Note in Figure 13-8 that the limits of the flexural strength $M_n$ are represented by $M_p$, which is the plastic bending moment that develops upon formation of the plastic hinge, and $M_r$, which is the moment at initial yielding and defines the end of the elastic range. The unbraced lengths that correspond to these limiting flexural strengths are designated, respectively, $L_p$ and $L_r$, where

$L_p$ = a maximum laterally unbraced length for the full plastic moment strength to develop. When $L_b > L_p$, the plastic moment strength $M_p$ cannot be attained.

$L_r$ = a maximum laterally unbraced length for inelastic lateral-torsional buckling limiting the flexural strength. When $L_b > L_r$, the section will buckle elastically before the yield stress is reached.

Note that an unbraced length designated $L_{pd}$ is shown in Figure 13-8. $L_{pd}$ represents a laterally unbraced length that would permit the use of a plastic analysis. When $L_{pd} < L_b < L_p$, the moment strength is still capable of reaching $M_p$, but the member lacks sufficient rotational capacity to redistribute moments and permit plastic analysis. The determination of $L_{pd}$ is not significant since both elastic and plastic analyses are permitted by the LRFDS, Section A5. Since elastic analysis is less complex and has fewer limitations than plastic analysis, it is the authors' preference to use only an elastic analysis procedure.

Note also in Figure 13-8 that applicable LRFDS expressions for the design flexural strength $M_n$ are shown on the curves where the unbraced length is in excess of $L_p$. These expressions have been established for the case of a uniform (constant) moment extending over the unbraced length $L_b$. Where the moment is *not* constant over the unbraced length, a bending coeffcient $C_b$ is used. $C_b$ is a liberalizing factor having a value between 1.0 and 2.3 and has the effect of increasing the flexural strength for the same unbraced length. It is calculated using the same expression as in ASD and is discussed in Section 4-6 of this text (also see the LRFDS, Chapter F). For our introductory treatment, we conservatively assume $C_b = 1.0$.

Chapter F of the LRFDS provides equations for the calculation of the nominal flexural strength $M_n$ based on the degree of lateral support of the compression flange of the member. Our discussion applies to hot-rolled, homogeneous, compact, doubly symmetric or singly symmetric sections used as beams (e.g., W, M, S, and C shapes as well as double angles and tees) loaded in a plane of symmetry or loaded and braced so that twisting of the member will not occur.

The various categories of the degree of lateral support as treated in the LRFDS are the following:

1. $L_b \le L_p$, where the full plastic moment may be developed.
2. $L_p < L_b \le L_r$, where the member buckles inelastically and $M_n$ varies linearly between $M_p$ and $M_r$.
3. $L_b > L_r$, where the member buckles elastically before the yield stress is reached.

In this section, we expand on only beams composed of compact shapes that have continuous lateral support for the entire span length or have an unbraced length that is very short such that $L_b \le L_p$. This category of a compact laterally supported beam bent about its strong axis and loaded in the plane of its weak axis represents the most common category of bending member as well as the simplest to analyze and design.

For the condition of $L_b \le L_p$, the nominal flexural strength is written as

$$M_n = M_p \qquad \text{LRFDS Eqn. (F1-1)}$$

where

$$M_p = F_y Z \le 1.5 M_y$$

where

$M_n$ = nominal flexural strength

$M_p$ = plastic moment that develops upon the formation of a plastic hinge

$F_y$ = specified minimum yield stress (ksi)

$Z$ = plastic section modulus

$M_y$ = moment corresponding to the onset of yielding at the extreme fiber from an elastic stress distribution

$L_p$ is as previously defined and may be calculated from

$$L_p = \frac{300 r_y}{\sqrt{F_y}}$$                          LRFDS Eqn. (F1-4)

where $r_y$ = radius of gyration about the minor ($y$-$y$) axis

The limit of $1.5 M_y$ for $M_p$ is to prevent excessive working load deformation. We may substitute $F_y Z$ for $M_p$. Likewise, we may substitute $F_y S$ for $M_y$, where $S$ = the elastic section modulus, in.[3] Rewriting the preceding expression results in

$$F_y Z \leq 1.5 F_y S$$

Rearranging this expression, it can be seen that the inequality is satisfied when

$$\frac{Z}{S} \leq 1.5$$

For the generally uncommon conditions of $L_p < L_b \leq L_r$ and $L_b > L_r$, the LRFDS, Chapter F, provides equations for the determination of $M_n$. These equations are somewhat involved and can be tedious to use. Fortunately for the designer, many design aids are available for the more common beam cross sections, and the calculations are much shorter than the complexity of the equations would indicate. There are three design aids furnished in the LRFDM, Part 4.

1.   The Load Factor Design Selection Table includes only W and M shapes used as bending members. The table is applicable to adequately braced beams with unbraced lengths $L_b$ not exceeding $L_r$. A sample of the Load Factor Design Selection Table from the LRFDM is shown in Figure 13-9. Note that its format is very similar to that of the Allowable Stress Design Selection Table in Part 2 of the AISC ASD Manual. Also note that this table contains a tabulated factor designated $BF$. This factor can be used to calculate the design flexural strength $\phi_b M_n$ for beams with unbraced lengths between the limiting bracing lengths of $L_p$ and $L_r$:

$$BF \, (\text{kips}) = \frac{\phi_b (M_p - M_r)}{L_r - L_p}$$

## LOAD FACTOR DESIGN SELECTION TABLE
### For shapes used as beams
### $\phi_b = 0.90$

$Z_x$

| Fy = 36 ksi | | | | | Zx | Shape | Fy = 50 ksi | | | | |
|---|---|---|---|---|---|---|---|---|---|---|---|
| BF | Lr | Lp | $\phi_b M_r$ | $\phi_b M_p$ | Zx | | $\phi_b M_p$ | $\phi_b M_r$ | Lp | Lr | BF |
| Kips | Ft | Ft | Kip-ft | Kip-ft | In.³ | | Kip-ft | Kip-ft | Ft | Ft | Kips |
| 12.7 | 16.6 | 5.6 | 222 | 362 | 134 | W24×55 | 503 | 342 | 4.7 | 12.9 | 19.6 |
| 8.08 | 23.2 | 7.0 | 228 | 359 | 133 | W18×65 | 499 | 351 | 6.0 | 17.1 | 13.3 |
| 2.90 | 56.4 | 12.8 | 230 | 356 | 132 | W12×87 | 495 | 354 | 10.9 | 38.4 | 5.12 |
| 2.00 | 77.4 | 11.0 | 218 | 351 | 130 | W10×100 | 488 | 336 | 9.4 | 50.8 | 3.66 |
| 5.57 | 32.3 | 10.3 | 228 | 351 | 130 | W16×67 | 488 | 351 | 8.7 | 23.8 | 9.02 |
| 11.3 | 17.3 | 5.6 | 216 | 348 | 129 | W21×57 | 484 | 333 | 4.8 | 13.1 | 18.0 |
| 4.10 | 40.0 | 10.3 | 218 | 340 | 126 | W14×74 | 473 | 336 | 8.8 | 28.0 | 7.12 |
| 7.91 | 22.4 | 7.0 | 211 | 332 | 123 | W18×60 | 461 | 324 | 6.0 | 16.7 | 12.8 |
| 2.88 | 51.8 | 12.7 | 209 | 321 | 119 | W12×79 | 446 | 321 | 10.8 | 35.7 | 5.03 |
| 4.05 | 37.3 | 10.3 | 201 | 311 | 115 | W14×68 | 431 | 309 | 8.7 | 26.4 | 6.91 |
| 1.97 | 68.4 | 11.0 | 192 | 305 | 113 | W10×88 | 424 | 296 | 9.3 | 45.1 | 3.58 |
| 7.65 | 21.4 | 7.0 | 192 | 302 | 112 | W18×55 | 420 | 295 | 5.9 | 16.1 | 12.2 |
| 10.5 | 16.2 | 5.4 | 184 | 297 | 110 | W21×50 | 413 | 284 | 4.6 | 12.5 | 16.4 |
| 2.87 | 48.2 | 12.7 | 190 | 292 | 108 | W12×72 | 405 | 292 | 10.7 | 33.6 | 4.93 |
| 6.43 | 22.8 | 6.7 | 180 | 284 | 105 | W16×57 | 394 | 277 | 5.7 | 16.6 | 10.7 |
| 3.91 | 34.7 | 10.2 | 180 | 275 | 102 | W14×61 | 383 | 277 | 8.7 | 24.9 | 6.51 |
| 7.31 | 20.5 | 6.9 | 173 | 273 | 101 | W18×50 | 379 | 267 | 5.8 | 15.6 | 11.5 |
| 1.95 | 60.1 | 10.8 | 168 | 264 | 97.6 | W10×77 | 366 | 258 | 9.2 | 39.9 | 3.53 |
| 2.80 | 44.7 | 12.6 | 171 | 261 | 96.8 | W12×65ᵇ | 358 | 264 | 11.8 | 31.7 | 4.72 |
| 9.68 | 15.4 | 5.3 | 159 | 258 | 95.4 | W21×44 | 358 | 245 | 4.5 | 12.0 | 14.9 |
| 6.18 | 21.3 | 6.6 | 158 | 248 | 92.0 | W16×50 | 345 | 243 | 5.6 | 15.8 | 10.1 |
| 8.13 | 16.6 | 5.4 | 154 | 245 | 90.7 | W18×46 | 340 | 236 | 4.6 | 12.6 | 13.0 |
| 4.17 | 28.0 | 8.0 | 152 | 235 | 87.1 | W14×53 | 327 | 233 | 6.8 | 20.1 | 7.02 |
| 2.91 | 38.4 | 10.5 | 152 | 233 | 86.4 | W12×58 | 324 | 234 | 8.9 | 27.0 | 4.96 |
| 1.93 | 53.7 | 10.8 | 148 | 230 | 85.3 | W10×68 | 320 | 227 | 9.2 | 36.0 | 3.46 |
| 5.91 | 20.2 | 6.5 | 142 | 222 | 82.3 | W16×45 | 309 | 218 | 5.6 | 15.2 | 9.43 |
| 7.51 | 15.7 | 5.3 | 133 | 212 | 78.4 | W18×40 | 294 | 205 | 4.5 | 12.1 | 11.7 |
| 4.06 | 26.3 | 8.0 | 137 | 212 | 78.4 | W14×48 | 294 | 211 | 6.8 | 19.2 | 6.70 |
| 2.85 | 35.8 | 10.3 | 138 | 210 | 77.9 | W12×53 | 292 | 212 | 8.8 | 25.6 | 4.77 |
| 1.91 | 48.1 | 10.7 | 130 | 201 | 74.6 | W10×60 | 280 | 200 | 9.1 | 32.6 | 3.38 |
| 5.54 | 19.3 | 6.5 | 126 | 197 | 72.9 | W16×40 | 273 | 194 | 5.6 | 14.7 | 8.67 |
| 3.06 | 30.8 | 8.2 | 126 | 195 | 72.4 | W12×50 | 272 | 194 | 6.9 | 21.7 | 5.25 |
| 1.30 | 64.0 | 8.8 | 118 | 190 | 70.2 | W8×67 | 263 | 181 | 7.5 | 41.9 | 2.38 |
| 3.91 | 24.7 | 7.9 | 122 | 188 | 69.6 | W14×43 | 261 | 188 | 6.7 | 18.2 | 6.32 |
| 1.89 | 43.9 | 10.7 | 117 | 180 | 66.6 | W10×54 | 250 | 180 | 9.1 | 30.2 | 3.30 |
| 6.95 | 14.8 | 5.1 | 112 | 180 | 66.5 | W18×35 | 249 | 173 | 4.3 | 11.5 | 10.7 |
| 3.01 | 28.5 | 8.1 | 113 | 175 | 64.7 | W12×45 | 243 | 174 | 6.9 | 20.3 | 5.07 |
| 5.23 | 18.3 | 6.3 | 110 | 173 | 64.0 | W16×36 | 240 | 170 | 5.4 | 14.1 | 8.08 |
| 4.41 | 20.0 | 6.5 | 106 | 166 | 61.5 | W14×38 | 231 | 164 | 5.5 | 14.9 | 7.07 |
| 1.88 | 40.7 | 10.6 | 106 | 163 | 60.4 | W10×49 | 227 | 164 | 9.0 | 28.3 | 3.25 |
| 1.27 | 56.0 | 8.8 | 101 | 161 | 59.8 | W8×58 | 224 | 156 | 7.4 | 36.8 | 2.32 |
| 2.92 | 26.5 | 8.0 | 101 | 155 | 57.5 | W12×40 | 216 | 156 | 6.8 | 19.3 | 4.82 |
| 1.96 | 35.1 | 8.4 | 95.7 | 148 | 54.9 | W10×45 | 206 | 147 | 7.1 | 24.1 | 3.45 |

ᵇIndicates noncompact shape; Fy = 50 ksi.

**FIGURE 13-9**  Load factor design selection table.

The design flexural strength may then be computed using all known or tabulated values:

$$\phi_b M_n = C_b[\phi_b M_p - BF(L_b - L_p)] \leq \phi_b M_p$$

Note that all terms have been previously defined.

2.   The Factored Uniform Load Tables are applicable for simple-span laterally supported beams and give the maximum uniformly distributed load in kips. They may also be used indirectly for several concentrated loading conditions. Limitations on the use of these tables are noted in the LRFDM, Part 4.

3.   The Design Flexural Strength of Beams with Unbraced Length $L_b > L_p$ charts show the design strength $\phi_b M_n$ for W and M shapes used as beams that have varying unbraced lengths. In general the unbraced lengths extend beyond most unbraced lengths frequently encountered in design practice. A sample of these charts is presented in this text as Figure 13-10. Note that it is very similar to the beam curves in the AISC ASD Manual. The LRFD charts have the general shape of the $M_n$ versus $L_b$ curve of Figure 13-8 but are specific for each particular section. Limitations on the use of these charts are noted in the LRFDM.

**Example 13-7** _____

Compute the design flexural strength $\phi_b M_n$ for a W18 × 40 beam of A36 steel for the following conditions (assume $C_b = 1.0$):

(a)   The beam has full lateral support ($L_b = 0$).
(b)   The unbraced length $L_b$ is 15 ft.

**Solution:**

(a)   The W18 × 40 is compact (all W shapes in A36 steel, except the W6 × 15, are compact), and $L_b = 0$. From Figure 13-9,

$$L_p = 5.3 \text{ ft}$$

$$Z_x = 78.4 \text{ in.}^3$$

From the LRFDM or the ASDM, $S_x = 68.4$ in.$^3$
      Since $L_b < L_p$,

$$\phi_b M_n = \phi_b M_p$$

From Figure 13-9, $\phi_b M_p = 212$ ft-kips. Therefore,

$$\phi_b M_n = 212 \text{ ft-kips}$$

Check to ensure that $M_p \leq 1.5 M_y$:

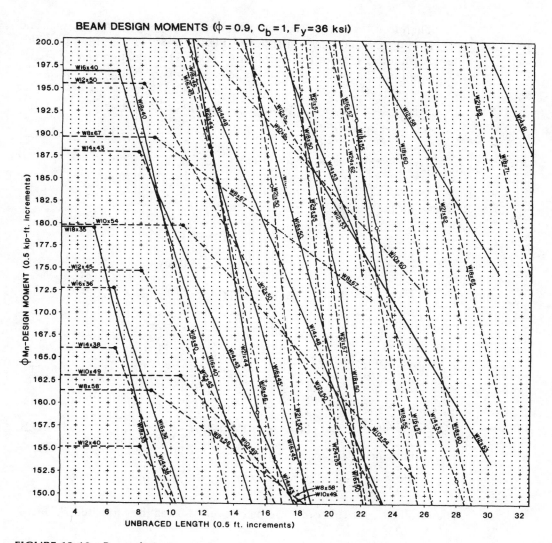

**FIGURE 13-10**   Beam design moment curves. (Courtesy of the American Institute of Steel Construction, Inc.)

$$\frac{Z}{S} = \frac{78.4}{68.4} = 1.15 < 1.5 \qquad\qquad \textbf{O.K.}$$

(b)   The unbraced length $L_b$ is 15 ft. From Figure 13-9, for the W18 × 40,

$$L_p = 5.3 \text{ ft}$$

$$L_r = 15.7 \text{ ft}$$

Therefore, $L_p < L_b < L_r$, and the value of $M_n$ is governed by inelastic lateral torsional buckling (refer to Figure 13-8). The LRFDS provides Equation (F1-2) for the calculation of $M_n$. This equation can be recognized as a linear equation of the form $y = mx + b$:

$$M_n = C_b\left[M_p - (M_p - M_r)\left(\frac{L_b - L_p}{L_r - L_p}\right)\right] \leq M_p$$

LRFDS Eqn. (F1-2)

The following data are furnished:

$$F_y = 36 \text{ ksi} \qquad C_b = 1.0 \qquad \phi_b = 0.90 \qquad Z_x = 78.4 \text{ in.}^3$$

From Figure 13-9,

$$\phi_b M_p = 212 \text{ ft-kips}$$

$$M_p = \frac{212}{0.90} = 236 \text{ ft-kips}$$

$$\phi_b M_r = 133 \text{ ft-kips}$$

$$M_r = \frac{133}{0.90} = 148 \text{ ft-kips}$$

Then, using LRFDS Equation (F1-2),

$$M_n = 1.0\left[236 - (236 - 148)\left(\frac{15 - 5.3}{15.7 - 5.3}\right)\right] \leq M_p$$

$$= 154 \text{ ft-kips} \leq 236 \text{ ft-kips} \qquad\qquad \textbf{O.K.}$$

from which

$$\phi_b M_n = 0.90(154) = 139 \text{ ft-kips}$$

The result for part (b) could also be computed using the tabulated $BF$ factor ($BF = 7.51$):

$$\phi_b M_n = C_b[\phi_b M_p - BF(L_b - L_p)] \leq \phi_b M_p$$

$$= 1.0[212 - 7.51(15 - 5.3)]$$

$$= 139 \text{ ft-kips} < 212 \text{ ft-kips} \qquad\qquad \textbf{O.K.}$$

## Design Shear Strength

LRFD is similar to ASD in the sense that bending member selection is based on moment requirements and is followed by a check for shear. The design shear strength criterion for unstiffened webs with $h/t_w \leq 260$ may be expressed as follows:

$$V_u \le \phi_v V_n$$

where

$V_u$ = required shear strength based on factored loads

$\phi_v$ = shear resistance factor (0.90)

$V_n$ = nominal shear strength

$\phi_v V_n$ = design shear strength furnished

Equations for the nominal shear strength are furnished in the LRFDS, Section F2.

1.   For

$$\frac{h}{t_w} \le \frac{418}{\sqrt{F_{yw}}}$$

shear yielding of the web occurs and

$$V_n = 0.6 F_{yw} A_w \qquad \text{LRFDS Eqn. (F2-1)}$$

2.   For

$$\frac{418}{\sqrt{F_{yw}}} < \frac{h}{t_w} \le \frac{523}{\sqrt{F_{yw}}}$$

inelastic shear buckling of the web occurs and

$$V_n = 0.6 F_{yw} A_w \left( \frac{418/\sqrt{F_{yw}}}{h/t_w} \right) \qquad \text{LRFDS Eqn. (F2-2)}$$

3.   For

$$\frac{523}{\sqrt{F_{yw}}} < \frac{h}{t_w} \le 260$$

elastic shear buckling of the web occurs and

$$V_n = A_w \frac{132{,}000}{(h/t_w)^2} \qquad \text{LRFDS Eqn. (F2-3)}$$

where

$h$ = clear distance between the web toe of fillets for rolled shapes and clear distance between flanges for welded sections

$t_w$ = thickness of web

$F_{yw}$ = specified minimum yield stress of the web

$A_w$ = web area (overall depth $d$ times the web thickness $t_w$)

If $h/t_w > 260$, web stiffeners are required and the provisions of Appendix F2 of the LRFDS must be taken into consideration.

It is useful to note that all rolled W, S, and HP shapes with $F_y$ up to 65 ksi will have their nominal shear strengths governed by shear yielding of the web.

**Example 13-8**

Using LRFD, compute the allowable uniformly distributed nominal live load that may be superimposed on W18 × 40 beams spaced at 8 ft–0 in. on center. The beams are simply supported on a span length of 30 ft and support a 6-in.-thick reinforced concrete slab. Assume A36 steel and full lateral support of the compression flange. Check shear. (Use a unit weight of reinforced concrete of 150 pcf.)

**Solution:**

1.   For the dead load, the weight of the concrete slab is calculated from

$$\frac{6}{12}(150) = 75 \text{ psf}$$

Note that each beam supports a load area 8 ft wide. Therefore, the weight of the slab per linear foot of beam is calculated from

$$75(8) = 600 \text{ lb/ft}$$

The beam weight is 40 lb/ft. Adding the slab dead load and the beam weight results in a total DL of 640 lb/ft.

2.   The beam is compact and has full lateral support ($L_b = 0$). Therefore, the plastic moment $M_p$ of the section may be developed, and the design flexural strength of the section $\phi_b M_n$ is calculated as

$$\phi_b M_n = \phi_b M_p$$

$$= \phi_b Z_x F_y$$

$$= \frac{0.9(78.4)(36)}{12} = 212 \text{ ft-kips}$$

where $Z_x$ is obtained from Figure 13-9. (Note that $\phi_b M_p$ can also be obtained directly from Figure 13-9.)

3.   Since, as a limit,

$$M_u = \phi_b M_n = \phi_b M_p$$

we can compute the total factored design load $w_u$ that will develop $M_u$:

$$M_u = \frac{w_u L^2}{8}$$

$$w_u = \frac{8M_u}{L^2} = \frac{8(212)}{30^2} = 1.88 \text{ kips/ft}$$

Solving for the nominal live load,

$$w_u = 1.2D + 1.6L$$

$$1.88 = 1.2(0.640) + 1.6L$$

From which

$$L = 0.695 \text{ kip/ft}$$

Check shear:

$$V_u = \frac{w_u L}{2} = \frac{1.88(30)}{2} = 28.2 \text{ kips}$$

Although it has been noted that all rolled W shapes with $F_y$ up to 65 ksi have their nominal shear strengths governed by shear yielding of the web, we check (for illustrative purposes):

$$\frac{h}{t_w} = \frac{15.5}{0.315} = 49.2$$

$$\frac{418}{\sqrt{F_{yw}}} = \frac{418}{\sqrt{36}} = 69.7$$

$$49.2 < 69.7$$

Therefore, shear yielding does govern, and

$$V_n = 0.6F_{yw}A_w$$

$$= 0.6(36)(17.9)(0.315) = 121.8 \text{ kips}$$

$$\phi_v V_n = 0.90(121.8) = 109.6 \text{ kips}$$

$$109.6 \text{ kips} > 28.2 \text{ kips} \qquad \qquad \textbf{O.K.}$$

The W18 × 40 is satisfactory in shear when subjected to a live load of 0.695 kip/ft.

---

**Example 13-9** _____

Using the LRFD method, select the lightest W shape for the beam shown in Figure 13-11. The given nominal loads are superimposed loads (they do not include the beam weight). Assume full lateral support ($L_b = 0$) and A36 steel. Consider flexure and shear.

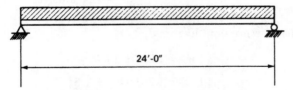

**FIGURE 13-11**   Beam load diagram.

**Solution:**

We follow the general procedure as given in Section 4-8 of this text. Assume a compact shape (recall that only the W6 × 15 was noncompact in A36 steel). Since $L_b = 0$, we also know that $M_n = Z_x F_y$ (see Figure 13-8). Determine the factored load (design load) and then the required flexural strength $M_u$ (initially neglect the beam weight):

$$w_u = 1.2D + 1.6L$$

$$= 1.2(0.5) + 1.6(1.1) = 2.36 \text{ kips/ft}$$

$$M_u = \frac{w_u L^2}{8} = \frac{2.36(24)^2}{8} = 169.9 \text{ ft-kips}$$

Since, as a limit,

$$M_u = \phi_b M_n = \phi_b Z_x F_y$$

we can write

$$\text{required } Z_x = \frac{M_u}{\phi_b F_y}$$

$$= \frac{169.9(12)}{0.9(36)} = 62.9 \text{ in.}^3$$

From Figure 13-9, select a W18 × 35 ($Z_x = 66.5 \text{ in.}^3$). Rework the calculations and include the beam weight:

$$w_u = 1.2(0.5 + 0.035) + 1.6(1.1) = 2.40 \text{ kips/ft}$$

$$M_u = \frac{2.40(24)^2}{8} = 172.8 \text{ ft-kips}$$

$$\text{required } Z_x = \frac{172.8(12)}{0.9(36)} = 64.0 \text{ in.}^3 \qquad\qquad \textbf{O.K.}$$

The W18 × 35 is satisfactory for flexure.
   Check shear:

$$V_u = \frac{w_u L}{2} = \frac{2.40(24)}{2} = 28.8 \text{ kips}$$

This is a W shape with $F_y < 65$ ksi. Therefore,

$$V_n = 0.6 F_{yw} A_w$$

$$= 0.6(36)(17.7)(0.300) = 114.7 \text{ kips}$$

$$\phi V_n = 0.90(114.7) + 103.2 \text{ kips}$$

$$103.2 \text{ kips} > 28.8 \text{ kips} \qquad \textbf{O.K.}$$

Use a W18 × 35.

Example 13-9 could also be solved using the curves of Figure 13-10. If we made a good estimate of 35 lbs/ft for the beam weight, the calculated design moment $M_u$ would be 172.8 ft-kips. Enter Figure 13-10 with $M_u$ of 172.8 ft-kips and unbraced length $L_b$ of 0 ft (left vertical axis). Select a W18 × 35 as the lightest W shape.

**Example 13-10** _____

Using the LRFD method, select the lightest W shape for the beam shown in Figure 13-12. Assume A36 steel. Lateral support exists at the reactions and at the point loads. The given loads are superimposed loads. Consider flexure and shear.

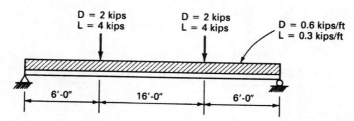

**FIGURE 13-12**  Beam load diagram.

**Solution:**

Determine the design point loads $P_u$ and the design uniformly distributed load $w_u$:

$$P_u = 1.2D + 1.6L$$

$$= 1.2(2) + 1.6(4) = 8.8 \text{ kips}$$

$$w_u = 1.2(0.6) + 1.6(0.3) = 1.20 \text{ kips/ft}$$

Determine the required design flexural strength $M_u$ (neglecting the beam weight):

$$M_u = \frac{w_u L^2}{8} + P_u a$$

$$= \frac{1.20(28)^2}{8} + 8.8(6) = 170.4 \text{ ft-kips}$$

Enter Figure 13-10 with $M_u$ of 170.4 ft-kips and unbraced length $L_b$ of 16 ft. Select a W14 × 48, and read $\phi M_n$ of 179.0 ft-kips (for $L_b = 16$ ft). Consider the additional moment due to the beam weight:

$$\text{additional } M_u = \frac{w_u L^2}{8} = \frac{1.2(0.048)(28)^2}{8} = 5.6 \text{ ft-kips}$$

$$\text{total } M_u = 170.4 + 5.6 = 176 \text{ ft-kips}$$

Since 176 ft-kips $<$ 179.0 ft-kips, the W14 × 48 is satisfactory for flexure. Check shear. The factored dead load due to the beam weight is

$$w_u = 1.2(0.048) = 0.058 \text{ kip/ft}$$

The required shear strength is then calculated from

$$V_u = \frac{(1.2 + 0.058)28}{2} + 8.8 = 26.4 \text{ kips}$$

This is a W shape with $F_y < 65$ ksi. Therefore,

$$V_n = 0.6 F_{yw} A_w$$

$$= 0.6(36)(13.79)(0.34) = 101.3 \text{ kips}$$

$$\phi V_n = 0.90(101.3) = 91.2 \text{ kips}$$

$$91.2 \text{ kips} > 26.2 \text{ kips} \qquad\qquad \textbf{(O.K.)}$$

Use a W14 × 48.

## Deflection

The consideration of deflection of beams is part of the design process. This is true whether designing with ASD or LRFD.

The checking of deflections when using the LRFD design method is identical to the ASD deflection computations. Deflections are computed using service loads for both methods. The LRFDS does not furnish precise criteria with respect to deflection limitations except that deflection should be checked and deflection should not impair the serviceability of the structure.

LRFD-designed structural members may be lighter and have a smaller moment of inertia than their ASD counterparts. Hence the deflection calculations may be thought of as being more significant when using LRFD as opposed to ASD.

## Block Shear

Block shear rupture strength is discussed in Section 13-3 (tension members) and in Chapters 2 and 7 where the ASD method is used. It must be considered at some beam connection locations where the end of the beam is coped as shown in Figure 13-13.

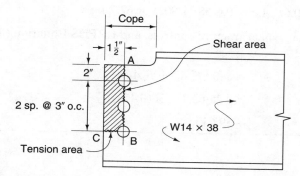

**FIGURE 13-13**  Block shear for coped beam.

The reader will recall that a block shear failure occurs by a combination of shear yielding and tension fracture or by shear fracture and tension yielding. The block shear strength check for beams is made using the two equations from LRFDS, Section J4.3. The governing equation is the one that has the larger fracture value where shear fracture is calculated from $0.6F_u A_{nv}$ and tensile fracture is calculated from $F_u A_{nt}$.

The resulting $\phi R_n$ is the maximum factored load reaction based on the block shear rupture strength.

## Example 13-11

Using the LRFDS, compute the block shear rupture strength for the beam shown in Figure 13-13 (this would be the maximum factored load reaction based on the block shear rupture strength). Use $\frac{3}{4}$-in.-diameter bolts and a W14 × 38 of A36 steel.

**Solution:**

For the W14 × 38, web $t_w = 0.310$ in. Use a hole diameter of $\frac{3}{4} + \frac{1}{8} = 0.875$ in. for the block shear calculations. Thus

$$A_{gv} = 8.00(0.310) = 2.48 \text{ in.}^2$$

$$A_{gt} = 1.5(0.310) = 0.465 \text{ in.}^2$$

$$A_{nv} = [8.00 - 2.5(0.875)](0.310) = 1.802 \text{ in.}^2$$

$$A_{nt} = [1.5 - 0.5(0.875)](0.310) = 0.329 \text{ in.}^2$$

Considering tensile fracture,

$$F_u A_{nt} = 58(0.329) = 19.08 \text{ kips}$$

Considering shear fracture,

$$0.6 F_u A_{nv} = 0.6(58)(1.802) = 62.7 \text{ kips}$$

Since $0.6 F_u A_{nv} > F_u A_{nt}$, shear fracture controls, and LRFDS Equation (J4-3b) is applicable:

$$\phi R_n = \phi [0.6 F_u A_{nv} + F_y A_{gt}]$$
$$= 0.75[62.7 + 36(0.465)]$$
$$= 59.6 \text{ kips}$$

## REFERENCES

[1] *Manual of Steel Construction Load and Resistance Factor Design,* 2nd ed., American Institute of Steel Construction, 1 East Wacker Drive, Suite 3100, Chicago, IL 60601.

[2] Abraham J. Rokach, *Structural Steel Design (Load and Resistance Factor Method)* (New York: Schaum's Outline Series, McGraw-Hill, Inc., 1991).

[3] J. C. Smith, *Structural Steel Design, LRFD Approach* (New York: John Wiley and Sons, Inc., 1991).

[4] Jack C. McCormac, *Structural Steel Design, LRFD Method* (New York: Harper & Row, 1989).

[5] Louis F. Geschwinder, Robert O. Disque, and Reidar Bjorhovde, *Load and Resistance Factor Design of Steel Structures* (Englewood Cliffs, NJ: Prentice Hall, Inc., 1994).

[6] William T. Segui, *LRFD Steel Design* (Boston, MA: PWS Publishing Company, 1994).

## PROBLEMS

*For the following problems, use the LRFD Specification as discussed in this chapter. All steel is A36.*

**13-1.** The nominal loads on a roof are 15 psf dead load, 45 psf snow load, 15 psf rain/ice load, and 20 psf live load. Determine the maximum design load (factored load).

**13-2.** The moments due to applied nominal loads acting on a beam are 40 ft-kips $D$ and 15 ft-kips $L$. Determine the required flexural strength $M_u$.

**13-3.** A C9 × 20 supports a nominal tensile dead load of 24 kips. The end connections are welded as shown. Determine the maximum simultaneous nominal tensile live load that this member can support. Neglect block shear. Assume that the welded connection itself is adequate.

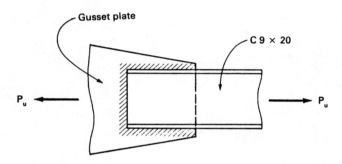

**PROBLEM 13-3**

**13-4.** An L4 × 4 × $\frac{1}{2}$ supports a nominal tensile load of 20 kips dead load and 40 kips live load. The connection is composed of one row of four $\frac{7}{8}$-in.-diameter bolts in one leg as shown. Is the member satisfactory? Assume that the connection itself is adequate.

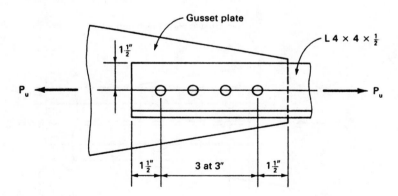

**PROBLEM 13-4**

**13-5.** Compute the design tensile strength $\phi_t P_n$ for a tension hanger composed of a bar 5 in. × $\frac{1}{2}$ in. in cross-sectional area. The hanger is connected with

$\frac{3}{4}$-in.-diameter bolts on one gage line as shown. Neglect block shear and assume the gusset plate is adequate for tension and the bolts adequate in shear and bearing.

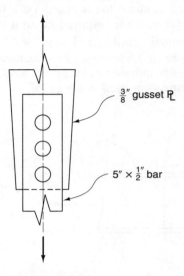

**PROBLEM 13-5**

**13-6.**  A tension member 18 ft–0 in. in length is to be composed of a wide-flange shape. End connections are assumed to be two rows of $\frac{7}{8}$-in.-diameter bolts in each flange as shown. The nominal loads are to be 90 kips dead load and 140 kips live load. Select the lightest W12 shape.

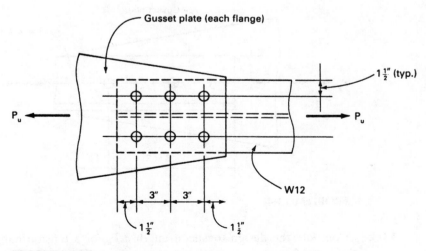

**PROBLEM 13-6**

**13-7.** Compute the design compressive strength $\phi_c P_n$ for a W10 × 45 column that has an unbraced length of 20 ft. The column is pin-connected at one end and fixed at the other.

**13-8.** Select the lightest W12 section for a column that will support an axial nominal load of 500 kips dead load and 500 kips live load. The unbraced length of the column is 16 ft, and the ends are assumed to be pin-connected.

**13-9.** Compute the design compressive strength $\phi_c P_n$ for an axial-loaded W12 × 65 column of A36 steel with an unbraced length of 24 ft with respect to the x-x axis. The column is pinned at both ends and braced in the weak direction at the third points. Assume a pin connection at the braced third points.

**13-10.** Compute the design compressive strength $\phi_c P_n$ for an axial-loaded W12 × 210 column of A36 steel. End connections and unbraced lengths are as follows:

**(a)** Pinned ends, $L = 25$ ft.

**(b)** One end pinned, one end fixed, $L = 30$ ft.

**(c)** Top end pinned, bottom end fixed, $L_x = 30$ ft, $L_y = 17$ ft.

**13-11.** Determine $\phi_b M_n$ for the following beams. Assume that $C_b = 1.0$.

**(a)** W12 × 53, $L_b = 26$ ft.

**(b)** W14 × 43, $L_b = 1.0$ ft.

**(c)** W8 × 67, $L_b = 16$ ft.

**13-12.** Calculate the maximum service live load $P$ that can be supported by the W21 × 50 shown. Assume A572 Grade 50 steel. The compressive flange has full lateral support. Neglect shear. Be sure to include the weight of the beam.

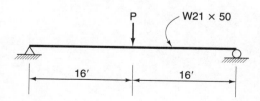

**PROBLEM 13-12**

**13-13.** Select a wide-flange beam for a span length of 25 ft. The beam is subjected to a factored bending moment of 275 ft-kips and is laterally supported at 5-ft intervals. Assume that $C_b = 1.0$ and that A572 Grade 50 steel is used.

**13-14.** The beam shown has full lateral bracing. The loads shown are factored design loads $(D + L)$. The uniformly distributed load does not include the weight of the beam. Select the lightest W shape. $C_b = 1.0$. Consider flexure and shear.

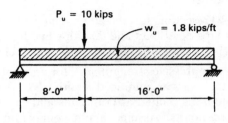

**PROBLEM 13-14**

**13-15.** The simply supported beam shown supports a uniformly distributed nominal dead load of 0.5 kip/ft (excluding the weight of the beam) and a point nominal live load of 3.75 kips. Lateral support exists at the reactions and at the point load. Select the lightest W shape. $C_b = 1.0$. Consider flexure and shear.

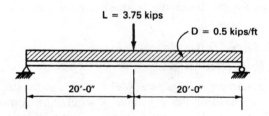

**PROBLEM 13-15**

# CHAPTER 14

# LRFD: Connections

14-1   INTRODUCTION

14-2   HIGH-STRENGTH BOLTED CONNECTIONS

14-3   FILLET WELDED CONNECTIONS

## 14-1

### INTRODUCTION

Much of the discussion of the first few sections of Chapter 7 (Bolted Connections) and Chapter 8 (Welded Connections) in this text is applicable to this chapter's discussion of the analysis and design of connections using the LRFD method. The basis for LRFD as discussed in Section 13-2 of this text is also applicable for this chapter. With respect to the limit states of connections, the LRFD method may be expressed in generic form by the equation

$$\Sigma \gamma_t Q_t \leq \phi R_n$$

where $\Sigma \gamma_t Q_t$ represents factored load effects and essentially is a required resistance or strength. (Each term is defined in Section 13-2.) Hence the preceding equation may be expressed as

$$\text{factored load effect} \leq \phi R_n$$

where $\phi$ is the appropriate resistance factor ($\leq 1.0$) and $R_n$ is the nominal resistance or strength of the member.

This chapter focuses on the evaluation of the right side of this expression, which we designate as the *design strength* of the member or connection.

For purposes of this chapter, references will be made to the LRFDM, Volume I, Part 6, which contains the following AISC criteria applicable to connections:

1.  The LRFDS, Chapter J, Connections, Joints, and Fasteners (also Appendix J and the Commentary for Chapter J)
2.  The LRFD Research Council on Structural Connections, Specification for Structural Joints Using ASTM A325 or A490 Bolts, referenced hereafter as *LRFD-SSJ*

Additionally, we consider only high-strength bolted connections and welded connections in this chapter.

## 14-2

## HIGH-STRENGTH BOLTED CONNECTIONS

A connection using high-strength bolts is classified as either a bearing-type connection (sometimes referred to as a shear-bearing connection) or a slip-critical connection.

For bearing-type connections not subject to tension loads, where slip can be permitted and where loosening or fatigue due to vibration or load fluctuations are not design considerations, the bolts need only be tightened to the snug-tight condition. Bolts tightened to only the snug condition must be clearly identified on the design and erection drawings.

For other bearing-type connections, such as those subject to direct tension loads, the bolts must be tightened to a bolt tension not less than that shown in Table 14-1. These bolts are installed and tightened using one of the methods described in Section 7-4 of this text.

Slip-critical connections are a type in which slip would be detrimental to the functioning of the connection. They are commonly designated for connections that are subject to fatigue or significant load reversal or for connections with oversized holes or slotted holes in which the applied force is approximately in the direction of the long dimension of the slots. In addition, a slip-critical connection must be

◼◼◼ TABLE 14-1   Minimum Bolt Tension, kips, and Bolt Cross-Sectional Area

| Bolt size (in.) | A325 bolts | A490 bolts | Area based on nominal diam. (in.$^2$) |
|:---:|:---:|:---:|:---:|
| $\frac{5}{8}$ | 19 | 24 | 0.3068 |
| $\frac{3}{4}$ | 28 | 35 | 0.4418 |
| $\frac{7}{8}$ | 39 | 49 | 0.6013 |
| 1 | 51 | 64 | 0.7854 |
| $1\frac{1}{8}$ | 56 | 80 | 0.9940 |

*Source:* Courtesy of AISC.

*Note:* Equal to 0.70 of minimum tensile strength of bolts, rounded to the nearest kip, as specified in ASTM specifications for A325 and A490 bolts with UNC threads.

used where welds and bolts share in transmitting shear load at a common faying surface. Bolts in slip-critical connections must be fully pretensioned to the minimum bolt tension specified in Table 14-1, which is taken from the LRFDS and the LRFD-SSJ. Slip-critical connections must be designated on contract plans and specifications. The bolts in slip-critical connections must be installed and tightened using one of the methods described in Section 7-4 of this text.

Despite the differences in assumed behavior between bearing-type connections and slip-critical connections, the analysis and design of both connections are similar. The bolts in each connection are assumed to be in shear and bearing. Slip-critical connections, however, must be checked for slip resistance to ensure that slip will not occur at specified load levels.

A simple single-shear lap connection of a tension member plate to a gusset plate is shown in Figure 14-1.

The three failure modes shown are as follows:

1. Figure 14-1c shows failure by single shear of the bolts in the plane between the plates.
2. Figure 14-1d shows that end tear-out is a failure of the end of the member that could occur if the end edge distance is too small. It is considered a shear failure due to bearing.
3. Figure 14-1e shows failure in the bearing resulting in hole distortion in the plate.

Additionally, the tension member plate could fail by rupture of the net section at the bolt or by yielding of the gross cross-sectional area. The two failure modes were discussed in Section 13-3 of this text.

Using the LRFD approach to analyze the type of connection shown in Figure 14-1 involves computing the various limit states applicable to this type of connection. These limit states are bolt shear, bearing on the plates, tension in the plates (both fracture and yield), and block shear (a combination of shear and tension in the

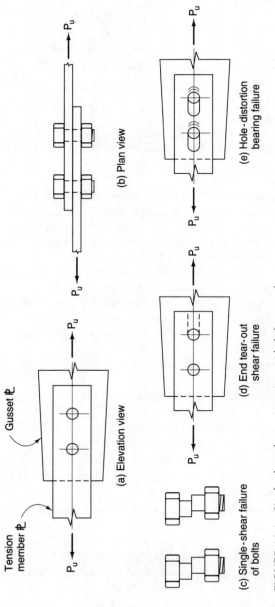

**FIGURE 14-1** Single-shear lap connection with failure modes.

(a) Elevation view

(b) Plan view

(c) Single-shear failure of bolts

(d) End tear-out shear failure

(e) Hole-distortion bearing failure

plates). In each case the design strength ($\phi R_n$) must be equal to or greater than the factored load effect (which in this case is designated $P_u$). The lowest design strength of the various failure modes will control.

## Shear Strength Limit State (High-Strength Bolts)

The design shear strength of both A325 and A490 bolts is $\phi R_n$, where the resistance factor $\phi$ is 0.75 and

$$R_n = F_n A_b \qquad \text{(LRFD-SSJ Eqn. 4.1)}$$

where

$R_n$ = nominal strength of a bolt subject to shear (kips)

$F_n$ = nominal strength of a bolt from Table 14-2 (ksi)

$A_b$ = area of bolt corresponding to nominal diameter (in.$^2$)

Therefore,

$$\phi R_n = \phi F_n A_b$$

The nominal strength ($F_n$) value is the same for both the bearing-type and slip-critical connections. The value is different based on whether or not the threads are in a plane of shear, however. Where threads are included in the shear plane, the bolt may be referred to as a type N bolt. For example, an A325 bolt of this type may be denoted as an A325N bolt. Where threads are excluded from the shear plane, the bolt may be referred to as a type X bolt (for example, A325X). Note that the N and X designations are applicable to both bearing-type and slip-critical

**TABLE 14-2   Nominal Strength of Fasteners ($F_n$)**

| Load condition | Nominal strength (ksi) | | Resistance factor $\phi$ |
|---|---|---|---|
| | A325 | A490 | |
| Applied Static Tension[a,b,c] | 90 | 113 | 0.75 |
| Shear in bolt with threads in the shear plane (N) | 48[d] | 60[d] | 0.75 |
| Shear in bolt with threads not in the shear plane (X) | 60[d] | 75[d] | 0.75 |

*Source:* From Table 2 of the SSJ. Courtesy of AISC.

[a]Bolts must be tensioned to requirements of Table 14-1.

[b]For bolts subject to tensile fatigue, reference is made to the LRFD-SSJ, Section 4(e).

[c]Except as required by the LRFD-SSJ, Section 4(b).

[d]In shear connections transmitting axial force whose length between extreme fasteners measured parallel to the line of force exceeds 50 in., tabulated values shall be reduced 20%.

connections. The use of the SC designation (refer to Section 7-6 of this text) is not valid for the LRFD method.

## Bearing Strength Limit State (High-Strength Bolts)

The design bearing strength is a function of many variables, among them bolt diameter, thickness of connected materials, and minimum tensile strength of the connected materials. End tear-out (Figure 14-1d) and hole deformation (Figure 14-1e) limit states depend on the hole type, bolt spacing, and edge distances. Design bearing strength does not depend on whether or not the threads of the bolt are in a plane of shear.

   In this chapter we consider only high-strength bolts in standard holes in a connection with two or more bolts in the line of force (unless noted otherwise).

   The design bearing strength is denoted by $\phi R_n$, where the resistance factor $\phi$ is 0.75. The nominal strength $R_n$ is calculated as follows, based on connection details:

1.   For all bolts in a connection with two or more bolts in the direction of the force, when the edge distance in the direction of the force is not less than 1.5$d$, the center-to-center bolt spacing is not less than 3$d$, and deformation around the bolt holes *is* a design consideration,

$$R_n = 2.4dtF_u \qquad \text{(LRFDS Eqn. J3-1a)}$$

2.   Same as step 1 preceding, except that deformation around the bolt holes *is not* a design consideration; for the bolt nearest the free edge in the direction of the force,

$$R_n = L_e tF_u \le 3.0\,dtF_u \qquad \text{(LRFDS Eqn. J3-1b)}$$

3.   Same as step 2 preceding, except that we now consider the remaining bolts:

$$R_n = \left(s - \frac{d}{2}\right)tF_u \le 3.0dtF_u \qquad \text{(LRFDS Eqn. J3-1c)}$$

4.   When $L_e < 1.5d$ or $s < 3d$, or for a single bolt or the bolt hole nearest the free edge when there are two or more bolts in the line of force,

$$R_n = L_e tF_u \le 2.4dtF_u \qquad \text{(LRFDS Eqn. J3-2a)}$$

5.   Same as step 4 preceding, except that we now consider the remaining bolts:

$$R_n = \left(s - \frac{d}{2}\right)tF_u \le 2.4dtF_u \qquad \text{(LRFDS Eqn. J3-2b)}$$

where

$R_n$ = nominal bearing strength of connected material (kips)

$F_u$ = specified minimum tensile strength of the connected part (ksi)

$L_e$ = distance in the direction of force from the center of a standard hole to the edge of the connected part (in.)

$d$ = nominal diameter of bolt (in.)

$t$ = thickness of connected material (in.)

$\phi$ = resistance factor (0.75)

$s$ = distance along line of force between centers of standard holes (in.)

The LRFDS directs in Section J3 that the distance between the centers of standard holes shall not be less than $2\frac{2}{3}$ times the nominal diameter of the bolt $d$ and that a distance of $3d$ is preferred. In addition, it is stipulated that the distance from the center of a standard hole to an edge of a connected part be not less than that furnished in the LRFDS, Table J3.4. This is the same as the ASDM, Table J3.5, of the ASDS.

## Block Shear Limit State

Block shear theory based on the AISC allowable stress design (ASD) method was discussed in Sections 2-2 and 7-6 of this text. Although the LRFD approach to block shear is more conservative than the ASD approach, the two methods are similar in their application to tension members, coped bending members, and various types of connections. The LRFD approach was discussed in Sections 13-3 and 13-5 of this text.

## Slip Resistance in Slip-Critical Connections

In theory, slip-critical connections are not subjected to shear and bearing. As discussed in the shear and bearing strength limit state paragraphs, however, the shear and bearing strength of the slip-critical connections must be computed due to the possibility that slip may occur as the result of an overload. To prevent slip, a design check for slip resistance is required by the LRFDS, Section J3.8.

This design check is a serviceability requirement, and it may be based on either service (nominal) loads or factored loads. If, for some reason, the consequences of slip are deemed critical, the limiting slip resistance should be based on the nominal loads. For the purpose of this discussion and consistent with Section 7-6 of this text, we compute the slip-resistance check based on the nominal load level. Further, we assume the common situation of a class A surface of contact that denotes an unpainted, clean, mill-scale surface or blast-cleaned surface with class A coatings and a mean slip coefficient of 0.33.

The LRFDS, Section J3, stipulates that the design resistance to shear of a bolt in a slip-critical connection is $\phi R_n$, where $R_n = F_v A_b$ and $\phi = 1.0$ for our condition of standard holes.

This design resistance to shear shall equal or exceed the shear on the bolt due to the nominal (service) loads. For our assumed surface condition and standard holes, $F_v = 17$ ksi for A325 bolts and $F_v = 21$ ksi for A490 bolts. Therefore,

$$\phi R_n = \phi F_v A_b$$

and since $\phi = 1.0$, this can be written as

$$\phi R_n = 17 A_b \qquad \text{for A325 bolts}$$

$$\phi R_n = 21 A_b \qquad \text{for A490 bolts}$$

**Example 14-1** _____

Compute the design strength $\phi R_n$ for the single-shear lap connection shown in Figure 14-2. Also compute the maximum service load permitted based on resistance to slip. The following apply:

1.  A36 steel plates ($F_u = 58$ ksi)
2.  Slip-critical connection
3.  $\frac{7}{8}$-in.-diameter A325 high-strength bolts
4.  Class A contact surface and standard holes
5.  Bolt threads in the shear plane (A325N)
6.  Deformation around the bolt hole *is* a design consideration

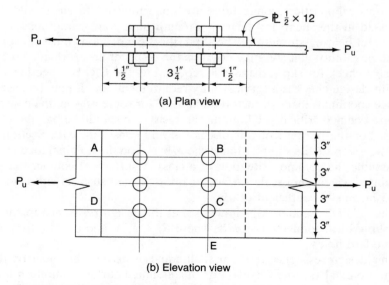

(a) Plan view

(b) Elevation view

**FIGURE 14-2**   Single-shear lap connection.

**Solution:**

The design strength limit states to be checked are bolt shear, bearing on the plates, tension in the plates (both fracture and yield), and block shear. The maximum service load will also be calculated based on slip resistance.

*Bolt shear:*   The design shear strength for one bolt is

$$\phi R_n = \phi F_n A_b$$

From Table 14-2,

$$F_n = 48 \text{ ksi}$$

$$\phi = 0.75$$

From Table 14-1,

$$A_b = 0.6013 \text{ in.}^2$$

From this,

$$R_n = F_n A_b = 48(0.6013) = 28.86 \text{ kips}$$

For six bolts, the design shear strength is calculated from

$$\phi R_n = 0.75(6)(28.86) = 129.9 \text{ kips}$$

Note that the design shear strength of one bolt may be obtained directly from the LRFDM, Volume II, Table 8-11.

*Bearing on the plates:*   As previously discussed, although this connection is designed as a slip-critical connection, it must nevertheless be checked for bearing in case, due to an overload, it should slip into bearing.

The design bearing strength depends on hole type, bolt spacing, and edge distance. For this connection,

$$1.5d = 1.5\left(\frac{7}{8}\right) = 1.313 \text{ in.}$$

$$3.0d = 3.0\left(\frac{7}{8}\right) = 2.63 \text{ in.}$$

The edge distance in the direction of the force is 1.5 in. ($> 1.5d$), the bolt spacing is 3.25 in. ($> 3.0d$), and there are two bolts in the direction of the load. Therefore, LRFDS Equation J3-1a is applicable, and

$$\phi R_n = \phi(2.4)dtF_u n$$

$$= 0.75(2.4)\left(\frac{7}{8}\right)\left(\frac{1}{2}\right)(58)(6)$$

$$= 274 \text{ kips}$$

Note that the design bearing strength for one bolt (based on 1-in.-thick material) may be obtained from the LRFDM, Volume II, Table 8-13.

*Tension in the plates:*   Based on yielding in the gross section (see Section 13-3 of this text),

$$\phi_t P_n = \phi_t F_y A_g$$

$$= 0.90(36)(12)\left(\frac{1}{2}\right)$$

$$= 194.4 \text{ kips}$$

Based on fracture in the net section (see Section 13-3 of this text),

$$\phi_t P_n = \phi_t F_u A_e$$

Since this is a flat-plate member,

$$A_e = A_n$$

where

$$A_n = A_g - A_{\text{holes}}$$

$$= \left(12 \times \frac{1}{2}\right) - 3\left(\frac{7}{8} + \frac{1}{8}\right)\left(\frac{1}{2}\right)$$

$$= 4.5 \text{ in.}^2$$

from which

$$\phi_t P_n = 0.75(58)(4.5)$$

$$= 195.8 \text{ kips}$$

*Block shear strength:*   As discussed in Section 13-3 of this text, since block shear is a rupture (fracture) limit state, two cases must be checked. The larger combination of fracture strength on one plane and yielding on a perpendicular plane will be the governing block shear strength. In this connection we apply this criterion to two possible failure paths (path *ABCD* and path *ABCE*).

Note that the hole diameter is taken as 1.0 in. (bolt diameter plus $\frac{1}{8}$ in.) and $\phi = 0.75$.

*Case I* (failure line *ABCD*; Figure 14-3):

$$A_{gv} = 2(0.50)(3.25 + 1.5) = 4.75 \text{ in}^2$$

$$A_{gt} = 0.50(6) = 3.00 \text{ in.}^2$$

$$A_{nv} = 2(0.50)[3.25 + 1.5 - 1.5(1.0)] = 3.25 \text{ in.}^2$$

$$A_{nt} = 0.50[6 - 2(1.0)] = 2.00 \text{ in.}^2$$

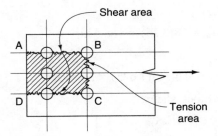

**FIGURE 14-3** Block shear.

Considering tensile fracture,

$$F_u A_{nt} = 58(2.00) = 116 \text{ kips}$$

Considering shear fracture,

$$0.6F_u A_{nv} = 0.6(58)(3.25) = 113.1 \text{ kips}$$

Since $F_u A_{nt} > 0.6F_u A_{nv}$, for block shear

$$\phi R_n = \phi[0.6F_y A_{gv} + F_u A_{nt}]$$
$$= 0.75[0.6(36)(4.75) + 116]$$
$$= 164.0 \text{ kips}$$

*Case II* (failure line *ABCE*; Figure 14-4):

$$A_{gv} = 0.50(3.25 + 1.5) = 2.38 \text{ in.}^2$$
$$A_{gt} = 0.50(9) = 4.50 \text{ in.}^2$$
$$A_{nv} = 0.50[3.25 + 1.5 - 1.5(1.0)] = 1.625 \text{ in.}^2$$
$$A_{nt} = 0.50[9 - 2.5(1.0)] = 3.25 \text{ in.}^2$$

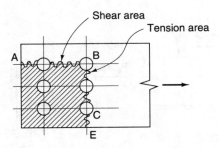

**FIGURE 14-4** Block shear.

Considering tensile fracture,

$$F_u A_{nt} = 58(3.25) = 188.5 \text{ kips}$$

Considering shear fracture,

$$0.6 F_u A_{nv} = 0.6(58)(1.625) = 56.5 \text{ kips}$$

Since $F_u A_{nt} > 0.6 F_u A_{nv}$, for block shear,

$$\phi R_n = \phi[0.6 F_y A_{gv} + F_u A_{nt}]$$
$$= 0.75[0.6(36)(2.38) + 188.5]$$
$$= 179.9 \text{ kips}$$

The governing block shear strength is 179.9 kips, the *larger* of the two cases. However, considering all the various limit states, bolt shear controls, and the design strength $\phi R_n$ of this connection is 129.9 kips.

*Slip resistance check:*   Since this is a slip-critical connection and no slippage is permitted, calculate the maximum nominal (service) load permitted by slip resistance.

The design slip resistance as per LRFDS is taken as

$$\phi R_n = \phi F_v A_b$$
$$= 1.0(17)(0.6013)$$
$$= 10.22 \text{ kips/bolt}$$

Since there are six bolts,

$$6(10.22) = 61.3 \text{ kips}$$

To prevent slip, the maximum service load must not exceed 61.3 kips.

**Example 14-2** _____

A tension member in a roof truss is composed of 2L4 × 3 × $\frac{1}{4}$ (LLBB) and connected to a gusset plate with one line of $\frac{7}{8}$-in.-diameter A325 bolts, as shown in Figure 14-5. Determine the required number of bolts and check

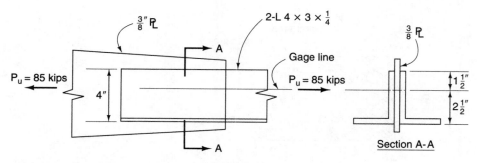

**FIGURE 14-5**   Tension member connection design.

block shear. The required design strength $P_u$ for the member is 85 kips. The following assumptions apply:

1. Gusset plate and angles are adequate for tension load
2. Bearing-type connection with threads in the shear plane (A325N)
3. Bolt spacing of 3 in. on center, with edge distance of $1\frac{1}{2}$ in.
4. Standard holes and A36 steel.
5. Deformation around holes *is not* a design consideration

**Solution:**

The tension member has been selected, and the required number of bolts will be determined by considering shear in the bolts and bearing on the $\frac{3}{8}$-in. gusset plate (the total thickness of the angles is $\frac{1}{2}$ in.). The block shear strength will also be checked, since this cannot be done until the bolts have been selected and the layout of the connection has been established.

1. The bolts are in double shear, and the design strength for one bolt is (refer to Tables 14-1 and 14-2)

$$\phi R_n = \phi F_n A_b$$
$$= 0.75(48)(2)(0.6013)$$
$$= 43.3 \text{ kips/bolt}$$

2. For bearing on the gusset plate, the design strength of one bolt (the bolt nearest the free edge) is

$$\phi R_n = \phi L_e t F_u \le \phi 3.0 dt F_u$$
$$= 0.75(1.5)(0.375)(58)$$
$$= 24.5 \text{ kips/bolt}$$

Check the upper limit:

$$\phi 3.0 dt F_u = 0.75(3.0)(0.875)(0.375)(58) = 42.8 \text{ kips/bolt}$$

$$24.5 \text{ kips} < 42.8 \text{ kips} \qquad \textbf{O.K.}$$

For the remaining bolts, the design strength for one bolt is

$$\phi R_n = \phi \left( s - \frac{d}{2} \right) t F_u \le \phi 3.0 dt F_u$$

$$= 0.75 \left( 3 - \frac{0.875}{2} \right)(0.375)(58)$$

$$= 41.8 \text{ kips/bolt}$$

$$41.8 \text{ kips} < 42.8 \text{ kips} \qquad \textbf{O.K.}$$

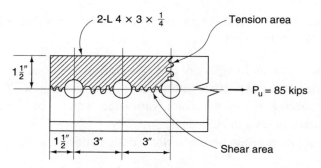

**FIGURE 14-6**   Connection detail.

Therefore, bearing controls, and the number of bolts $N_b$ required is calculated from

$$\text{required } N_b = 1 + \frac{85 - 24.5}{41.8} = 2.45 \text{ bolts} \qquad \text{(try three bolts)}$$

3.  Check block shear. The detail of the connection is shown in Figure 14-6 with the block shear failure path indicated.

$$A_{gv} = 2(0.25)(7.5) = 3.75 \text{ in.}^2$$

$$A_{gt} = 2(0.25)(1.5) = 0.75 \text{ in.}^2$$

$$A_{nv} = 2(0.25)[7.5 - 2.5(1.0)] = 2.50 \text{ in.}^2$$

$$A_{nt} = 2(0.25)[1.5 - 0.5(1.0)] = 0.50 \text{ in.}^2$$

Considering tensile fracture,

$$F_u A_{nt} = 58(0.50) = 29.0 \text{ kips}$$

Considering shear fracture,

$$0.6 F_u A_{nv} = 0.6(58)(2.50) = 87.0 \text{ kips}$$

Since $0.6 F_u A_{nv} > F_u A_{nt}$, shear fracture controls, and LRFDS Equation (J4-3b) is applicable:

$$\phi R_n = \phi[0.6 F_u A_{nv} + F_y A_{gt}]$$

$$= 0.75[87.0 + 36(0.75)]$$

$$= 85.5 \text{ kips}$$

$$85.5 \text{ kips} > 85 \text{ kips} \qquad\qquad \textbf{O.K.}$$

## 14-3

### FILLET WELDED CONNECTIONS

The discussions of Sections 8-1 through 8-3 of this text concerning the welding process, types of welds and joints, and the behavior of fillet welded connections are pertinent to our discussion of the LRFD method of fillet welded connection design and analysis. The LRFDS requires that the factored load effect not exceed the design strength of the weld metal or the base material, whichever controls. This may be expressed as

$$\text{factored load effect} \leq \phi R_n$$

where $\phi$ is the appropriate resistance factor and $R_n$ is the nominal resistance or strength of either the weld metal or the base material. The nominal strength of fillet welds is determined from the effective throat area (see Figure 8-10), whereas the nominal strength of the base materials is governed by their respective thicknesses. Figure 14-7 illustrates the shear planes for fillet welds and base material:

(a)   Plane 1–1 in which the resistance is governed by the shear strength for material A

(b)   Plane 2–2 in which the resistance is governed by the shear strength of the weld metal

(c)   Plane 3–3 in which the resistance is governed by the shear strength of the material B

The strength or resistance of the welded connection is the lowest of the resistance calculated in each plane of shear transfer.

With respect to the weld metal, failure is assumed to occur in shear on a plane through the throat of the weld. When using the shielded metal-arc process, the throat distance is perpendicular to the face of the weld and is equal to 0.707 times the leg size of the weld. Note that the effective throat distance for a weld made

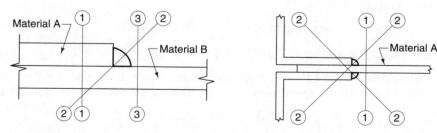

**FIGURE 14-7**   Shear planes for fillet welds loaded in longitudinal shear. (Direction of load is into the page.) (Courtesy of the American Institute of Steel Construction.)

with the submerged arc welding process is larger (see Section 8-3). For our purpose, the shielded metal arc welding (SMAW) process are specified here.

Since failure of the weld metal occurs on the throat area, the nominal resistance or strength of the weld metal can be written as

$$R_n = A_w F_w$$

and the design strength can be written as

$$\phi R_n = \phi A_w F_w$$

where

$\phi$ = resistance factor (0.75)

$A_w$ = effective cross-sectional area of the weld (0.707 × leg size) (in.$^2$)

$F_w$ = nominal strength of the weld electrode (ksi)

The strength of the fillet weld is based on the type of electrode used. It is common to use an E70XX electrode with an ultimate tensile strength of 70 ksi in combination with steels having a yield stress in the range between 36 ksi and 60 ksi.

Since, according to the LRFDS, the nominal shear strength $F_w$ is taken as 0.6 times the tensile strength of the weld metal (where the tensile strength is also defined as the classification strength of the weld metal and denoted $F_{EXX}$), the design strength previously written as $\phi A_w F_w$ may now be written as

$$\phi R_n = \phi A_w (0.6) F_{EXX}$$

Substituting $\phi = 0.75$ and $F_{EXX} = 70$ ksi (E70 electrode) results in

$$\phi R_n = 0.75 A_w (0.6)(70) = 31.5 A_w \text{ (kips)}$$

In addition to the weld metal considerations, the factored load effect on the base material must not exceed the design strength of $\phi F_{BM}$. This can be written as

$$\phi R_n \le \phi F_{BM} A_{BM}$$

where

$\phi$ = resistance factor (0.90)

$F_{BM}$ = nominal shear strength of the base metal (ksi)

$A_{BM}$ = cross-sectional area of the base metal subject to shear (in.$^2$)

According to the LRFDS, $F_{BM}$ may be taken as $0.60F_y$. Additionally, $A_{BM}$ may be designated as $A_g$. Therefore, the preceding expression may be rewritten as

$$\phi R_n = \phi(0.60F_y)A_g$$
$$= 0.90(0.60)F_y A_g$$
$$= 0.54 F_y A_g \text{ (kips)}$$

Hence, the design strength of the welded connection is taken as the lower of $\phi F_{BM} A_{BM}$ and $\phi F_w A_w$ when applicable. These expressions, in turn, using our assumptions, may be written as $31.5 A_w$ and $0.54 F_y A_g$.

As in our discussion in Section 8-3, it is convenient to calculate the design strength of a 1-in.-long fillet weld having a leg size of $\frac{1}{16}$ in. based on shear through the effective throat area. For an E70XX electrode and using $A_w = 0.707(\frac{1}{16})(1.0)$, we write

$$\phi R_n = 31.5 A_w$$

$$= 31.5(0.707)\left(\frac{1}{16}\right)(1.0)$$

$$= 1.392 \text{ kips/in. for a } \tfrac{1}{16}\text{-in. weld}$$

The shear strength for any size fillet weld for an E70XX electrode can then be calculated by multiplying the size of the weld in sixteenths of an inch by 1.392 kips/in. Table 14-3 provides these values for fillet welds between $\frac{1}{16}$ in. and $\frac{7}{8}$ in.

We can similarly calculate the design strength of the base material for a connection such as that shown in Figure 14-7. The thinner of the two connected materials would govern. Assuming a $\frac{1}{16}$-in.-thick material "A" and A36 steel,

$$\phi R_n = \phi F_{BM} A_{BM}$$

$$= 0.54 F_y A_g$$

$$= 0.54(36)\left(\frac{1}{16}\right)(1.0)$$

1.215 kips/in. per $\frac{1}{16}$-in. material thickness

The commentary to the LRFDS, Section J4, states that the block shear failure mode should be checked around the periphery of welded connections. The value of the resistance factor $\phi$ should be taken as 0.75 for both the fracture and the yielding planes.

**TABLE 14-3   Strength of Welds $\phi R_n$ (kips per linear inch)**

| Weld size, in. | Strength | Weld size, in. | Strength |
|---|---|---|---|
| 1/16 | 1.392 | 1/2 | 11.14 |
| 1/8 | 2.78 | 9/16 | 12.53 |
| 3/16 | 4.18 | 5/8 | 13.92 |
| 1/4 | 5.57 | 11/16 | 15.31 |
| 5/16 | 6.96 | 3/4 | 16.70 |
| 3/8 | 8.35 | 13/16 | 18.10 |
| 7/16 | 9.74 | 7/8 | 19.49 |

**Example 14-3** _____

Determine the design strength of the welded connection shown in Figure 14-8. The steel is A36, and the electrode used was E70 (SMAW). The weld is a $\frac{7}{16}$-in. fillet weld. Neglect the design strength of the plates.

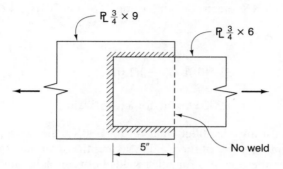

**FIGURE 14-8**   Welded lap connection.

**Solution:**

*Strength of weld metal:*   From Table 14-3, the strength of a $\frac{7}{16}$-in. fillet weld is 9.74 kips/in. For 16 in. of weld, the design strength of the connection based on the weld metal is

$$\phi R_n = 16(9.74) = 155.8 \text{ kips}$$

*Strength of base metal:*   For a $\frac{3}{4}$-in.-thick plate, using our previously determined constant of $\phi R_n = 1.215$ kips/in. per $\frac{1}{16}$-in. material thickness, we calculate

$$\phi R_n = 1.215(12) = 14.58 \text{ kips/in.}$$

and for a 16-in. length of weld,

$$\phi R_n = 14.58(16) = 233 \text{ kips}$$

*Check block shear strength:*   For this check, the *gross* areas in shear and tension are used:

$$A_{gv} = A_{nv} = 5(2)(0.75) = 7.50 \text{ in.}^2$$

$$A_{gt} = A_{nt} = 6(0.75) = 4.50 \text{ in.}^2$$

Considering tensile fracture,

$$F_u A_{nt} = 58(4.50) = 261 \text{ kips}$$

Considering shear fracture,

$$0.6 F_u A_{nv} = 0.6(58)(7.50) = 261 \text{ kips}$$

Since the two cases of fracture are equally critical, the block shear design strength is calculated from LRFDS Equation (J4-3a):

$$\phi R_n = \phi[0.6F_y A_{gv} + F_u A_{nt}]$$

$$= 0.75[0.6(36)(7.50) + 58(4.50)]$$

$$= 317 \text{ kips}$$

The design strength of the connection is therefore governed by the strength of the fillet weld metal:

$$\phi R_n = 155.8 \text{ kips}$$

---

**Example 14-4** _____

Design an end connection using longitudinal welds and an end transverse weld to develop the full tensile capacity of the hanger shown in Figure 14-9. Use A36 steel and E70XX electrodes (SMAW). The hanger is a flat 1-in.-thick plate.

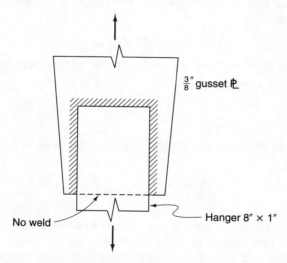

$\frac{3}{8}''$ gusset ℄

No weld

Hanger 8″ × 1″

**FIGURE 14-9**  Welded lap connection.

**Solution:**

Determine the design strength of the hanger plate. Based on yielding of the gross section,

$$\phi_t P_n = \phi F_y A_g$$

$$= 0.90(36)(8)(1) = 259 \text{ kips}$$

Based on fracture of the net section,

$$\phi_t P_n = \phi F_u A_e$$

Since the member is a flat plate with longitudinal and transverse welds,

$$A_e = A_g$$

and

$$\phi_t P_n = 0.75(58)(8)(1) = 348 \text{ kips}$$

Therefore, the connection must be designed for a factored load of 259 kips. The maximum weld size is calculated from

$$1 \text{ in.} - \frac{1}{16} \text{ in.} = \frac{15}{16} \text{ in.}$$

We will try a $\frac{3}{8}$-in. E70 fillet weld. From Table 14-3, the design strength of this weld is 8.35 kips/in. The design strength of the base material using the thinner of the connected members ($\frac{3}{8}$ in.), and recalling our previously determined constant of 1.215 kips/in. per $\frac{1}{16}$ in. thickness for the given conditions, is calculated from

$$1.215(6) = 7.29 \text{ kips/in.}$$

The design strength of the base material controls and the required length of weld $l$ is calculated from

$$l = \frac{259}{7.29} = 35.5 \text{ in.}$$

With an 8-in.-long transverse weld, the required length of longitudinal weld ($l_l$) on each side is

$$l_l = \frac{35 - 8}{2} \doteq 13.75 \text{ in.}$$

The minimum length of weld on each side must not be less than the perpendicular distance between them (8 in.). Therefore, use a 14-in. longitudinal weld on each side and an 8-in. transverse weld (the total length of weld is 36-in.).

Last, check block shear for the designed connection:

$$A_{gv} = A_{nv} = 0.375(28) = 10.50 \text{ in.}^2$$

$$A_{gt} = A_{nt} = 0.375(8) = 3.00 \text{ in.}^2$$

Considering tensile fracture,

$$F_u A_{nt} = 58(3.00) = 174 \text{ kips}$$

Considering shear fracture,

$$0.6F_u A_{nv} = 0.6(58)(10.50) = 365 \text{ kips}$$

$$\phi R_n = \phi(0.6F_u A_{nv} + F_y A_{gt})$$

$$= 0.75[365 + 36(3.00)] = 355 \text{ kips}$$

$$355 \text{ kips} > 259 \text{ kips} \qquad \qquad \textbf{O.K.}$$

## PROBLEMS

*Note: In all problems for this chapter, use the LRFDS. The following assumptions apply, unless otherwise noted:*

1.  All plates and structural shapes are A36 steel.
2.  For slip-critical connections, hole deformation is a consideration.
3.  For bearing-type connections, hole deformation is not a consideration.
4.  All bolts are high-strength bolts in standard holes.
5.  Use class A contact surface with mean slip coefficient of 0.33, as applicable.
6.  Gusset plates are adequate for tension.
7.  For welded connections, use E70XX electrodes (SMAW.)

**14-1.**  Determine the design strength for the single-angle tension member shown. Bolts are $\frac{3}{4}$-in.-diameter A490N in a bearing-type connection. Assume that hole deformation is a consideration for this connection.

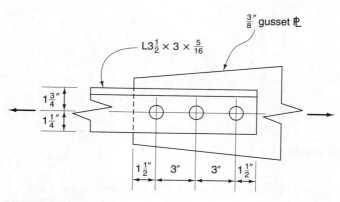

**PROBLEM 14-1**

**14-2.**     Determine the design strength of the wide-flange tension member shown. Bolts are $\frac{7}{8}$-in.-diameter A325X in a bearing-type connection. Assume that hole deformation is a consideration for this connection.

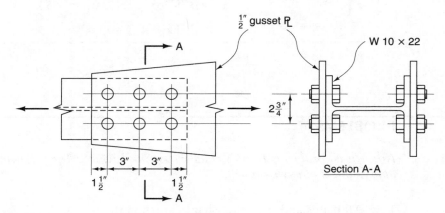

**PROBLEM 14-2**

**14-3.**     Determine the design strength for the single-shear lap connection shown. The bolts are 1-in.-diameter A325N in a slip-critical connection. Threads are in the shear plane. Calculate the maximum permitted service load based on slip resistance.

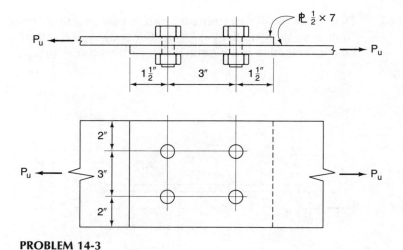

**PROBLEM 14-3**

**14-4.**     Rework Problem 14-3 using $\frac{7}{8}$-in.-diameter A490N bolts in a bearing-type connection.

**14-5.** Determine the design strength for the connection shown. Bolts are $\frac{3}{4}$-in.-diameter A325X in a bearing-type connection. Assume A588 steel.

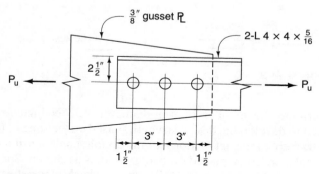

**PROBLEM 14-5**

**14-6.** Determine the number of $\frac{3}{4}$-in.-diameter A490 bolts required for the slip-critical splice shown. Consider only bolt shear and bearing on the plates. Assume that bolts are placed on two gage lines. Also assume an edge distance in the line of force of 1.5 in. and a spacing of 3 in. State any other assumptions.

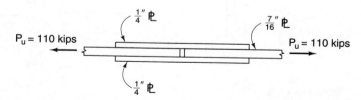

**PROBLEM 14-6**

**14-7.** Rework Problem 14-6 using $\frac{3}{4}$-in.-diameter A325 bolts in a bearing-type connection. The applied factored tensile load $P_u$ is 150 kips. Consider only bolt shear and bearing on the plates.

**(a)** Threads are excluded from the shear plane (bolts are A325X).

**(b)** Threads are in the shear plane (bolts are A325N).

**14-8.** Determine the required number of $\frac{7}{8}$-in.-diameter A325 bolts required for the single-shear lap connection shown. The connection is subjected to a factored load of 200 kips and is to be a bearing-type connection with threads assumed to be in the shear plane (A325N). Consider only bolt shear and bearing in the plates. Neglect block shear. Assume bolts are placed on two gage lines. Also assume an edge distance in the line of force of 1.5 in. and spacing of 3 in. along the gage lines. Check the tension capacity of the plates after detailing the connection.

**PROBLEM 14-8**

**14-9.** Determine the number of $\frac{3}{4}$-in.-diameter A325N high-strength bolts required to develop the design strength (based on yielding of the gross section and fracture of the net section) for the double-angle tension member shown. The bolts are to be placed on two gage lines as shown. Specify the spacing and edge distance to be used. Be sure to check block shear. This is a slip-critical connection. Calculate the maximum permitted service load based on slip resistance.

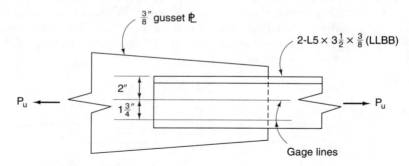

**PROBLEM 14-9**

**14-10.** Compute the design strength for the double-shear butt connection shown. The bolts are $\frac{7}{8}$-in.-diameter A325 high-strength bolts. Assume a bearing-type connection.

(a) Threads are excluded from the shear plane (A325X).

(b) Threads are in the shear plane (A325N).

**14-11.** Determine the design strength of the welded connections shown. The welds are $\frac{5}{16}$-in. fillet welds. Neglect block shear.

**14-12.** Determine the maximum value of $P_u$ based on the information shown in the diagram. Neglect block shear.

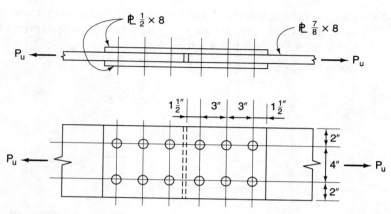

**PROBLEM 14-10**

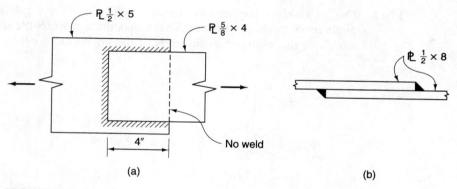

(a)

(b)

**PROBLEM 14-11**

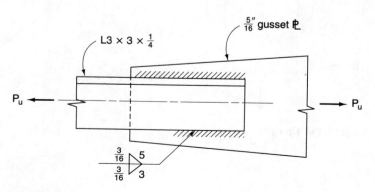

**PROBLEM 14-12**

**14-13.** Determine the length of weld required for the connection shown. Use a $\frac{5}{16}$-in. fillet weld, and assume 1-in. end returns. The factored tensile load $P_u$ is 108 kips. Neglect block shear.

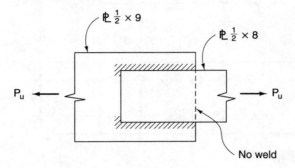

**PROBLEM 14-13**

**14-14.** Design a welded connection to resist the full design strength of the double-angle tension hanger shown. Use the maximum permissible size fillet weld. Continue the weld across the ends of the angles as shown.

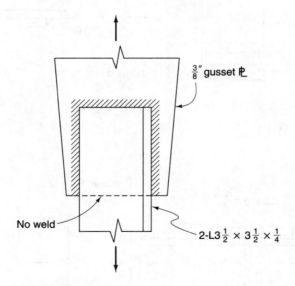

**PROBLEM 14-14**

# Appendices

# Design Tables
# for Open Web Steel Joists

# TABLE A-1   Standard Load Table/Open Web Steel Joists, K-Series

| Joist Designation | 8K1 | 10K1 | 12K1 | 12K3 | 12K5 | 14K1 | 14K3 | 14K4 | 14K6 | 16K2 | 16K3 | 16K4 | 16K5 | 16K6 | 16K7 | 16K9 |
|---|---|---|---|---|---|---|---|---|---|---|---|---|---|---|---|---|
| Depth (in.) | 8 | 10 | 12 | 12 | 12 | 14 | 14 | 14 | 14 | 16 | 16 | 16 | 16 | 16 | 16 | 16 |
| Approx. Wt. (lb/ft) | 5.1 | 5.0 | 5.0 | 5.7 | 7.1 | 5.2 | 6.0 | 6.7 | 7.7 | 5.5 | 6.3 | 7.0 | 7.5 | 8.1 | 8.6 | 10.0 |
| Span (ft) | | | | | | | | | | | | | | | | |
| 8 | 550/550 | | | | | | | | | | | | | | | |
| 9 | 550/550 | | | | | | | | | | | | | | | |
| 10 | 550/480 | 550/550 | | | | | | | | | | | | | | |
| 11 | 532/377 | 550/542 | | | | | | | | | | | | | | |
| 12 | 444/288 | 550/455 | 550/550 | 550/550 | 550/550 | | | | | | | | | | | |
| 13 | 377/225 | 479/363 | 550/510 | 550/510 | 550/510 | | | | | | | | | | | |
| 14 | 324/179 | 412/289 | 500/425 | 550/463 | 550/463 | 550/550 | 550/550 | 550/550 | 550/550 | | | | | | | |
| 15 | 281/145 | 358/234 | 434/344 | 543/428 | 550/434 | 511/475 | 550/507 | 550/507 | 550/507 | | | | | | | |
| 16 | 246/119 | 313/192 | 380/282 | 476/351 | 550/396 | 448/390 | 550/467 | 550/467 | 550/467 | 550/550 | 550/550 | 550/550 | 550/550 | 550/550 | 550/550 | 550/550 |
| 17 | | 277/159 | 336/234 | 420/291 | 550/366 | 395/324 | 495/404 | 550/443 | 550/443 | 512/488 | 550/526 | 550/526 | 550/526 | 550/526 | 550/526 | 550/526 |
| 18 | | 246/134 | 299/197 | 374/245 | 507/317 | 352/272 | 441/339 | 530/397 | 550/408 | 456/409 | 508/456 | 550/490 | 550/490 | 550/490 | 550/490 | 550/490 |
| 19 | | 221/113 | 268/167 | 335/207 | 454/269 | 315/230 | 395/287 | 475/336 | 550/383 | 408/347 | 455/386 | 547/452 | 550/455 | 550/455 | 550/455 | 550/455 |
| 20 | | 199/97 | 241/142 | 302/177 | 409/230 | 284/197 | 356/246 | 428/287 | 525/347 | 368/297 | 410/330 | 493/386 | 550/426 | 550/426 | 550/426 | 550/426 |
| 21 | | | 218/123 | 273/153 | 370/198 | 257/170 | 322/212 | 388/248 | 475/299 | 333/255 | 371/285 | 447/333 | 503/373 | 548/405 | 550/406 | 550/406 |
| 22 | | | 199/106 | 249/132 | 337/172 | 234/147 | 293/184 | 353/215 | 432/259 | 303/222 | 337/247 | 406/289 | 458/323 | 498/351 | 550/385 | 550/385 |
| 23 | | | 181/93 | 227/116 | 308/150 | 214/128 | 268/160 | 322/188 | 395/226 | 277/194 | 308/216 | 371/252 | 418/282 | 455/307 | 507/339 | 550/363 |
| 24 | | | 166/81 | 208/101 | 282/132 | 196/113 | 245/141 | 295/165 | 362/199 | 254/170 | 283/189 | 340/221 | 384/248 | 418/269 | 465/298 | 550/346 |
| 25 | | | | | | 180/100 | 226/124 | 272/145 | 334/175 | 234/150 | 260/167 | 313/195 | 353/219 | 384/238 | 428/263 | 514/311 |
| 26 | | | | | | 166/88 | 209/110 | 251/129 | 308/156 | 216/133 | 240/148 | 289/173 | 326/194 | 355/211 | 395/233 | 474/276 |
| 27 | | | | | | 154/79 | 193/98 | 233/115 | 285/139 | 200/119 | 223/132 | 268/155 | 302/173 | 329/188 | 366/208 | 439/246 |
| 28 | | | | | | 143/70 | 180/88 | 216/103 | 265/124 | 186/106 | 207/118 | 249/138 | 281/155 | 306/168 | 340/186 | 408/220 |
| 29 | | | | | | | | | | 173/95 | 193/106 | 232/124 | 261/139 | 285/151 | 317/167 | 380/198 |
| 30 | | | | | | | | | | 161/86 | 180/96 | 216/112 | 244/126 | 266/137 | 296/151 | 355/178 |
| 31 | | | | | | | | | | 151/78 | 168/87 | 203/101 | 228/114 | 249/124 | 277/137 | 332/161 |
| 32 | | | | | | | | | | 142/71 | 158/79 | 190/92 | 214/103 | 233/112 | 259/124 | 311/147 |

Reprinted courtesy of the Steel Joist Institute.

Notes for Table A-1:

1. Based on a maximum allowable tensile stress of 30,000 psi.

2. The approximate joist weights per linear foot shown in these tables do not include accessories.

3. Where the joist span exceeds the unshaded area of the table, the row of bridging nearest the midspan shall be diagonal bridging with bolted connections at chords and midspan.

**■■■ TABLE A-1   Standard Load Table/Open Web Steel Joists, K-Series (cont.)**

| Joist Designation | 18K3 | 18K4 | 18K5 | 18K6 | 18K7 | 18K9 | 18K10 | 20K3 | 20K4 | 20K5 | 20K6 | 20K7 | 20K9 | 20K10 | 22K4 | 22K5 | 22K6 | 22K7 | 22K9 | 22K10 | 22K11 |
|---|---|---|---|---|---|---|---|---|---|---|---|---|---|---|---|---|---|---|---|---|---|
| Depth (in.) | 18 | 18 | 18 | 18 | 18 | 18 | 18 | 20 | 20 | 20 | 20 | 20 | 20 | 20 | 22 | 22 | 22 | 22 | 22 | 22 | 22 |
| Approx. Wt. (lb/ft) | 6.6 | 7.2 | 7.7 | 8.5 | 9.0 | 10.2 | 11.7 | 6.7 | 7.6 | 8.2 | 8.9 | 9.3 | 10.8 | 12.2 | 8.0 | 8.8 | 9.2 | 9.7 | 11.3 | 12.6 | 13.8 |
| **Span (ft) ↓** | | | | | | | | | | | | | | | | | | | | | |
| 18 | 550 | 550 | 550 | 550 | 550 | 550 | 550 | | | | | | | | | | | | | | |
|  | 550 | 550 | 550 | 550 | 550 | 550 | 550 | | | | | | | | | | | | | | |
| 19 | 514 | 550 | 550 | 550 | 550 | 550 | 550 | | | | | | | | | | | | | | |
|  | 494 | 523 | 523 | 523 | 523 | 523 | 523 | | | | | | | | | | | | | | |
| 20 | 463 | 550 | 550 | 550 | 550 | 550 | 550 | 517 | 550 | 550 | 550 | 550 | 550 | 550 | | | | | | | |
|  | 423 | 490 | 490 | 490 | 490 | 490 | 490 | 517 | 550 | 550 | 550 | 550 | 550 | 550 | | | | | | | |
| 21 | 420 | 506 | 550 | 550 | 550 | 550 | 550 | 468 | 550 | 550 | 550 | 550 | 550 | 550 | | | | | | | |
|  | 364 | 426 | 460 | 460 | 460 | 460 | 460 | 453 | 520 | 520 | 520 | 520 | 520 | 520 | | | | | | | |
| 22 | 382 | 460 | 518 | 550 | 550 | 550 | 550 | 426 | 514 | 550 | 550 | 550 | 550 | 550 | 550 | 550 | 550 | 550 | 550 | 550 | 550 |
|  | 316 | 370 | 414 | 438 | 438 | 438 | 438 | 393 | 461 | 490 | 490 | 490 | 490 | 490 | 548 | 548 | 548 | 548 | 548 | 548 | 548 |
| 23 | 349 | 420 | 473 | 516 | 550 | 550 | 550 | 389 | 469 | 529 | 550 | 550 | 550 | 550 | 518 | 550 | 550 | 550 | 550 | 550 | 550 |
|  | 276 | 323 | 362 | 393 | 418 | 418 | 418 | 344 | 402 | 451 | 468 | 468 | 468 | 468 | 491 | 518 | 518 | 518 | 518 | 518 | 518 |
| 24 | 320 | 385 | 434 | 473 | 526 | 550 | 550 | 357 | 430 | 485 | 518 | 550 | 550 | 550 | 475 | 536 | 550 | 550 | 550 | 550 | 550 |
|  | 242 | 284 | 318 | 345 | 382 | 396 | 396 | 302 | 353 | 396 | 430 | 448 | 448 | 448 | 431 | 483 | 495 | 495 | 495 | 495 | 495 |
| 25 | 294 | 355 | 400 | 435 | 485 | 550 | 550 | 329 | 396 | 446 | 486 | 541 | 550 | 550 | 438 | 493 | 537 | 550 | 550 | 550 | 550 |
|  | 214 | 250 | 281 | 305 | 337 | 377 | 377 | 266 | 312 | 350 | 380 | 421 | 426 | 426 | 381 | 427 | 464 | 474 | 474 | 474 | 474 |
| 26 | 272 | 328 | 369 | 402 | 448 | 538 | 550 | 304 | 366 | 412 | 449 | 500 | 550 | 550 | 404 | 455 | 496 | 550 | 550 | 550 | 550 |
|  | 190 | 222 | 249 | 271 | 299 | 354 | 361 | 236 | 277 | 310 | 337 | 373 | 405 | 405 | 338 | 379 | 411 | 454 | 454 | 454 | 454 |
| 27 | 252 | 303 | 342 | 372 | 415 | 498 | 550 | 281 | 339 | 382 | 416 | 463 | 550 | 550 | 374 | 422 | 459 | 512 | 550 | 550 | 550 |
|  | 169 | 198 | 222 | 241 | 267 | 315 | 347 | 211 | 247 | 277 | 301 | 333 | 389 | 389 | 301 | 337 | 367 | 406 | 432 | 432 | 432 |
| 28 | 234 | 282 | 318 | 346 | 385 | 463 | 548 | 261 | 315 | 355 | 386 | 430 | 517 | 550 | 348 | 392 | 427 | 475 | 550 | 550 | 550 |
|  | 151 | 177 | 199 | 216 | 239 | 282 | 331 | 189 | 221 | 248 | 269 | 298 | 353 | 375 | 270 | 302 | 328 | 364 | 413 | 413 | 413 |
| 29 | 218 | 263 | 296 | 322 | 359 | 431 | 511 | 243 | 293 | 330 | 360 | 401 | 482 | 550 | 324 | 365 | 398 | 443 | 532 | 550 | 550 |
|  | 136 | 159 | 179 | 194 | 215 | 254 | 298 | 170 | 199 | 223 | 242 | 268 | 317 | 359 | 242 | 272 | 295 | 327 | 387 | 399 | 399 |
| 30 | 203 | 245 | 276 | 301 | 335 | 402 | 488 | 227 | 274 | 308 | 336 | 374 | 450 | 533 | 302 | 341 | 371 | 413 | 497 | 550 | 550 |
|  | 123 | 144 | 161 | 175 | 194 | 229 | 269 | 153 | 179 | 201 | 218 | 242 | 286 | 336 | 219 | 245 | 266 | 295 | 349 | 385 | 385 |
| 31 | 190 | 229 | 258 | 281 | 313 | 376 | 446 | 212 | 256 | 289 | 314 | 350 | 421 | 499 | 283 | 319 | 347 | 387 | 465 | 550 | 550 |
|  | 111 | 130 | 146 | 158 | 175 | 207 | 243 | 138 | 162 | 182 | 198 | 219 | 259 | 304 | 198 | 222 | 241 | 267 | 316 | 369 | 369 |
| 32 | 178 | 215 | 242 | 264 | 294 | 353 | 418 | 199 | 240 | 271 | 295 | 328 | 395 | 468 | 265 | 299 | 326 | 363 | 436 | 517 | 549 |
|  | 101 | 118 | 132 | 144 | 159 | 188 | 221 | 126 | 147 | 165 | 179 | 199 | 235 | 276 | 180 | 201 | 219 | 242 | 287 | 337 | 355 |
| 33 | 168 | 202 | 228 | 248 | 276 | 332 | 393 | 187 | 226 | 254 | 277 | 309 | 371 | 440 | 249 | 281 | 306 | 341 | 410 | 486 | 532 |
|  | 92 | 108 | 121 | 131 | 145 | 171 | 201 | 114 | 134 | 150 | 163 | 181 | 214 | 251 | 164 | 183 | 199 | 221 | 261 | 307 | 334 |
| 34 | 158 | 190 | 214 | 233 | 260 | 312 | 370 | 176 | 212 | 239 | 261 | 290 | 349 | 414 | 235 | 265 | 288 | 321 | 386 | 458 | 516 |
|  | 84 | 98 | 110 | 120 | 132 | 156 | 184 | 105 | 122 | 137 | 149 | 165 | 195 | 229 | 149 | 167 | 182 | 202 | 239 | 280 | 314 |
| 35 | 149 | 179 | 202 | 220 | 245 | 294 | 349 | 166 | 200 | 226 | 246 | 274 | 329 | 390 | 221 | 249 | 272 | 303 | 364 | 432 | 494 |
|  | 77 | 90 | 101 | 110 | 121 | 143 | 168 | 96 | 112 | 126 | 137 | 151 | 179 | 210 | 137 | 153 | 167 | 185 | 219 | 257 | 292 |
| 36 | 141 | 169 | 191 | 208 | 232 | 278 | 330 | 157 | 189 | 213 | 232 | 259 | 311 | 369 | 209 | 236 | 257 | 286 | 344 | 408 | 467 |
|  | 70 | 82 | 92 | 101 | 111 | 132 | 154 | 88 | 103 | 115 | 125 | 139 | 164 | 193 | 126 | 141 | 153 | 169 | 201 | 236 | 269 |
| 37 | | | | | | | | 148 | 179 | 202 | 220 | 245 | 294 | 349 | 198 | 223 | 243 | 271 | 325 | 386 | 442 |
|  | | | | | | | | 81 | 95 | 106 | 115 | 128 | 151 | 178 | 116 | 130 | 141 | 156 | 185 | 217 | 247 |
| 38 | | | | | | | | 141 | 170 | 191 | 208 | 232 | 279 | 331 | 187 | 211 | 230 | 256 | 308 | 366 | 419 |
|  | | | | | | | | 74 | 87 | 98 | 106 | 118 | 139 | 164 | 107 | 119 | 130 | 144 | 170 | 200 | 228 |
| 39 | | | | | | | | 133 | 161 | 181 | 198 | 220 | 265 | 314 | 178 | 200 | 218 | 243 | 292 | 347 | 397 |
|  | | | | | | | | 69 | 81 | 90 | 98 | 109 | 129 | 151 | 98 | 110 | 120 | 133 | 157 | 185 | 211 |
| 40 | | | | | | | | 127 | 153 | 172 | 188 | 209 | 251 | 298 | 169 | 190 | 207 | 231 | 278 | 330 | 377 |
|  | | | | | | | | 64 | 75 | 84 | 91 | 101 | 119 | 140 | 91 | 102 | 111 | 123 | 146 | 171 | 195 |
| 41 | | | | | | | | | | | | | | | 161 | 181 | 197 | 220 | 264 | 314 | 359 |
|  | | | | | | | | | | | | | | | 85 | 95 | 103 | 114 | 135 | 159 | 181 |
| 42 | | | | | | | | | | | | | | | 153 | 173 | 188 | 209 | 252 | 299 | 342 |
|  | | | | | | | | | | | | | | | 79 | 88 | 96 | 106 | 126 | 148 | 168 |
| 43 | | | | | | | | | | | | | | | 146 | 165 | 179 | 200 | 240 | 285 | 326 |
|  | | | | | | | | | | | | | | | 73 | 82 | 89 | 99 | 117 | 138 | 157 |
| 44 | | | | | | | | | | | | | | | 139 | 157 | 171 | 191 | 229 | 272 | 311 |
|  | | | | | | | | | | | | | | | 68 | 76 | 83 | 92 | 109 | 128 | 146 |

**TABLE A-1   Standard Load Table/Open Web Steel Joists, K-Series (cont.)**

| Joist Designation | 24K4 | 24K5 | 24K6 | 24K7 | 24K8 | 24K9 | 24K10 | 24K12 | 26K5 | 26K6 | 26K7 | 26K8 | 26K9 | 26K10 | 26K12 | 28K6 | 28K7 | 28K8 | 28K9 | 28K10 | 28K12 |
|---|---|---|---|---|---|---|---|---|---|---|---|---|---|---|---|---|---|---|---|---|---|
| Depth (in.) | 24 | 24 | 24 | 24 | 24 | 24 | 24 | 24 | 26 | 26 | 26 | 26 | 26 | 26 | 26 | 28 | 28 | 28 | 28 | 28 | 28 |
| Approx. Wt. (lb/ft) | 8.4 | 9.3 | 9.7 | 10.1 | 11.5 | 12.0 | 13.1 | 16.0 | 9.8 | 10.6 | 10.9 | 12.1 | 12.2 | 13.8 | 16.6 | 11.4 | 11.8 | 12.7 | 13.0 | 14.3 | 17.1 |
| Span (ft) ↓ | | | | | | | | | | | | | | | | | | | | | |
| 24 | 520/516 | 550/544 | 550/544 | 550/544 | 550/544 | 550/544 | 550/544 | 550/544 | | | | | | | | | | | | | |
| 25 | 479/456 | 540/511 | 550/520 | 550/520 | 550/520 | 550/520 | 550/520 | 550/520 | | | | | | | | | | | | | |
| 26 | 442/405 | 499/453 | 543/493 | 550/499 | 550/499 | 550/499 | 550/499 | 550/499 | 542/535 | 550/541 | 550/541 | 550/541 | 550/541 | 550/541 | 550/541 | | | | | | |
| 27 | 410/361 | 462/404 | 503/439 | 550/479 | 550/479 | 550/479 | 550/479 | 550/479 | 502/477 | 547/519 | 550/522 | 550/522 | 550/522 | 550/522 | 550/522 | | | | | | |
| 28 | 381/323 | 429/362 | 467/393 | 521/436 | 550/456 | 550/456 | 550/456 | 550/456 | 466/427 | 508/464 | 550/501 | 550/501 | 550/501 | 550/501 | 550/501 | 548/541 | 550/543 | 550/543 | 550/543 | 550/543 | 550/543 |
| 29 | 354/290 | 400/325 | 435/354 | 485/392 | 536/429 | 550/436 | 550/436 | 550/436 | 434/384 | 473/417 | 527/463 | 550/479 | 550/479 | 550/479 | 550/479 | 511/486 | 550/522 | 550/522 | 550/522 | 550/522 | 550/522 |
| 30 | 331/262 | 373/293 | 406/319 | 453/353 | 500/387 | 544/419 | 550/422 | 550/422 | 405/346 | 441/377 | 492/417 | 544/457 | 550/459 | 550/459 | 550/459 | 477/439 | 531/486 | 550/500 | 550/500 | 550/500 | 550/500 |
| 31 | 310/237 | 349/266 | 380/289 | 424/320 | 468/350 | 510/379 | 550/410 | 550/410 | 379/314 | 413/341 | 460/378 | 509/413 | 550/444 | 550/444 | 550/444 | 446/397 | 497/440 | 550/466 | 550/480 | 550/480 | 550/480 |
| 32 | 290/215 | 327/241 | 357/262 | 397/290 | 439/318 | 478/344 | 549/393 | 549/393 | 356/285 | 387/309 | 432/343 | 477/375 | 519/407 | 549/431 | 549/431 | 418/361 | 466/400 | 515/438 | 549/463 | 549/463 | 549/463 |
| 33 | 273/196 | 308/220 | 335/239 | 373/265 | 413/289 | 449/313 | 532/368 | 532/368 | 334/259 | 364/282 | 406/312 | 448/342 | 488/370 | 532/404 | 532/404 | 393/329 | 438/364 | 484/399 | 527/432 | 532/435 | 532/435 |
| 34 | 257/179 | 290/201 | 315/218 | 351/242 | 388/264 | 423/286 | 502/337 | 516/344 | 315/237 | 343/257 | 382/285 | 422/312 | 459/338 | 516/378 | 516/378 | 370/300 | 412/333 | 456/364 | 496/395 | 516/410 | 516/410 |
| 35 | 242/164 | 273/184 | 297/200 | 331/221 | 366/242 | 399/262 | 473/308 | 501/324 | 297/217 | 323/236 | 360/261 | 398/286 | 433/310 | 501/356 | 501/356 | 349/275 | 389/305 | 430/333 | 468/361 | 501/389 | 501/389 |
| 36 | 229/150 | 258/169 | 281/183 | 313/203 | 346/222 | 377/241 | 447/283 | 487/306 | 280/199 | 305/216 | 340/240 | 376/263 | 409/284 | 486/334 | 487/334 | 330/252 | 367/280 | 406/306 | 442/332 | 487/366 | 487/366 |
| 37 | 216/138 | 244/155 | 266/169 | 296/187 | 327/205 | 356/222 | 423/260 | 474/290 | 265/183 | 288/199 | 322/221 | 356/242 | 387/262 | 460/308 | 474/315 | 312/232 | 348/257 | 384/282 | 418/305 | 474/344 | 474/344 |
| 38 | 205/128 | 231/143 | 252/156 | 281/172 | 310/189 | 338/204 | 401/240 | 461/275 | 251/169 | 274/184 | 305/204 | 337/223 | 367/241 | 436/284 | 461/299 | 296/214 | 329/237 | 364/260 | 396/282 | 461/325 | 461/325 |
| 39 | 195/118 | 219/132 | 239/144 | 266/159 | 294/174 | 320/189 | 380/222 | 449/261 | 238/156 | 260/170 | 289/188 | 320/206 | 348/223 | 413/262 | 449/283 | 280/198 | 313/219 | 346/240 | 376/260 | 447/306 | 449/308 |
| 40 | 185/109 | 208/122 | 227/133 | 253/148 | 280/161 | 304/175 | 361/206 | 438/247 | 227/145 | 247/157 | 275/174 | 304/191 | 331/207 | 393/249 | 438/269 | 266/183 | 297/203 | 328/222 | 357/241 | 424/284 | 438/291 |
| 41 | 176/101 | 198/114 | 216/124 | 241/137 | 266/150 | 290/162 | 344/191 | 427/235 | 215/134 | 235/146 | 262/162 | 289/177 | 315/192 | 374/225 | 427/256 | 253/170 | 283/189 | 312/206 | 340/224 | 404/263 | 427/277 |
| 42 | 168/94 | 189/106 | 206/115 | 229/127 | 253/139 | 276/151 | 327/177 | 417/224 | 205/125 | 224/136 | 249/150 | 275/164 | 300/178 | 356/210 | 417/244 | 241/158 | 269/175 | 297/192 | 324/208 | 384/245 | 417/264 |
| 43 | 160/88 | 180/98 | 196/107 | 219/118 | 242/130 | 263/140 | 312/165 | 406/213 | 196/116 | 213/126 | 238/140 | 263/153 | 286/166 | 339/195 | 407/232 | 230/147 | 257/163 | 284/179 | 309/194 | 367/228 | 407/252 |
| 44 | 153/82 | 172/92 | 187/100 | 209/110 | 231/121 | 251/131 | 298/154 | 387/199 | 187/108 | 204/118 | 227/131 | 251/143 | 273/155 | 324/182 | 398/222 | 220/137 | 245/152 | 271/167 | 295/181 | 350/212 | 398/240 |
| 45 | 146/76 | 164/86 | 179/93 | 199/103 | 220/113 | 240/122 | 285/144 | 370/185 | 179/101 | 194/110 | 217/122 | 240/133 | 261/145 | 310/170 | 389/212 | 210/128 | 234/142 | 259/156 | 282/169 | 334/198 | 389/229 |
| 46 | 139/71 | 157/80 | 171/87 | 191/97 | 211/106 | 230/114 | 272/135 | 354/174 | 171/95 | 186/103 | 207/114 | 229/125 | 250/135 | 296/159 | 380/203 | 201/120 | 224/133 | 248/146 | 270/158 | 320/186 | 380/219 |
| 47 | 133/67 | 150/75 | 164/82 | 183/90 | 202/99 | 220/107 | 261/126 | 339/163 | 164/89 | 178/96 | 199/107 | 219/117 | 239/127 | 284/149 | 369/192 | 192/112 | 214/125 | 237/136 | 258/148 | 306/174 | 372/210 |
| 48 | 128/63 | 144/70 | 157/77 | 175/85 | 194/93 | 211/101 | 250/118 | 325/153 | 157/83 | 170/90 | 190/100 | 210/110 | 229/119 | 272/140 | 353/180 | 184/105 | 206/117 | 227/128 | 247/139 | 294/163 | 365/201 |
| 49 | | | | | | | | | 150/78 | 164/85 | 183/94 | 202/103 | 220/112 | 261/131 | 339/169 | 177/99 | 197/110 | 218/120 | 237/130 | 282/153 | 357/193 |
| 50 | | | | | | | | | 144/73 | 157/80 | 175/89 | 194/97 | 211/105 | 250/124 | 325/159 | 170/93 | 189/103 | 209/113 | 228/123 | 270/144 | 350/185 |
| 51 | | | | | | | | | 139/69 | 151/75 | 168/83 | 186/91 | 203/99 | 241/116 | 313/150 | 163/88 | 182/97 | 201/106 | 219/115 | 260/136 | 338/175 |
| 52 | | | | | | | | | 133/65 | 145/71 | 162/79 | 179/86 | 195/93 | 231/110 | 301/142 | 157/83 | 175/92 | 193/100 | 210/109 | 250/128 | 325/165 |
| 53 | | | | | | | | | | | | | | | | 151/78 | 168/87 | 186/95 | 203/103 | 240/121 | 313/156 |
| 54 | | | | | | | | | | | | | | | | 145/74 | 162/82 | 179/89 | 195/97 | 232/114 | 301/147 |
| 55 | | | | | | | | | | | | | | | | 140/70 | 156/77 | 173/85 | 188/92 | 223/108 | 290/139 |
| 56 | | | | | | | | | | | | | | | | 135/66 | 151/73 | 166/80 | 181/87 | 215/102 | 280/132 |

**TABLE A-1** Standard Load Table/Open Web Steel Joists, K-Series (cont.)

| Joist Designation | 30K7 | 30K8 | 30K9 | 30K10 | 30K11 | 30K12 |
|---|---|---|---|---|---|---|
| Depth (in.) | 30 | 30 | 30 | 30 | 30 | 30 |
| Approx. Wt. (lb/ft) | 12.3 | 13.2 | 13.4 | 15.0 | 16.4 | 17.6 |
| Span (ft) ↓ | | | | | | |
| 30 | 550 / 543 | 550 / 543 | 550 / 543 | 550 / 543 | 550 / 543 | 550 / 543 |
| 31 | 534 / 508 | 550 / 520 | 550 / 520 | 550 / 520 | 550 / 520 | 550 / 520 |
| 32 | 501 / 461 | 549 / 500 | 549 / 500 | 549 / 500 | 549 / 500 | 549 / 500 |
| 33 | 471 / 420 | 520 / 460 | 532 / 468 | 532 / 468 | 532 / 468 | 532 / 468 |
| 34 | 443 / 384 | 490 / 420 | 516 / 441 | 516 / 441 | 516 / 441 | 516 / 441 |
| 35 | 418 / 351 | 462 / 384 | 501 / 415 | 501 / 415 | 501 / 415 | 501 / 415 |
| 36 | 395 / 323 | 436 / 353 | 475 / 383 | 487 / 392 | 487 / 392 | 487 / 392 |
| 37 | 373 / 297 | 413 / 325 | 449 / 352 | 474 / 374 | 474 / 374 | 474 / 374 |
| 38 | 354 / 274 | 391 / 300 | 426 / 325 | 461 / 353 | 461 / 353 | 461 / 353 |
| 39 | 336 / 253 | 371 / 277 | 404 / 300 | 449 / 333 | 449 / 333 | 449 / 333 |
| 40 | 319 / 234 | 353 / 256 | 384 / 278 | 438 / 315 | 438 / 315 | 438 / 315 |
| 41 | 303 / 217 | 335 / 238 | 365 / 258 | 427 / 300 | 427 / 300 | 427 / 300 |
| 42 | 289 / 202 | 320 / 221 | 348 / 240 | 413 / 282 | 417 / 284 | 417 / 284 |
| 43 | 276 / 188 | 305 / 206 | 332 / 223 | 394 / 263 | 407 / 270 | 407 / 270 |
| 44 | 263 / 176 | 291 / 192 | 317 / 208 | 376 / 245 | 398 / 258 | 398 / 258 |
| 45 | 251 / 164 | 278 / 179 | 303 / 195 | 359 / 229 | 389 / 246 | 389 / 246 |
| 46 | 241 / 153 | 266 / 168 | 290 / 182 | 344 / 214 | 380 / 236 | 380 / 236 |
| 47 | 230 / 144 | 255 / 157 | 277 / 171 | 329 / 201 | 372 / 226 | 372 / 226 |
| 48 | 221 / 135 | 244 / 148 | 266 / 160 | 315 / 188 | 362 / 215 | 365 / 216 |
| 49 | 212 / 127 | 234 / 139 | 255 / 150 | 303 / 177 | 347 / 202 | 357 / 207 |
| 50 | 203 / 119 | 225 / 130 | 245 / 141 | 291 / 166 | 333 / 190 | 350 / 199 |
| 51 | 195 / 112 | 216 / 123 | 235 / 133 | 279 / 157 | 320 / 179 | 343 / 192 |
| 52 | 188 / 106 | 208 / 116 | 226 / 126 | 268 / 148 | 308 / 169 | 336 / 184 |
| 53 | 181 / 100 | 200 / 109 | 218 / 119 | 258 / 140 | 296 / 159 | 330 / 177 |
| 54 | 174 / 94 | 192 / 103 | 209 / 112 | 249 / 132 | 285 / 150 | 324 / 170 |
| 55 | 168 / 89 | 185 / 98 | 202 / 106 | 240 / 125 | 275 / 142 | 312 / 161 |
| 56 | 162 / 84 | 179 / 92 | 195 / 100 | 231 / 118 | 265 / 135 | 301 / 153 |
| 57 | 156 / 80 | 173 / 88 | 188 / 95 | 223 / 112 | 256 / 128 | 290 / 145 |
| 58 | 151 / 76 | 167 / 83 | 181 / 90 | 215 / 106 | 247 / 121 | 280 / 137 |
| 59 | 146 / 72 | 161 / 79 | 175 / 86 | 208 / 101 | 239 / 115 | 271 / 130 |
| 60 | 141 / 69 | 156 / 75 | 169 / 81 | 201 / 96 | 231 / 109 | 262 / 124 |

## ▮ TABLE A-2   K-Series Economy Table

| Joist Designation | 10K1 | 12K1 | 8K1 | 14K1 | 16K2 | 12K3 | 14K3 | 16K3 | 18K3 | 14K4 | 20K3 | 16K4 | 12K5 | 18K4 | 16K5 | 20K4 |
|---|---|---|---|---|---|---|---|---|---|---|---|---|---|---|---|---|
| Depth (in.) | 10 | 12 | 8 | 14 | 16 | 12 | 14 | 16 | 18 | 14 | 20 | 16 | 12 | 18 | 16 | 20 |
| Approx. Wt. (lb/ft) | 5.0 | 5.0 | 5.1 | 5.2 | 5.5 | 5.7 | 6.0 | 6.3 | 6.6 | 6.7 | 6.7 | 7.0 | 7.1 | 7.2 | 7.5 | 7.6 |
| **Span (ft) ↓** | | | | | | | | | | | | | | | | |
| 8 | | | 550/550 | | | | | | | | | | | | | |
| 9 | | | 550/550 | | | | | | | | | | | | | |
| 10 | 550/550 | | 550/480 | | | | | | | | | | | | | |
| 11 | 550/542 | | 532/377 | | | | | | | | | | | | | |
| 12 | 550/455 | 550/550 | 444/288 | | | 550/550 | | | | | | | 550/550 | | | |
| 13 | 479/363 | 550/510 | 377/225 | | | 550/510 | | | | | | | 550/510 | | | |
| 14 | 412/289 | 500/425 | 324/179 | 550/550 | | 550/463 | 550/550 | | | 550/550 | | | 550/463 | | | |
| 15 | 358/234 | 434/344 | 281/145 | 511/475 | | 543/428 | 550/507 | | | 550/507 | | | 550/434 | | | |
| 16 | 313/192 | 380/282 | 246/119 | 448/390 | 550/550 | 476/351 | 550/467 | 550/550 | | 550/467 | | 550/550 | 550/396 | | 550/550 | |
| 17 | 277/159 | 336/234 | | 395/324 | 512/488 | 420/291 | 495/404 | 550/526 | | 550/443 | | 550/526 | 550/366 | | 550/526 | |
| 18 | 246/134 | 299/197 | | 352/272 | 456/409 | 374/245 | 441/339 | 508/456 | 550/550 | 530/397 | | 550/490 | 507/317 | 550/550 | 550/490 | |
| 19 | 221/113 | 268/167 | | 315/230 | 408/347 | 335/207 | 395/287 | 455/386 | 514/494 | 475/336 | | 547/452 | 454/269 | 550/523 | 550/455 | |
| 20 | 199/97 | 241/142 | | 284/197 | 368/297 | 302/177 | 356/246 | 410/330 | 463/423 | 428/287 | 517/517 | 493/386 | 409/230 | 550/490 | 550/426 | 550/550 |
| 21 | | 218/123 | | 257/170 | 333/255 | 273/153 | 322/212 | 371/285 | 420/364 | 388/248 | 468/447 | 447/333 | 370/198 | 506/426 | 503/373 | 550/550 |
| 22 | | 199/106 | | 234/147 | 303/222 | 249/132 | 293/184 | 337/247 | 382/316 | 353/215 | 426/393 | 406/289 | 337/172 | 460/370 | 458/323 | 514/461 |
| 23 | | | | 214/128 | 277/194 | 227/116 | 268/160 | 308/216 | 349/276 | 322/188 | 389/344 | 371/252 | 308/150 | 420/323 | 418/282 | 469/402 |
| 24 | | | | 196/113 | 254/170 | 208/101 | 245/141 | 283/189 | 320/242 | 295/165 | 357/302 | 340/221 | 282/132 | 385/284 | 384/248 | 430/353 |
| 25 | | | | 180/100 | 234/150 | | 226/124 | 260/167 | 294/214 | 272/145 | 329/266 | 313/195 | | 355/250 | 353/219 | 396/312 |
| 26 | | | | 166/88 | 216/133 | | 209/110 | 240/148 | 272/190 | 251/129 | 304/236 | 289/173 | | 328/222 | 326/194 | 366/277 |
| 27 | | | | 154/79 | 200/119 | | 193/98 | 223/132 | 252/169 | 233/115 | 281/211 | 268/155 | | 303/198 | 302/173 | 339/247 |
| 28 | | | | 143/70 | 186/106 | | 180/88 | 207/118 | 234/151 | 216/103 | 261/189 | 249/138 | | 282/177 | 281/155 | 315/221 |
| 29 | | | | | 173/95 | | | 193/106 | 218/136 | | 243/170 | 232/124 | | 263/159 | 261/139 | 293/199 |
| 30 | | | | | 161/86 | | | 180/96 | 203/123 | | 227/153 | 216/112 | | 245/144 | 244/126 | 274/179 |
| 31 | | | | | 151/78 | | | 168/87 | 190/111 | | 212/138 | 203/101 | | 229/130 | 228/114 | 256/162 |
| 32 | | | | | 142/71 | | | 158/79 | 178/101 | | 199/126 | 190/92 | | 215/118 | 214/103 | 240/147 |
| 33 | | | | | | | | | 168/92 | | 187/114 | | | 202/108 | | 226/134 |
| 34 | | | | | | | | | 158/84 | | 176/105 | | | 190/98 | | 212/122 |
| 35 | | | | | | | | | 149/77 | | 166/96 | | | 179/90 | | 200/112 |
| 36 | | | | | | | | | 141/70 | | 157/88 | | | 169/82 | | 189/103 |
| 37 | | | | | | | | | | | 148/81 | | | | | 179/95 |
| 38 | | | | | | | | | | | 141/74 | | | | | 170/87 |
| 39 | | | | | | | | | | | 133/69 | | | | | 161/81 |
| 40 | | | | | | | | | | | 127/64 | | | | | 153/75 |

### ▰ TABLE A-2   K-Series Economy Table (cont.)

| Joist Designation | 14K6 | 18K5 | 22K4 | 16K6 | 20K5 | 24K4 | 18K6 | 16K7 | 22K5 | 20K6 | 18K7 | 22K6 | 20K7 | 24K5 | 22K7 | 24K6 |
|---|---|---|---|---|---|---|---|---|---|---|---|---|---|---|---|---|
| Depth (in.) | 14 | 18 | 22 | 16 | 20 | 24 | 18 | 16 | 22 | 20 | 18 | 22 | 20 | 24 | 22 | 24 |
| Approx. Wt. (lb/ft) | 7.7 | 7.7 | 8.0 | 8.1 | 8.2 | 8.4 | 8.5 | 8.6 | 8.8 | 8.9 | 9.0 | 9.2 | 9.3 | 9.3 | 9.7 | 9.7 |
| **Span (ft) ↓** | | | | | | | | | | | | | | | | |
| 14 | 550/550 | | | | | | | | | | | | | | | |
| 15 | 550/507 | | | | | | | | | | | | | | | |
| 16 | 550/467 | | | 550/550 | | | | 550/550 | | | | | | | | |
| 17 | 550/443 | | | 550/526 | | | | 550/526 | | | | | | | | |
| 18 | 550/408 | 550/550 | | 550/490 | | | 550/550 | 550/490 | | | 550/550 | | | | | |
| 19 | 550/383 | 550/523 | | 550/455 | | | 550/523 | 550/455 | | | 550/523 | | | | | |
| 20 | 525/347 | 550/490 | | 550/426 | 550/550 | | 550/490 | 550/426 | | 550/550 | 550/490 | | 550/550 | | | |
| 21 | 475/299 | 550/460 | | 548/405 | 550/520 | | 550/460 | 550/406 | | 550/520 | 550/460 | | 550/520 | | | |
| 22 | 432/259 | 518/414 | 550/548 | 498/351 | 550/490 | | 550/438 | 550/385 | 550/548 | 550/490 | 550/438 | 550/548 | 550/490 | | 550/548 | |
| 23 | 395/226 | 473/362 | 518/491 | 455/307 | 529/451 | | 516/393 | 507/339 | 550/518 | 550/468 | 550/418 | 550/518 | 550/468 | | 550/518 | |
| 24 | 362/199 | 434/318 | 475/431 | 418/269 | 485/396 | 520/516 | 473/345 | 465/298 | 536/483 | 528/430 | 526/382 | 550/495 | 550/448 | 550/544 | 550/495 | 550/544 |
| 25 | 334/175 | 400/281 | 438/381 | 384/238 | 446/350 | 479/456 | 435/305 | 428/263 | 493/427 | 486/380 | 485/337 | 537/464 | 541/421 | 540/511 | 550/474 | 550/520 |
| 26 | 308/156 | 369/249 | 404/338 | 355/211 | 412/310 | 442/405 | 402/271 | 395/233 | 455/379 | 449/337 | 448/299 | 496/411 | 500/373 | 499/453 | 550/454 | 543/493 |
| 27 | 285/139 | 342/222 | 374/301 | 329/188 | 382/277 | 410/361 | 372/241 | 366/208 | 422/337 | 416/301 | 415/267 | 459/367 | 463/333 | 462/404 | 512/406 | 503/439 |
| 28 | 265/124 | 318/199 | 348/270 | 306/168 | 355/248 | 381/323 | 346/216 | 340/186 | 392/302 | 386/269 | 385/239 | 427/328 | 430/298 | 429/362 | 475/364 | 467/393 |
| 29 | | 296/179 | 324/242 | 285/151 | 330/223 | 354/290 | 322/194 | 317/167 | 365/272 | 360/242 | 359/215 | 398/295 | 401/268 | 400/325 | 443/327 | 435/354 |
| 30 | | 276/161 | 302/219 | 266/137 | 308/201 | 331/262 | 301/175 | 296/151 | 341/245 | 336/218 | 335/194 | 371/266 | 374/242 | 373/293 | 413/295 | 406/319 |
| 31 | | 258/146 | 283/198 | 249/124 | 289/182 | 310/237 | 281/158 | 277/137 | 319/222 | 314/198 | 313/175 | 347/241 | 350/219 | 349/266 | 387/267 | 380/289 |
| 32 | | 242/132 | 265/180 | 233/112 | 271/165 | 290/215 | 264/144 | 259/124 | 299/201 | 295/179 | 294/159 | 326/219 | 328/199 | 327/241 | 363/242 | 357/262 |
| 33 | | 228/121 | 249/164 | | 254/150 | 273/196 | 248/131 | | 281/183 | 277/163 | 276/145 | 306/199 | 309/181 | 308/220 | 341/221 | 335/239 |
| 34 | | 214/110 | 235/149 | | 239/137 | 257/179 | 233/120 | | 265/167 | 261/149 | 260/132 | 288/182 | 290/165 | 290/201 | 321/202 | 315/218 |
| 35 | | 202/101 | 221/137 | | 226/126 | 242/164 | 220/110 | | 249/153 | 246/137 | 245/121 | 272/167 | 274/151 | 273/184 | 303/185 | 297/200 |
| 36 | | 191/92 | 209/126 | | 213/115 | 229/150 | 208/101 | | 236/141 | 232/125 | 232/111 | 257/153 | 259/139 | 258/169 | 286/169 | 281/183 |
| 37 | | | 198/116 | | 202/106 | 216/138 | | | 223/130 | 220/115 | | 243/141 | 245/128 | 244/155 | 271/156 | 266/169 |
| 38 | | | 187/107 | | 191/98 | 205/128 | | | 211/119 | 208/106 | | 230/130 | 232/118 | 231/143 | 256/144 | 252/156 |
| 39 | | | 178/98 | | 181/90 | 195/118 | | | 200/110 | 198/98 | | 218/120 | 220/109 | 219/132 | 243/133 | 239/144 |
| 40 | | | 169/91 | | 172/84 | 185/109 | | | 190/102 | 188/91 | | 207/111 | 209/101 | 208/122 | 231/123 | 227/133 |
| 41 | | | 161/85 | | | 176/101 | | | 181/95 | | | 197/103 | | 198/114 | 220/114 | 216/124 |
| 42 | | | 153/79 | | | 168/94 | | | 173/88 | | | 188/96 | | 189/106 | 209/106 | 206/115 |
| 43 | | | 146/73 | | | 160/88 | | | 165/82 | | | 179/89 | | 180/98 | 200/99 | 196/107 |
| 44 | | | 139/68 | | | 153/82 | | | 157/76 | | | 171/83 | | 172/92 | 191/92 | 187/100 |
| 45 | | | | | | 146/76 | | | | | | | | 164/86 | | 179/93 |
| 46 | | | | | | 139/71 | | | | | | | | 157/80 | | 171/87 |
| 47 | | | | | | 133/67 | | | | | | | | 150/75 | | 164/82 |
| 48 | | | | | | 128/63 | | | | | | | | 144/70 | | 157/77 |

| Joist Designation | 26K5 | 16K9 | 24K7 | 18K9 | 26K6 | 20K9 | 26K7 | 22K9 | 28K6 | 24K8 | 18K10 | 28K7 | 24K9 | 26K8 | 20K10 | 26K9 |
|---|---|---|---|---|---|---|---|---|---|---|---|---|---|---|---|---|
| Depth (in.) | 26 | 16 | 24 | 18 | 26 | 20 | 26 | 22 | 28 | 24 | 18 | 28 | 24 | 26 | 20 | 26 |
| Approx. Wt. (lb/ft) | 9.8 | 10.0 | 10.1 | 10.2 | 10.6 | 10.8 | 10.9 | 11.3 | 11.4 | 11.5 | 11.7 | 11.8 | 12.0 | 12.1 | 12.2 | 12.2 |
| Span (ft) ↓ | | | | | | | | | | | | | | | | |
| 16 | | 550/550 | | | | | | | | | | | | | | |
| 17 | | 550/526 | | | | | | | | | | | | | | |
| 18 | | 550/490 | | 550/550 | | | | | | | 550/550 | | | | | |
| 19 | | 550/455 | | 550/523 | | | | | | | 550/523 | | | | | |
| 20 | | 550/426 | | 550/490 | 550/550 | | | | | | 550/490 | | | | 550/550 | |
| 21 | | 550/406 | | 550/460 | 550/520 | | | | | | 550/460 | | | | 550/520 | |
| 22 | | 550/385 | | 550/438 | 550/490 | | | 550/548 | | | 550/438 | | | | 550/490 | |
| 23 | | 550/363 | | 550/418 | 550/468 | | | 550/518 | | | 550/418 | | | | 550/468 | |
| 24 | | 550/346 | 550/544 | 550/396 | 550/448 | | | 550/495 | | 550/544 | 550/396 | | 550/544 | | 550/448 | |
| 25 | | 514/311 | 550/520 | 550/377 | 550/426 | | | 550/474 | | 550/520 | 550/377 | | 550/520 | | 550/426 | |
| 26 | 542/535 | 474/276 | 550/499 | 538/354 | 550/541 | 550/405 | 550/541 | 550/454 | | 550/499 | 550/361 | | 550/499 | 550/541 | 550/405 | 550/541 |
| 27 | 502/477 | 439/246 | 550/479 | 498/315 | 547/519 | 550/389 | 550/522 | 550/432 | | 550/479 | 550/347 | | 550/479 | 550/522 | 550/389 | 550/522 |
| 28 | 466/427 | 408/220 | 521/436 | 463/282 | 508/464 | 517/353 | 550/501 | 550/413 | 548/541 | 550/456 | 548/331 | 550/543 | 550/456 | 550/501 | 550/375 | 550/501 |
| 29 | 434/384 | 380/198 | 485/392 | 431/254 | 473/417 | 482/317 | 527/463 | 532/387 | 511/486 | 536/429 | 511/298 | 550/522 | 550/436 | 550/479 | 550/359 | 550/479 |
| 30 | 405/346 | 355/178 | 453/353 | 402/229 | 441/377 | 450/286 | 492/417 | 497/349 | 477/439 | 500/387 | 477/269 | 531/486 | 544/419 | 544/457 | 533/336 | 550/459 |
| 31 | 379/314 | 332/161 | 424/320 | 376/207 | 413/341 | 421/259 | 460/378 | 465/316 | 446/397 | 468/350 | 446/243 | 497/440 | 510/379 | 509/413 | 499/304 | 550/444 |
| 32 | 356/285 | 311/147 | 397/290 | 353/188 | 387/309 | 395/235 | 432/343 | 436/287 | 418/361 | 439/318 | 418/221 | 466/400 | 478/344 | 477/375 | 468/276 | 519/407 |
| 33 | 334/259 | | 373/265 | 332/171 | 364/282 | 371/214 | 406/312 | 410/261 | 393/329 | 413/289 | 393/201 | 438/364 | 449/313 | 448/342 | 440/251 | 488/370 |
| 34 | 315/237 | | 351/242 | 312/156 | 343/257 | 349/195 | 382/285 | 386/239 | 370/300 | 388/264 | 370/184 | 412/333 | 423/286 | 422/312 | 414/229 | 459/338 |
| 35 | 297/217 | | 331/221 | 294/143 | 323/236 | 329/179 | 360/261 | 364/219 | 349/275 | 366/242 | 349/168 | 389/305 | 399/286 | 398/312 | 390/210 | 433/310 |
| 36 | 280/199 | | 313/203 | 278/132 | 305/216 | 311/164 | 340/240 | 344/201 | 330/252 | 346/222 | 330/154 | 367/280 | 377/241 | 376/263 | 369/193 | 409/284 |
| 37 | 265/183 | | 296/187 | | 289/199 | 294/151 | 322/221 | 325/185 | 312/232 | 327/205 | | 348/257 | 356/222 | 356/242 | 349/178 | 387/262 |
| 38 | 251/169 | | 281/172 | | 274/184 | 279/139 | 305/204 | 308/170 | 296/214 | 310/189 | | 329/237 | 338/204 | 337/223 | 331/164 | 367/241 |
| 39 | 238/156 | | 266/159 | | 260/170 | 265/129 | 289/188 | 292/157 | 280/198 | 294/174 | | 313/219 | 320/206 | 320/206 | 314/151 | 348/223 |
| 40 | 227/145 | | 253/148 | | 247/157 | 251/119 | 275/174 | 278/146 | 266/183 | 280/161 | | 297/203 | 304/175 | 304/191 | 298/140 | 331/207 |
| 41 | 215/134 | | 241/137 | | 235/146 | | 262/162 | 264/135 | 253/170 | 266/150 | | 283/189 | 290/162 | 289/177 | | 315/192 |
| 42 | 205/125 | | 229/127 | | 224/136 | | 249/150 | 252/126 | 241/158 | 253/139 | | 269/175 | 276/151 | 275/164 | | 300/178 |
| 43 | 196/116 | | 219/118 | | 213/126 | | 238/140 | 240/117 | 230/147 | 242/130 | | 257/163 | 263/140 | 263/153 | | 286/166 |
| 44 | 187/108 | | 209/110 | | 204/118 | | 227/131 | 229/109 | 220/137 | 231/121 | | 245/152 | 251/131 | 251/143 | | 273/155 |
| 45 | 179/101 | | 199/103 | | 194/110 | | 217/122 | | 210/128 | 220/113 | | 234/142 | 240/122 | 240/133 | | 261/145 |
| 46 | 171/95 | | 191/97 | | 186/103 | | 207/114 | | 201/120 | 211/106 | | 224/133 | 230/114 | 229/125 | | 250/135 |
| 47 | 164/89 | | 183/90 | | 178/96 | | 199/107 | | 192/112 | 202/99 | | 214/125 | 220/107 | 219/117 | | 239/127 |
| 48 | 157/83 | | 175/85 | | 171/90 | | 190/100 | | 184/105 | 194/93 | | 206/117 | 211/101 | 210/110 | | 229/119 |
| 49 | 150/78 | | | | 164/85 | | 183/94 | | 177/99 | | | 197/110 | | 202/103 | | 220/112 |
| 50 | 144/73 | | | | 157/80 | | 175/89 | | 170/93 | | | 189/103 | | 194/97 | | 211/105 |
| 51 | 139/69 | | | | 151/75 | | 168/83 | | 163/88 | | | 182/97 | | 186/91 | | 203/99 |
| 52 | 133/65 | | | | 145/71 | | 162/79 | | 157/83 | | | 175/92 | | 179/86 | | 195/93 |
| 53 | | | | | | | | | 151/78 | | | 168/87 | | | | |
| 54 | | | | | | | | | 145/74 | | | 162/82 | | | | |
| 55 | | | | | | | | | 140/70 | | | 156/77 | | | | |
| 56 | | | | | | | | | 135/66 | | | 151/73 | | | | |

| Joist Designation | 30K7 | 22K10 | 28K8 | 28K9 | 24K10 | 30K8 | 30K9 | 22K11 | 26K10 | 28K10 | 30K10 | 24K12 | 30K11 | 26K12 | 28K12 | 30K12 |
|---|---|---|---|---|---|---|---|---|---|---|---|---|---|---|---|---|
| Depth (in.) | 30 | 22 | 28 | 28 | 24 | 30 | 30 | 22 | 26 | 28 | 30 | 24 | 30 | 26 | 28 | 30 |
| Approx. Wt. (lb/ft) | 12.3 | 12.6 | 12.7 | 13.0 | 13.1 | 13.2 | 13.4 | 13.8 | 13.8 | 14.3 | 15.0 | 16.0 | 16.4 | 16.6 | 17.1 | 17.6 |
| **Span (ft) ↓** | | | | | | | | | | | | | | | | |
| 22 | | 550/548 | | | | | | 550/548 | | | | | | | | |
| 23 | | 550/518 | | | | | | 550/518 | | | | | | | | |
| 24 | | 550/495 | | | 550/544 | | | 550/495 | | | | 550/544 | | | | |
| 25 | | 550/474 | | | 550/520 | | | 550/474 | | | | 550/520 | | | | |
| 26 | | 550/454 | | | 550/499 | | | 550/454 | 550/541 | | | 550/499 | | 550/541 | | |
| 27 | | 550/432 | | | 550/479 | | | 550/432 | 550/522 | | | 550/479 | | 550/522 | | |
| 28 | | 550/413 | 550/543 | 550/543 | 550/456 | | | 550/413 | 550/501 | 550/543 | | 550/456 | | 550/501 | 550/543 | |
| 29 | | 550/399 | 550/522 | 550/522 | 550/436 | | | 550/399 | 550/479 | 550/522 | | 550/436 | | 550/479 | 550/522 | |
| 30 | 550/543 | 550/385 | 550/500 | 550/500 | 550/422 | 550/543 | 550/543 | 550/385 | 550/459 | 550/500 | 550/543 | 550/422 | 550/543 | 550/459 | 550/543 | 550/543 |
| 31 | 534/508 | 550/369 | 550/480 | 550/480 | 550/410 | 550/520 | 550/520 | 550/369 | 550/444 | 550/480 | 550/520 | 550/410 | 550/520 | 550/444 | 550/480 | 550/520 |
| 32 | 501/461 | 517/337 | 515/438 | 549/463 | 549/393 | 549/500 | 549/500 | 549/355 | 549/431 | 549/463 | 549/500 | 549/393 | 549/500 | 549/431 | 549/463 | 549/500 |
| 33 | 471/420 | 486/307 | 484/399 | 527/432 | 532/368 | 520/460 | 532/468 | 532/334 | 532/404 | 532/435 | 532/468 | 532/368 | 532/468 | 532/404 | 532/435 | 532/468 |
| 34 | 443/384 | 458/280 | 456/364 | 496/395 | 502/337 | 490/420 | 516/441 | 516/314 | 516/378 | 516/410 | 516/441 | 516/344 | 516/441 | 516/378 | 516/410 | 516/441 |
| 35 | 418/351 | 432/257 | 430/333 | 468/361 | 473/308 | 462/384 | 501/415 | 494/292 | 501/356 | 501/389 | 501/415 | 501/324 | 501/415 | 501/356 | 501/389 | 501/415 |
| 36 | 395/323 | 408/236 | 406/306 | 442/332 | 447/283 | 436/353 | 475/383 | 467/269 | 486/334 | 487/366 | 487/392 | 487/306 | 487/392 | 487/334 | 487/366 | 487/392 |
| 37 | 373/297 | 386/217 | 384/282 | 418/305 | 423/260 | 413/325 | 449/352 | 442/247 | 460/308 | 474/344 | 474/374 | 474/290 | 474/374 | 474/315 | 474/344 | 474/374 |
| 38 | 354/274 | 366/200 | 364/260 | 396/282 | 401/240 | 391/300 | 426/325 | 419/228 | 436/284 | 461/325 | 461/353 | 461/275 | 461/353 | 461/299 | 461/325 | 461/353 |
| 39 | 336/253 | 347/185 | 346/240 | 376/260 | 380/222 | 371/277 | 404/300 | 397/211 | 413/262 | 447/306 | 449/333 | 449/261 | 449/333 | 449/283 | 449/308 | 449/333 |
| 40 | 319/234 | 330/171 | 328/222 | 357/241 | 361/206 | 353/256 | 384/278 | 377/195 | 393/243 | 424/284 | 438/315 | 438/247 | 438/315 | 438/269 | 438/291 | 438/315 |
| 41 | 303/217 | 314/159 | 312/206 | 340/224 | 344/191 | 335/238 | 365/258 | 359/181 | 374/225 | 404/263 | 427/300 | 427/235 | 427/300 | 427/256 | 427/277 | 427/300 |
| 42 | 289/202 | 299/148 | 297/192 | 324/208 | 327/177 | 320/221 | 348/240 | 342/168 | 356/210 | 384/245 | 413/282 | 417/224 | 417/284 | 417/244 | 417/264 | 417/284 |
| 43 | 276/188 | 285/138 | 284/179 | 309/194 | 312/165 | 305/206 | 332/223 | 326/157 | 339/195 | 367/228 | 394/263 | 406/213 | 407/270 | 407/232 | 407/252 | 407/270 |
| 44 | 263/176 | 272/128 | 271/167 | 295/181 | 298/154 | 291/192 | 317/208 | 311/146 | 324/182 | 350/212 | 376/245 | 387/199 | 398/258 | 398/222 | 398/240 | 398/258 |
| 45 | 251/164 | | 259/156 | 282/169 | 285/144 | 278/179 | 303/195 | | 310/170 | 334/198 | 359/229 | 370/185 | 389/246 | 389/212 | 389/229 | 389/246 |
| 46 | 241/153 | | 248/146 | 270/158 | 272/135 | 266/168 | 290/182 | | 296/159 | 320/186 | 344/214 | 354/174 | 380/236 | 380/203 | 380/219 | 380/236 |
| 47 | 230/144 | | 237/136 | 258/148 | 261/126 | 255/157 | 277/171 | | 284/149 | 306/174 | 329/201 | 339/163 | 372/226 | 369/192 | 372/210 | 372/226 |
| 48 | 221/135 | | 227/128 | 247/139 | 250/118 | 244/148 | 266/160 | | 272/140 | 294/163 | 315/188 | 325/153 | 362/215 | 353/180 | 365/201 | 362/216 |
| 49 | 212/127 | | 218/120 | 237/130 | | 234/139 | 255/150 | | 261/131 | 282/153 | 303/177 | | 347/202 | 339/169 | 357/193 | 357/207 |
| 50 | 203/119 | | 209/113 | 228/123 | | 225/130 | 245/141 | | 250/124 | 270/144 | 291/166 | | 333/190 | 325/159 | 350/185 | 350/199 |
| 51 | 195/112 | | 201/106 | 219/115 | | 216/123 | 235/133 | | 241/116 | 260/136 | 279/157 | | 320/179 | 313/150 | 338/175 | 343/192 |
| 52 | 188/106 | | 193/100 | 210/109 | | 208/116 | 226/126 | | 231/110 | 250/128 | 268/148 | | 308/169 | 301/142 | 325/165 | 336/184 |
| 53 | 181/100 | | 186/95 | 203/103 | | 200/109 | 218/119 | | | 240/121 | 258/140 | | 296/159 | | 313/156 | 330/177 |
| 54 | 174/94 | | 179/89 | 195/97 | | 192/103 | 209/112 | | | 232/114 | 249/132 | | 285/150 | | 301/147 | 324/170 |
| 55 | 168/89 | | 173/85 | 188/92 | | 185/98 | 202/106 | | | 223/108 | 240/125 | | 275/142 | | 290/139 | 312/161 |
| 56 | 162/84 | | 166/80 | 181/87 | | 179/92 | 195/100 | | | 215/102 | 231/118 | | 265/135 | | 280/132 | 301/153 |
| 57 | 156/80 | | | | | 173/88 | 188/95 | | | | 223/112 | | 256/128 | | | 290/145 |
| 58 | 151/76 | | | | | 167/83 | 181/90 | | | | 215/106 | | 247/121 | | | 280/137 |
| 59 | 146/72 | | | | | 161/79 | 175/86 | | | | 208/101 | | 239/115 | | | 271/130 |
| 60 | 141/69 | | | | | 156/75 | 169/81 | | | | 201/96 | | 231/109 | | | 262/124 |

# APPENDIX B

# Metrication

## THE INTERNATIONAL SYSTEM OF UNITS

The U.S. Customary System of weights and measures has been used as the primary unit system in this book. This system was developed from the English system (British), which was introduced in the original 13 colonies when they were under British rule. Although the English system spread to many parts of the world during the past three centuries, it was widely recognized that there was a need for a single international coordinated measurement system. As a result, a second system of weights and measures, known as the metric system, was developed by a commission of French scientists and was adopted by France as the legal system of weights and measures in 1799.

Although the metric system was not accepted with enthusiasm at first, adoption by other nations occurred steadily after France made its use compulsory in 1840. In the United States, an Act of Congress in 1866 made it lawful throughout the land to employ the weights and measures of the metric system in all contracts, dealings, or court proceedings.

*Note:* Much of this material has been taken from Leonard Spiegel and George F. Limbrunner, *Reinforced Concrete Design,* 3rd ed., 1992, pp. 447–51. Reprinted by permission of Prentice Hall, Englewood Cliffs, NJ.

By 1900 a total of 35 nations, including the major nations of continental Europe and most of South America, had officially accepted the metric system. Because of the many versions of the metric system that developed, an International General Conference on Weights and Measures in 1960 adopted an extensive revision and simplification of the system. The name *Le Système International d' Unités* (International System of Units), with the international abbreviation SI, was adopted for this modernized metric system. The American Society of Civil Engineers (ASCE) resolved in 1970 to actively support conversion to SI and to adopt the revised edition of the American Society for Testing and Materials (ASTM) *Metric Practice* guide.

In 1971 the U.S. Secretary of Commerce, in transmitting to Congress the results of a three-year study authorized by the Metric Study Act of 1968, recommended that the United States change to predominant use of the metric system through a coordinated national program. Congress responded by enacting the Metric Conversion Act of 1975, which established a U.S. Metric Board to carry out the planning, coordination, and public education that would facilitate a voluntary conversion to a modern metric system.

In 1988 Congress passed the Omnibus Trade and Competitiveness Act, which, among other things, mandated that by the end of fiscal 1992, the federal government would require metric specifications on all the goods it purchased. In 1991 then-President George Bush signed Executive Order 12770, *Metric Usage in Federal Government Programs,* which required federal agencies to develop specific timetables and milestones for the transition to metric. At this writing, many federal construction programs are fully converted, and most of the others will be converted in the near term. Federal construction constitutes a large portion of the nation's construction industry. The authors continue to believe that in the near future virtually the entire construction industry will "go metric."

Metric conversion involves two distinct aspects. On one hand, design calculations and detailing practices will be affected by the new units. Many publications, such as specifications, building codes, and design handbooks (and their associated software), are being produced using SI units. On the other hand, the physical sizes of some products will be affected (for instance, plywood will change from a 4-ft width to a slightly smaller 1200-mm width). For structural steel, physical sizes of shapes will remain the same, but designations will change. Inches will change to millimeters, and pounds per foot will change to kilograms per meter (a current W18 × 60 will be known as a W460 × 89). Properties will undergo a "soft conversion" (meaning that physical properties will remain unchanged, but be converted mathematically to the new units). For example, section modulus will be listed in millimeters cubed, and moment of inertia will be listed in millimeters to the fourth power. Bolts, however, will change to metric diameters and threads as per ASTM A325M and A490M.

As a means of introducing the SI metric system, this appendix furnishes various tables as well as examples purely for familiarization purposes.

## B-2

### SI STYLE AND USAGE

The SI consists of a limited number of *base* units, which correspond to fundamental quantities, and a large number of *derived* units (derived from the base units) that describe other quantities. The SI base units and derived units pertinent to structural steel design are listed in Tables B-1 and B-2, respectively.

**TABLE B-1   SI Base Units and Symbols**

| Quantity | Unit | SI symbol |
|----------|------|-----------|
| Length | meter | m |
| Mass | kilogram | kg |
| Time | second | s |
| Angle[a] | radian | rad |
| Temperature[b] | kelvin | K |

[a]It is also permissible to use the arc degree *and its decimal submultiples* when the radian is not convenient.

[b]It is also permissible to use degrees Celsius (°C).

**TABLE B-2   SI Derived Units**

| Quantity | Unit | SI symbol | Formula |
|----------|------|-----------|---------|
| Acceleration | meter per second squared | — | $m/s^2$ |
| Area | square meter | — | $m^2$ |
| Density (mass per unit volume) | kilogram per cubic meter | — | $kg/m^3$ |
| Force | newton | N | $kg \cdot m/s^2$ |
| Pressure or stress | pascal | Pa | $N/m^2$ |
| Volume | cubic meter | — | $m^3$ |
| Section modulus | millimeter to third power | — | $mm^3$ |
| Moment of inertia | millimeter to fourth power | — | $mm^4$ |
| Moment of force, torque | newton meter | — | $N \cdot m$ |
| Force per unit length | newton per meter | — | $N/m$ |
| Mass per unit length | kilogram per meter | — | $kg/m$ |
| Mass per unit area | kilogram per square meter | — | $kg/m^2$ |

Although specified in SI, the pascal is not universally accepted as the unit of stress. Because section properties are expressed in millimeters, it is more convenient to express stress in newtons per square millimeter ($1 \, N/mm^2 = 1 \, Mpa$). This is the practice followed in recent international structural design standards, including the International Standards Organization (ISO) Draft International Standard for Steel Design [1].

■■■■■ TABLE B-3    SI Prefixes

| Prefix | SI symbol | Factor | Example |
|--------|-----------|--------|---------|
| mega | M | $1\ 000\ 000 = 10^6$ | Mm = 1 000 000 meters |
| kilo | k | $1000 = 10^3$ | km = 1000 meters |
| milli | m | $0.001 = 10^{-3}$ | mm = 0.001 meter |

Since the orders of magnitude of many quantities cover wide ranges of numerical values, SI *prefixes* have been established to deal with decimal point placement. SI prefixes representing steps of 1000 are recommended to indicate orders of magnitude. Those recommended for use in structural steel design are listed in Table B-3.

From Table B-3, it is clear that there is a choice of ways to represent numbers. It is preferable to use numbers between 1 and 1000, whenever possible, by selecting the appropriate prefix. For example, 18 m is preferred to 0.018 km or 18 000 mm. A brief note must be made at this point concerning the presentation of numbers with many digits. It is common practice in the United States to separate digits into groups of three by means of commas. To avoid confusion with the widespread European practice of using a comma on the line as the decimal marker, the method of setting off groups of three digits with a gap, as shown in Table B-3, is recommended international practice. Note that this method is used on both the left and right sides of the decimal marker for any string of five or more digits. A group of four digits on either side of the decimal marker need not be separated.

A significant difference between SI and other measurement systems is the use of explicit and distinctly separate units for mass and force. The SI base unit kilogram (kg) denotes the base unit of mass, which is the quantity of matter of an object. This is a constant quantity that is independent of gravitational attraction. The derived SI unit newton (N) denotes the absolute derived unit of force (mass times acceleration: $kg \cdot m/s^2$). The term *weight* should be avoided because it may be confused with mass and also because it describes a particular force that is related solely to gravitational acceleration, which varies on the surface of the earth. For the conversion of mass to force, the recommended value for the acceleration of gravity in the United States may be taken as $g = 9.81$ m/s$^2$.

Since weight is defined as the force of gravity on a body and, since gravity may vary, the weight of a body may theoretically vary. Weight may be computed using Newton's second law of motion, which may be stated mathematically as

$$F = Ma$$

where

$F$ = force

$M$ = mass

$a$ = acceleration

Since weight represents a force of gravity, it may be expressed as

weight = force of gravity on a body

= (mass of the body) × (acceleration of gravity)

or

$$\text{weight} = Mg$$

where $g$ is the acceleration of gravity.

As an example, if a member supports 500 kg (kilograms) of dead load, the weight of the dead load may be obtained from

$$\text{weight} = Mg = (500 \text{ kg})(9.81 \text{ m/s}^2)$$

$$= 4910 \text{ kg·m/s}^2$$

$$= 4910 \text{ N} = 4.91 \text{ kN}$$

Note that the preceding reflects that 1 kg·m/s² is defined as one newton (N).

**B-3**

## CONVERSION FACTORS

Table B-4 contains conversion factors for the conversion of U.S. Customary System units to SI units for quantities that may be frequently used in structural steel design.

**Example B-1** _____

A steel beam is designated in the U.S. Customary System as a W14 × 48. Calculate its weight per unit length in the SI System.

**Solution:**

The weight of lb/ft in the U.S. Customary System will be converted to N/m (newtons per meter) in the SI System. In the following calculations the second term will convert lb to kg and the third term will convert ft to m, which will furnish mass per unit length. The fourth term will convert the mass to force (weight) using $F = Mg$. All conversion factors are from Table B-4. Thus

$$48 \frac{\text{lb}}{\text{ft}} \left( \frac{0.454 \text{ kg}}{1 \text{ lb}} \right) \left( \frac{1 \text{ ft}}{0.3048 \text{ m}} \right) \left( 9.81 \frac{\text{m}}{\text{s}^2} \right) = 701 \frac{\text{kg·m}}{\text{m·s}^2} = 701 \text{ N/m}$$

or, using a direct conversion factor,

$$48(14.594) = 701 \text{ N/m}$$

■■■  **TABLE B-4   Conversion Factors: U.S. Customary to SI Units**

| | Multiply | | By | To obtain |
|---|---|---|---|---|
| Length | inches | × | 25.4 | = millimeters |
| | feet | × | 0.3048 | = meters |
| | yards | × | 0.9144 | = meters |
| | miles (statute) | × | 1.609 | = kilometers |
| Area | square inches | × | 645.2 | = square millimeters |
| | square feet | × | 0.0929 | = square meters |
| | square yards | × | 0.8361 | = square meters |
| Volume | cubic inches | × | 16 387. | = cubic millimeters |
| | cubic feet | × | 0.028 32 | = cubic meters |
| | cubic yards | × | 0.7646 | = cubic meters |
| | gallons (U.S. liquid) | × | 0.003 785 | = cubic meters |
| Force | pounds | × | 4.448 | = newtons |
| | kips | × | 4448. | = newtons |
| Force per unit length | pounds per foot | × | 14.594 | = newtons per meter |
| | kips per foot | × | 14 594. | = newtons per meter |
| Load per unit volume | pounds per cubic foot | × | 0.157 14 | = kilonewtons per cubic meter |
| Bending moment or torque | inch-pounds | × | 0.1130 | = newton meters |
| | foot-pounds | × | 1.356 | = newton meters |
| | inch-kips | × | 113.0 | = newton meters |
| | foot-kips | × | 1356. | = newton meters |
| | inch-kips | × | 0.1130 | = kilonewton meters |
| | foot-kips | × | 1.356 | = kilonewton meters |
| Stress, pressure, loading (force per unit area) | pounds per square inch | × | 6895. | = pascals |
| | pounds per square inch | × | 6.895 | = kilopascals |
| | pounds per square inch | × | 0.006 895 | = megapascals |
| | kips per square inch | × | 6.895 | = megapascals |
| | pounds per square foot | × | 47.88 | = pascals |
| | pounds per square foot | × | 0.047 88 | = kilopascals |
| | kips per square foot | × | 47.88 | = kilopascals |
| | kips per square foot | × | 0.047 88 | = megapascals |
| Mass | pounds | × | 0.454 | = kilograms |
| Mass per unit volume (density) | pounds per cubic foot | × | 16.02 | = kilograms per cubic meter |
| | pounds per cubic yard | × | 0.5933 | = kilograms per cubic meter |
| Moment of inertia | inches$^4$ | × | 416 231. | = millimeters$^4$ |
| Mass per unit length | pounds per foot | × | 1.488 | = kilograms per meter |
| Mass per unit area | pounds per square foot | × | 4.882 | = kilograms per square meter |

**Example B-2**

Using the ASD method, compute the allowable axial compressive load for a W250 × 80 (W10 × 54) column that has an unbraced length of 6 m. Use A36 steel, and assume that the ends are pin-connected.

**Solution:**

The following properties for the W250 × 80 are obtained from ASDM, Part 1, and converted to SI:

$$A = (15.8 \text{ in.}^2)(645.2 \text{ mm}^2/\text{in.}^2) = 10\ 190 \text{ mm}^2 = 10.19 \times 10^{-3} \text{ m}^2$$

$$r_y = (2.56 \text{ in.})(25.4 \text{ mm/in.}) = 65.0 \text{ mm} = 65.0 \times 10^{-3} \text{ m}$$

Note that the preceding properties may also be obtained from the AISC publication "Metric Properties of Structural Shapes" (see Reference 2).

The slenderness ratio with respect to the member $Y$ axis is

$$\frac{K\ell}{r_y} = \frac{1.0(6 \text{ m})}{65.0 \times 10^{-3} \text{ m}} = 92.3 \qquad (\text{use } 92)$$

The values for $F_a$ (in ksi) from ASDM, Part 3, Table C-36, may be converted to N/mm² in the SI system by multiplying the tabulated values by 6.895. Therefore,

$$F_a = 13.97(6.895) = 96.3 \text{ N/mm}^2 = 96.3 \times 10^6 \text{ N/m}^2$$

and

$$P_a = F_a A = (96.3 \times 10^6 \text{ N/m}^2)(10.19 \times 10^{-3} \text{ m}^2)$$

$$= 981 \times 10^3 \text{ N}$$

$$= 981 \text{ kN}$$

**Example B-3**

Calculate the allowable span length for a W460 × 74 (W18 × 50) simply supported beam subjected to a uniformly distributed load of 29.2 kN/m (which includes the weight of the beam). Assume A36 steel and full lateral support. Consider both moment and shear.

**Solution:**

$$F_y = 250 \text{ N/mm}^2$$

$$F_b = 0.66F_y = 165 \text{ N/mm}^2 = 165 \times 10^6 \text{ N/m}^2$$

Note that this member is compact whether in the U.S. Customary System or the SI system:

$$F_v = 0.40F_y = 100 \text{ N/mm}^2$$

The following properties are obtained from the ASDM, Part 1, and converted to SI:

$$d = (17.99 \text{ in.})(25.4 \text{ mm/in.}) = 457 \text{ mm}$$

$$t_w = (0.355 \text{ in.})(25.4 \text{ mm/in.}) = 9.02 \text{ mm}$$

$$S_x = (88.9 \text{ in.}^3)(25.4^3 \text{ mm}^3/\text{in.}^3) = 1457 \times 10^3 \text{ mm}^3 = 1.457 \times 10^{-3} \text{ m}^3$$

1.   The allowable bending moment is

$$M_R = F_b S_x$$

$$= (165 \times 10^6 \text{ N/m}^2)(1.457 \times 10^{-3} \text{ m}^3)$$

$$= 240 \times 10^3 \text{ N·m}$$

2.   The allowable span length based on moment only is

$$M = \frac{wL^2}{8}$$

Therefore,

$$L^2 = \frac{8M}{w}$$

$$= \frac{8(240 \times 10^3 \text{ N·m})}{(29.2 \times 10^3 \text{ N/m})}$$

$$= 65.8 \text{ m}^2$$

$$L = 8.11 \text{ m}$$

3.   Check shear:

$$\text{allowable shear } V = F_v \, dt_w$$

$$= (100 \text{ N/mm}^2)(457 \text{ mm})(9.02 \text{ mm})$$

$$= 412 \times 10^3 \text{ N}$$

$$V = \frac{wL}{2}$$

Therefore,

$$L = \frac{2V}{w}$$

$$= \frac{2(412 \times 10^3 \text{ N})}{(29.2 \times 10^3 \text{ N/m})}$$

$$= 28.2 \text{ m}$$

The controlling value is the smaller of the two span lengths (indicating that moment controls). Therefore, the maximum allowable span length is 8.11 meters.

## Example B-4

With reference to the lap joint shown in Figure 8-14 (Chapter 8 of this text), use a plate 150 mm × 10 mm thick connected to a 10-mm-thick gusset plate. Design the longitudinal fillet welds to develop the capacity of the 150 mm × 10 mm plate. Use the ASD approach, A36 steel, and E-70 electrodes (SMAW). Neglect block shear.

**Solution:**

$$F_y = 250 \text{ N/mm}^2$$

$$(\text{Plate}) \, F_u = (58 \text{ ksi})(6.895) = 400 \text{ N/mm}^2$$

$$(\text{Electrode}) \, F_u = (70 \text{ ksi})(6.895) = 480 \text{ N/mm}^2$$

1.  Calculate the tensile capacity of the plate. Based on gross area,

$$P_t = A_g F_t = A_g 0.60 F_y$$

$$= (150 \text{ mm})(10 \text{ mm})(0.60)(250 \text{ N/mm}^2)$$

$$= 225 \times 10^3 \text{ N} = 225 \text{ kN}$$

Based on effective net area (ASDS, Section B3) (assume $1.5w > \ell > w$ and $U = 0.75$),

$$P_t = A_e F_t = U A_g F_t = U A_g 0.5 F_u$$

$$= 0.75(150 \text{ mm})(10 \text{ mm})(0.5)(400 \text{ N/mm}^2)$$

$$= 225 \times 10^3 \text{ N} = 225 \text{ kN}$$

Use a plate capacity of 225 kN.

2.  The maximum weld size is

$$10 \text{ mm} - 2 \text{ mm} = 8 \text{ mm} \qquad (\text{use an 8-mm fillet weld})$$

3.  The capacity of 8-mm weld per linear mm is

$$F_v = 0.3 F_u = 0.3(480 \text{ N/mm}^2) = 144 \text{ N/mm}^2$$

$$P = F_v(0.707)(\text{leg size})$$

$$= (144 \text{ N/mm}^2)(0.707)(8 \text{ mm}) = 814 \text{ N/mm}$$

4. The length of weld required is

$$\frac{P_t}{P} = \frac{225 \times 10^3\,\text{N}}{814\,\text{N/mm}} = 276\,\text{mm}$$

$$\frac{276}{2} = 138\,\text{mm each side of plate}$$

5. Use end returns with a minimum length of $2\times$ (leg size).

$$2(8\,\text{mm}) = 16\,\text{mm}$$

Use 25-mm end returns. Therefore, length of longitudinal welds required is

$$138\,\text{mm} - 25\,\text{mm} = 113\,\text{mm each side of plate}$$

The minimum length of longitudinal weld allowed by specification (ASDS) is equal to the perpendicular distance between longitudinal welds. Therefore, use a longitudinal weld length of 150 mm on each side of the plate with 25-mm end returns.

6. Check the assumed $U$ value:

$$w = 150\,\text{mm}$$

$$1.5w = 225\,\text{mm}$$

$$\ell = 150\,\text{mm}$$

Therefore,

$$1.5w > \ell \geq w$$

O.K., since $U$ was assumed to be 0.75.

## REFERENCES

[1] American Institute of Steel Construction, Inc., "SI Units for Structural Steel Design," *AISC Engineering Journal,* 2nd Qtr., 1993.

[2] *Metric Properties of Structural Shapes,* 1992, American Institute of Steel Construction, Inc., 1 East Wacker Drive, Suite 3100, Chicago, IL 60601.

# APPENDIX C

# Flowcharts

Flowcharts for some of the basic analysis and design procedures are provided in this appendix. These flowcharts parallel the step-by-step procedures given in the text. They may be used to guide the reader through the steps of a particular analysis or design problem. They also depict graphically the types of decision-making processes that accompany the development of computer program solutions for these problems.

Figure C-5, Beam Design, includes a step wherein the beam weight is estimated *prior* to initially computing shear and moment. This is the approach used in text Example 4-8. Although the inclusion of an estimated beam weight is usually desirable, it is not mandatory, provided that the beam weight is included before the final selection is made. The latter approach is used in text Examples 4-7 and 4-9 and is, essentially, equivalent to assuming an initial beam weight of zero.

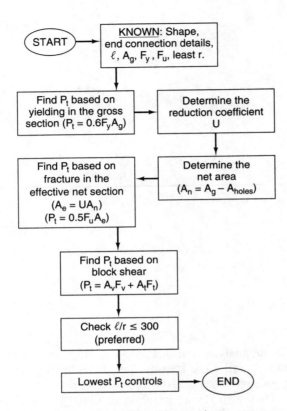

**FIGURE C-1**   Tension member analysis.

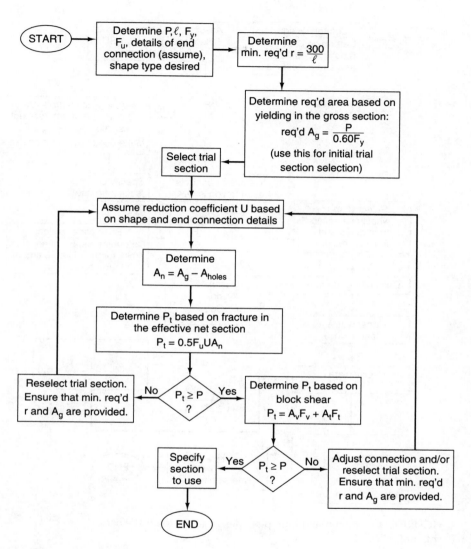

**FIGURE C-2**    Tension member design.

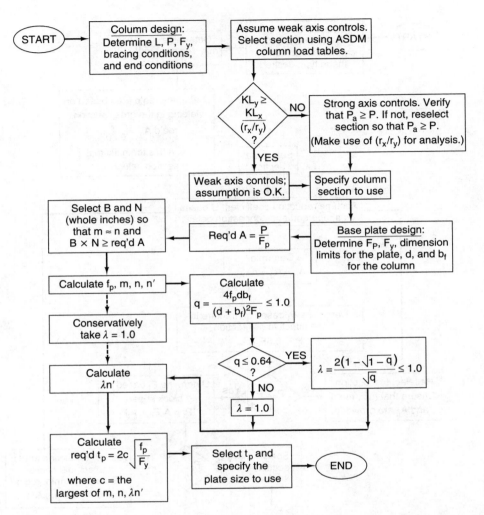

**FIGURE C-3**   Column and base plate design (axial load; W-shape column).

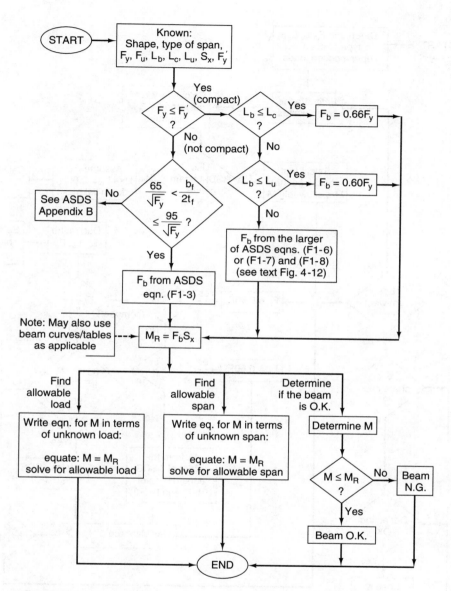

**FIGURE C-4**   Beam analysis (W shape, moment only, strong-axis bending).

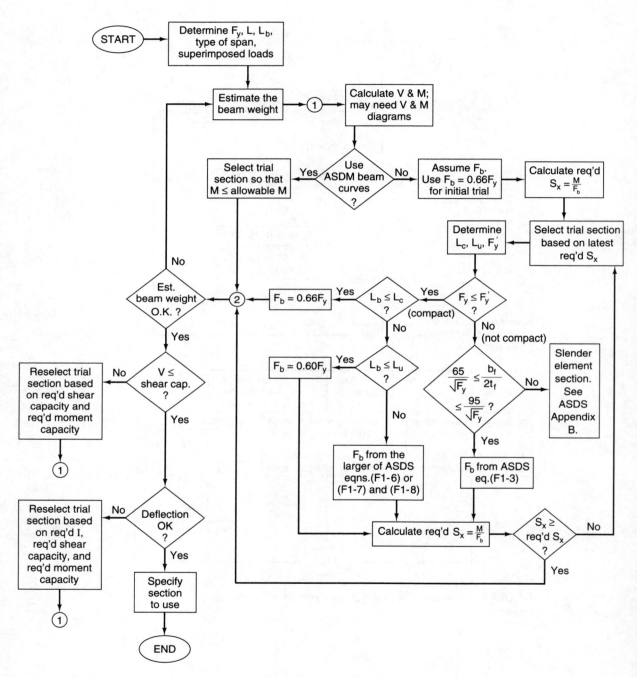

**FIGURE C-5**   Beam design (W and M; moment, shear, and deflection).

# Answers to Selected Problems

**CHAPTER 1**

**1-4.** (a) 62    (b) 25    (c) 88
    (d) 37.4    (e) 89.68
**1-6.** 101.7 lb
**1-7.** (a) 18    (b) 36
    (c) 13    (d) 59

**CHAPTER 2**

**2-1.** 47.1 kips
**2-3.** (a) 48.9 kips    (b) 101.9 kips
**2-5.** 97.2 kips
**2-7.** 243 kips
**2-9.** 203 kips
**2-11.** 146.7 kips
**2-13.** 162 kips
**2-15.** (a) 29.5 ft    (b) 106.5 ft
**2-17.** 58.4 kips > 58.1 kips, DE is O.K.
**2-19.** W10 × 22
**2-21.** W8 × 40

**2-23.** (a) $2L3\frac{1}{2} \times 2\frac{1}{2} \times \frac{1}{4}$ (LLBB)
    (b) $2L3 \times 3 \times \frac{3}{16}$
**2-25.** $TS5 \times 5 \times \frac{3}{16}$
**2-27.** $2C15 \times 33.9$
**2-29.** $1\frac{3}{4}$-in.-diam. rod: $1\frac{3}{4} - 5UNC2A$
**2-31.** $1\frac{1}{8}$-in.-diam. rod: $1\frac{1}{8} - 7UNC2A$

**CHAPTER 3**

**3-1.** 467 kips
**3-3.** 119.3 kips
**3-5.** (a) $f_e = 41.4$ ksi (N.G.)
    (b) $P_a = 13.8$ kips
**3-7.** (a) 253 kips    (b) 355 kips
    (c) 184.8 kips
**3-9.** (a) 136 kips    (b) 69 kips
    (c) 178 kips
**3-11.** (a) 296 kips    (b) 440 kips
    (c) 1326 kips    (d) 477 kips
    (e) 442 kips    (f) 1227 kips

(g) 193 kips    (h) 275 kips

(i) 83 kips

**3-13.** (a) 1341 kips    (b) 1629 kips

**3-15.** (a) W12 × 106

(b) W12 × 79    (c) W8 × 24

(d) W10 × 49    (e) W10 × 49

(f) Pipe 4 STD

(g) TS5 × 5 × $\frac{3}{16}$

(h) W14 × 500

(i) W14 × 426

**3-17.** W14 × 61 and W14 × 109

**3-19.** Use W10 × 112 with 2 $\frac{1}{4}$-in. plates welded across the flange tips parallel to the web.

**3-21.** 100 kips, two intermediate connectors, 32-in. spacing.

**3-23.** (a) 2L5 × 3$\frac{1}{2}$ × $\frac{1}{2}$ (LLBB) with two intermediate connectors at maximum spacing of 45.6 in.

(b) 2L5 × 3$\frac{1}{2}$ × $\frac{3}{8}$ (LLBB) with two intermediate connectors at maximum spacing of 89.3 in.

(c) 2L4 × 3$\frac{1}{2}$ × $\frac{1}{4}$ (LLBB) with two intermediate connectors at maximum spacing of 41.6 in.

**3-25.** (a) Plate 9 × $\frac{5}{8}$ × 1′–1 (20.7 lb)

(b) Plate 24 × 2$\frac{1}{8}$ × 2′–2 (376 lb)

(c) Plate 22 × 2$\frac{1}{4}$ × 2′–7 (435 lb)

## CHAPTER 4

**4-3.** $f_b$ = 21.1 ksi

**4-5.** Compact: W36 × 300, W18 × 46; Noncompact: W12 × 65, W14 × 99

**4-7.** $f_b$ = 20.5 ksi < 22 ksi; therefore, O.K.

**4-9.** $M$ = 162.6 ft-kips; $M_R$ = 163.2 ft-kips; therefore, O.K.

**4-11.** No allowable load $P$ is permitted.

**4-13.** (a) 24 ksi    (b) 17.48 ksi

(c) 8.10 ksi

**4-15.** 1.20 kips/ft

**4-17.** W24 × 76

**4-19.** W30 × 108

**4-21.** W18 × 35

**4-23.** Beam: W33 × 118; girder: W36 × 300

**4-25.** W18 × 50

**4-27.** Max. $f_v$ = 12.47 ksi; average web shear = 11.04 ksi

**4-29.** W14 × 48

**4-31.** W30 × 99

**4-33.** W14 × 22

**4-35.** W14 × 74 (N.G.)

**4-37.** w = 10.69 kips/ft

**4-39.** W12 × 19

**4-41.** N.G. for deflection; use a W27 × 178.

**4-43.** W18 × 55

**4-45.** B1: W30 × 90; G1: W40 × 215

**4-47.** W18 × 35

**4-49.** W21 × 44

**4-51.** Minimum $N$ = 5.54 in.

**4-53.** Plate: 1$\frac{1}{2}$ × 10 × 1′–8

**4-55.** No plate needed.

**4-57.** Maximum reaction = 23.3 kips

## CHAPTER 5

**5-1.** (a) Use 2L3$\frac{1}{2}$ × 2$\frac{1}{2}$ × $\frac{1}{4}$ (SLBB)

(b) Use 2L3$\frac{1}{2}$ × 3 × $\frac{1}{4}$ (SLBB)

**5-3.** Use WT8 × 22.5 and L6 × 3$\frac{1}{2}$ × $\frac{3}{8}$

**5-5.** (a) 19.3 ft-kips    (b) 765 lb/ft

(c) Shear (O.K.)

**5-7.** Steel: 56.4%; wood: 43.6%

**5-9.** (a) Use plate: $\frac{5}{8}$ × 11$\frac{1}{4}$

(b) Use plate: 1 × 9$\frac{1}{4}$

**5-11.** $f_b$ = 12.3 ksi

**5-13.** Use two plates: $\frac{5}{8}$ × 9

**5-15.** (a) 4.72 kips/ft

(b) 3.28 kips/ft

**5-17.** Beam is N.G.

**5-19.** Beam is O.K.

**5-21.** 420 ft-kips

**5-23.** Web plate: $\frac{1}{2}$ × 46; flange plates: 1 × 18

## CHAPTER 6

**6-1.** Member is O.K.

**6-3.** Member is N.G.

**6-5.** Member is O.K.

**6-7.** 84.9 kips

**6-9.** W12 × 53

**6-11.** W14 × 90

**6-13.** W14 × 145

**6-15.** W12 × 106

**6-17.** W14 × 132

**6-19.** Column AB: $K = 1.81$; column DC: $K = 1.88$

## CHAPTER 7

**7-1.** **(a)** 90.1 kips **(b)** 113.4 kips **(c)** 111.6 kips

**7-3.** **(a)** 60 kips **(b)** 74.4 kips

**7-5.** **(a)** 135 kips **(b)** 227 kips **(c)** 167.4 kips

**7-7.** Connection is satisfactory.

**7-9.** Connection (14 bolts) is satisfactory.

**7-11.** Use 6 bolts and $1\frac{1}{4}$-in. edge distance.

**7-13.** Use 24 bolts, 3-in. pitch, and $1\frac{1}{2}$-in. edge distance.

**7-15.** Use 10 bolts.

**7-17.** Capacity = 190.4 kips

**7-19.** 36.1 kips

**7-21.** 30.0 kips

**7-23.** Top and bottom angles: L8 × 4 × 1 × 0′−6

**7-25.** **(a)** 2598 lb **(b)** 1976 lb

**7-27.** Connection is O.K.

## CHAPTER 8

**8-1.** 111 kips

**8-3.** 95.9 kips

**8-5.** $8\frac{1}{2}$ in. long, $\frac{5}{16}$ in. weld (each side)

**8-7.** Use $\frac{11}{16}$ in. weld; 11 in. at the heel of the angle, 6 in. at the connected leg.

**8-9.** Use $\frac{7}{16}$ in. weld as shown, with $3\frac{1}{2}$ in. × 1 in. slot weld.

**8-11.** Use 9 in. plate length with $\frac{3}{16}$ in. weld each side of the web.

**8-13.** 2.30 kips/in.

**8-15.** Use a $\frac{1}{4}$-in. fillet weld.

**8-17.** 35.4 kips

## CHAPTER 9

**9-1.** **(a)** 8K1 **(b)** 30K12 **(c)** 11.5 lb/ft **(d)** 9728 lb

**(e)** 218 lb/ft

**9-3.** 166.5 lb/ft

**9-5.** 28K7

**9-7.** 12K1

**9-9.** 22K6

## CHAPTER 10

**10-1.** W21 × 44

**10-3.** W24 × 76

**10-5.** W18 × 35

## CHAPTER 11

**11-1.** 2L4 × $3\frac{1}{2}$ × $\frac{1}{4}$ × 0′−$8\frac{1}{2}$; three rows of bolts

**11-3.** 2L4 × $3\frac{1}{2}$ × $\frac{3}{8}$ × 1′−1; four rows of bolts

**11-7.** Seat angle: L4 × 4 × $\frac{3}{4}$ × 0′−8 (Type A pattern)

## CHAPTER 13

**13-1.** 90 psf

**13-3.** 101.1 kips

**13-5.** 81.0 kips

**13-7.** 251 kips

**13-9.** 499 kips

**13-11.** **(a)** 165.5 ft-kips **(b)** 188.0 ft-kips **(c)** 180.3 ft-kips

**13-13.** W18 × 40

## CHAPTER 14

**14-1.** 53.9 kips

**14-3.** 103.5 kips; 53.4 kips

**14-5.** 118.1 kips

**14-7.** **(a)** Four bolts each side of splice **(b)** Six bolts each side of splice

**14-9.** Six bolts; bolt spacing = $3\frac{1}{2}$ in.; edge distance = $1\frac{1}{2}$ in.; service load = 90.1 kips

**14-11.** **(a)** 83.5 kips **(b)** 111.4 kips

**14-13.** 16 in.

# Index